This book is to be returned on or before
the last date stamped below.

Towards
A Healthy Baby

Congenital Disorders and The New Genetics in Primary Care

BERNADETTE MODELL

Consultant in Perinatal Medicine, Wellcome Principle Fellow, and Director of the WHO Collaborating Centre for Control of Hereditary Diseases, University College and Middlesex School of Medicine, London

and

MICHAEL MODELL

General Practitioner, London Borough of Camden, and Senior Clinical Lecturer in Primary Health Care, University College and Middlesex School of Medicine, London

OXFORD NEW YORK TOKYO
OXFORD UNIVERSITY PRESS
1992

Oxford University Press, Walton Street, Oxford OX2 6DP
Oxford   New York   Toronto
Delhi   Bombay   Calcutta   Madras   Karachi
Petaling Jaya   Singapore   Hong Kong   Tokyo
Nairobi   Dar es Salaam   Cape Town
Melbourne   Auckland
and associated companies in
Berlin   Ibadan

Oxford is a trade mark of Oxford University Press

Published in the United States
by Oxford University Press, New York

A catalogue record for this book is available from the British Library

Library of Congress Cataloging in Publication Data
Modell, Bernadette.
Towards a healthy baby: congenital disorders and the new genetics in
primary care/Bernadette Modell and Michael Modell.
(Oxford medical publications)
Includes bibliographical references and index.
1. Genetic counselling.   2. Genetic disorders—Prevention.
3. Human chromosome abnormalities—Diagnosis.
I. Modell, Michael.   II. Title.   III. Series.
[DNLM: 1. Abnormalities—prevention & control—Europe.
2. Genetic Counseling.   3. Genetic Screening.
4. Hereditary Diseases-prevention & control—Europe.
5. Prenatal Diagnosis.   6. Primary Health Care.   QS 675 M689t]
RB155.7.M63   1992   616'.042—dc20   91-32222
ISBN 0-19-262234-X
ISBN 0-19-261486-X (pbk.)

Set by
Colset (Private) Ltd, Singapore
Printed in Great Britain by
Bookcraft (Bath) Ltd
Midsomer Norton, Avon

Preface

The objective of medical genetics is to help people with a genetic disadvantage to live and reproduce as normally as possible.

WHO Advisory Group 1985

This book is primarily for doctors, nurses, and counsellors working in primary care in the United Kingdom and the rest of Europe, but we hope that senior medical students will also find it useful. Our objectives are to summarize recent advances in the understanding of congenital and genetic disorders, and to discuss their relevance to the care of people in the community. We aim to help family doctors, health visitors, midwives, and nurses to explain the benefits and limitations of screening programmes to their patients, to counsel and support people undergoing screening tests, and to provide initial advice for couples at risk of producing a child with one of the commoner genetic or congenital disorders. We also discuss the indications for referring families for further advice to specialists such as clinical geneticists.

Immunization and a higher standard of living have almost eliminated many previously lethal childhood infections and the risks associated with poor sanitation and malnutrition. Consequently, in most of Europe a baby boy can on average expect to live for 70 and a girl for 76 years, an improvement of more than 20 years since 1900. The residual infant and childhood mortality and morbidity is largely due to genetic and congenital disorders and the consequences of prematurity; these problems are more intractable.

In the Western world, the main causes of death in adult life are cancer, ischaemic heart disease, and lung disease. Advances in adult medicine have as yet made relatively little impact on these diseases; the life expectancy of people aged 50 has increased by only a few years during this century. Prevention of the common chronic diseases of middle age may prove easier than cure. Survival to old age and quality of life are largely dependent on economic and social circumstances and the characteristics inherited from our parents.

In addition, new problems have emerged related to contamination of food and degradation of the environment, and effects of radiation from natural and man-made sources. New viruses are causing new infectious diseases; for example, the human immunodeficiency virus (HIV), which works by becoming incorporated into the human genome.

At the same time, progress in molecular biology is creating new understanding, new tools, and new approaches to these difficult problems. It is becoming possible to identify individuals or couples who may have a child with a severe genetic or congenital disorder, or who may themselves suffer from a genetically determined disease in later life. Methods are being developed to examine the developing fetus, to identify the fundamental causes of genetic disease, and to understand how interaction of genes and environment can lead to disease. It is becoming necessary for primary care workers to understand these advances.

Some of the concepts of molecular biology are quite complicated, so Part I includes a concise description of basic genetics. Our intention is to help doctors and nurses to explain the relevant concepts more confidently to their patients, and to keep abreast of developments in the future.

The new technology may benefit almost everyone, but equitable delivery of the services to the community will depend on population screening. It is recognized that these services must be based in primary health care if they are to be offered to the whole population (WHO 1985b; Royal College of Physicians 1989; Modell *et al.* 1991). The forms of screening mentioned below are now well established in maternal and child health services, but they are still not optimally delivered. We also need to learn as much as we can from these existing 'community genetics services', to be prepared for further applications of genetic screening in the future. It may soon be possible to detect familial predispositions to many common diseases, including cancer, and advice will be needed on how to minimize such risks: but screening for genetic risk for diseases of adult life presents many problems.

Pre-conception counselling is now accepted as an important concept. It can help to avoid fetal damage from infection or from cigarettes and alcohol, and should include a brief family history as a guide to some genetic risks. It enables an 'elderly' mother-to-be to benefit from early prenatal diagnostic tests, and allows screening to be offered for haemoglobin disorders and Tay–Sachs disease. These involve specific ethnic groups, but screening for carriers of cystic fibrosis is becoming a reality, and will ultimately involve the whole European population.

Screening during pregnancy can identify many women at risk of having a child with a serious disability in time to allow them an informed choice amongst the available options. These range from selective abortion of an affected fetus to optimal management of an affected infant from the outset.

Comprehensive *neonatal screening* ensures that some of the commoner conditions caused by the accumulation of toxic metabolites or lack of essential hormones are detected early, so that treatment can be started before symptoms develop.

We have placed particular emphasis on conditions that are current targets of screening, because doctors and nurses must know the implications of having a child with a neural tube defect, Down syndrome, cystic fibrosis, or thalassaemia, in order to inform their patients. It is impossible to discuss every congenital or genetic disorder, so we have selected some that occur in more than one in 10 000 births and used them to illustrate some general principles of medical genetics. A handful of families with such conditions is likely to be present within the group of 10 000–15 000 people registered with an urban group practice in the UK, or a polyclinic population in many other countries.

Genetic and congenital disease raise important ethical issues, largely related to our ability to decide who shall be permitted to survive. We briefly examine these issue as they arise.

In the references we have limited ourselves as far as possible to outstanding books that should be available in any medical or scientific library, inexpensive monographs on some specific disorders, and articles in journals that are easily accessible to primary care workers. This has inevitably led to a limited and local selection which omits many outstanding papers. We are sorry if, by chance, our selection unintentionally causes any offence.

We would like to thank the colleagues and friends who have generously spent time reading and commenting on the text. These particularly include Dr Anver Kuliev, former director of the Hereditary Diseases Programme of the WHO, and Dr Marsden Wagner, director of the Maternal and Child Health unit at the WHO Regional Office for Europe, Professor Rodney Harris, Gill Denniss and Dr Elizabeth Anionwu. Dr Gerald Corney and Mr Humphry Ward reviewed individual sections. Their advice and criticisms have been most useful in shaping the final form of the book and reducing the number of avoidable errors. This book would not have been written without the secretarial help of Christine Connolly, Veronica Fernie, and Madeleine Waller. We are very grateful.

We dedicate the book to Rachel and David — our contribution to the human gene pool.

London   B.M.
February 1991   M.M.

Contents

Part 1. Basic genetics

Part 2. Genetic diagnosis and counselling

Part 3. Specific congenital and genetic disorders

Part 4. Healthy pregnancy

Part 5. Long-term implications

Abbreviations

AFP	alphafetoprotein
ATP	adenosine triphosphate
cDNA	complementary DNA
DNA	deoxyribonucleic acid
G6PD	glucose-6-phosphate dehydrogenase
GTP	guanosine triphosphate
HLA	human leucocyte antigen
Ig	immune globulin
Kb	kilobase (1000 base pairs of DNA)
LDL	low density lipoprotein
MHC	major histocompatibility complex
mRNA	messenger RNA
NTD	neural tube defect
PCR	polymerase chain reaction
PND	prenatal diagnosis
RFLP	restriction fragment length polymorphism
Rh	rhesus blood group
RNA	ribonucleic acid
rRNA	ribosomal RNA
tRNA	transfer RNA
UTP	uridine triphosphate

1. Genetics and primary health care

The practical implications of basic medical genetics are of increasing relevance to family doctors and community health workers. There is an important genetic element in common disorders of adult life and old age, while over 10 per cent of the population carry one of the common inherited disorders. Individual inherited diseases are rare (cystic fibrosis, which affects about one in 2000 infants, is one of the most common), yet collectively they are quite common, and make a disproportionately high impact on the family and the medical services because they tend to be chronic and severe. A congenital or genetic disorder in one family member often involves a risk for relatives, and it is increasingly possible to detect and advise carriers of inherited diseases; however, most health workers have had little training in genetics, and so often either fail to inform or misinform people about their genetic risks.

Information is the main clinical tool in medical genetics. Avoiding genetic risk often depends on personal choice, for instance whether or not to alter life-style or diet, or to request prenatal diagnosis and selective abortion to avoid the birth of a handicapped child. The 'right' choice differs for each person depending on their individual circumstances, and ideas of right and wrong. In genetic counselling, information is therefore given in a relatively neutral way, and time is spent in helping people to work out the best decision for themselves. This to some extent differentiates the practice of medical genetics from that of acute medicine.

The main emphasis in this book is on congenital and genetically determined disorders of childhood, because this is the area in which we have most experience of how genetic knowledge can be used to prevent disease. Several types of screening programmes are well established, and primary care workers play an important role in them. Table 1.1 shows the main categories of congenital and genetic disorders, together with common examples selected because of their importance in primary care, that are used throughout the book to illustrate the basic principles of medical genetics and screening.

In developed countries, a high proportion of deaths in infancy and childhood is due to intractable chronic disorders that are often congenital or genetic in origin. Figure 1.1 compares the typical survival curves of the USA and Malawi (which closely resembles that in Europe 100 years ago). In both cases, mortality is highest in infancy and childhood and in late middle and old age, but in developing countries there is vastly disproportionate early

Table 1.1 Examples of congenital disorders discussed in this book

'Multifactorial'	Congenital malformations	Neural tube defects, congenital heart disease, urinary tract abnormalities, congenital dislocation of hip
	Cerebral palsy	
	Mental retardation	
	Coronary heart disease	
Chromosomal	Miscarriage	
	Down Syndrome	
	Disorders of sex chromosomes	
Inherited diseases	Recessively inherited	Cystic fibrosis, haemoglobin disorders, phenylketonuria, lactase non-persistence, Tay–Sachs disease, haemochromatosis
	Dominantly inherited	Huntington's chorea, adult polycystic kidney disease, achondroplasia, familial hypercholesterolaemia
	X-linked	Haemophilia, Duchenne muscular dystrophy, G6PD deficiency, fragile X mental retardation
Environmental hazards	Intra-uterine infections	Rubella, CMV, toxoplasmosis, syphilis, HIV, herpes simplex
	Maternal diseases	Diabetes,
	Maternal behaviour	Alcohol, smoking,
	Drugs	e.g. anti-epileptics.
	Radiation	

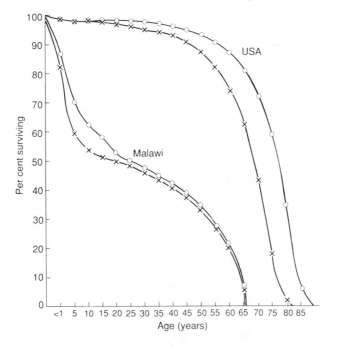

Fig. 1.1 Comparison of survival in a typical Western and a typical African country. Males = ×, Females = ○. (Data from United Nations Demographic Yearbook 1978).

mortality, largely due to infections aggravated by malnutrition. Most infants with congenital or genetic disorders die in this group without ever being diagnosed. When improved public health, family planning, and nutrition reduce infant mortality, children with intractable chronic disorders become an important problem.

Even in developed countries, changes due to improved health care are still in progress. Paediatric surgery for congenital malformations is increasingly successful, and as medical treatment improves, more children with chronic disorders are surviving into adult life. As a result, the cumulative number of affected people is rising, and chronic disease in young adults is increasingly of genetic origin. To take only one example, in the UK since the late 1960s, about 60 newborns with phenylketonuria (PKU) are detected and started on treatment every year. At the time of writing about 2500 patients are known, and ultimately, assuming a normal average survival of 70 years, there will be about 4200 healthy individuals with PKU requiring surveillance and support in the community at any one time.

The situation is similar for chronic disorders of later life. As life expectancy increases, genetic predispositions have more time to show themselves, and affected people survive longer, and often in better health, than previously. This is equally true for disorders characteristic of early adult life (e.g. diabetes mellitus, rheumatoid arthritis, ankylosing spondylitis, schizophrenia, and thyroid disease), of middle age (e.g. coronary heart disease, hypertension, and cancers) and of old age (e.g. senile dementia — Alzheimer's disease). Despite their importance in primary care, we touch on these conditions only briefly, because there is as yet very little experience of how people use information about genetic predisposition to common disorders.

Incidence and prevalence of congenital disorders

In discussing the frequency of genetic and congenital disorders, it is important to distinguish between incidence and prevalence.

Incidence is the rate at which new cases appear. For example, for infectious diseases it is usually expressed as new cases diagnosed annually per 100 000 population. For congenital disorders, whether detected at birth or not, it may be expressed as the number affected per 1000 or 10 000 births; this is the **birth incidence**. This is a useful measure of the frequency with which individuals with abnormalities enter the population.

Prevalence is the proportion of the whole population affected at any one time, expressed as the number per 1000–100 000. The prevalence of people with congenital or genetic disorders is lower than the birth incidence, because they usually do not live as long as the general population. Prevalence also depends on factors such as the survival of untreated patients, age at onset of the disease, the life expectancy of the general population, the effectiveness of treatment, and the number of years for which treatment has been available. For example, Alzheimer's disease (senile dementia) is a genetic disease that usually affects people only after the age of 70. In populations where life expectancy is around 50 years, the condition is rare, but as life expectancy rises towards 80 years, both the proportion of elderly in the population and the proportion of the population affected increase.

Table 1.2 shows the approximate birth incidence of the major categories of congenital disorder, and gives an indication of what can be done for them. The commonest congenital disorders, the *malformations*, are due to a disturbance of embryonic development. They can be single or multiple, internal or external, major or trivial, may involve any organ system, and may present at birth or much later. Most occur sporadically without any obvious explanation and are considered of 'multifactorial' origin (p. 106). Relatively few can be traced to a single environmental or genetic cause. The main approaches for prevention are avoiding environmental causes, prenatal screening in order to allow the option of selective abortion, and neonatal surgery. Up to 50 per cent can be successfully corrected (Czeizel and Sankanarayanan 1984).

Table 1.2 Congenital anomalies: classification, and estimate of total disability generated annually in Europe

Category of anomaly	Births/1000	Annual births (number)	Early mortality		Chronic problems		Successful treatment		Main therapeutic needs
			%	Number	%	Number	%	Number	
Congenital malformations	30	408 000	22	89 800	24	98 000	54	220 500	Paediatric surgery
Chromosomal disorders	3.2	43 500	34	14 800	64	28 000	2	900	Social support
Inherited diseases (single gene defects)	7.0	95 200	58	55 200	31	29 500	11	10 500	Medical treatment and support
Total	40	546 700	29	160 000	28	155 500	43	232 900	

Total annual births ≈ 13.6 × 10^6. (Calculations based on Czeizel and Sankanarayanan 1984 and Modell *et al.* 1990)
Treatment is considered successful if it permits an approximately normal length and quality of life, including education, work, and the ability to have a family.

The commonest *chromosomal disorders* are aneuploidies. Most embryos with chromosomal abnormalities miscarry, making miscarriage the most common genetic problem in pregnancy (Alberman and Creasey 1977). The birth incidence of infants with chromosomal aneuploidies depends on maternal age distribution, which fluctuates with time and place (p. 63). Chromosomal abnormalities in the newborn fall into two main groups. *Autosomal aneuploidies* (such as Down syndrome, which accounts for over half of all severe chromosomal abnormalities) are common, severe, and untreatable. They can be detected in the first or second trimester of pregnancy by examining the chromosomes of the fetus, and are an important target of prenatal screening. By contrast, *sex chromosome abnormalities* (like Turner's or Klinefelter's syndromes) cause from moderate to very mild handicap. Some chromosomal abnormalities have almost no physical effect at all (p. 65).

McKusick's catalogue (1988) lists over 4000 *inherited diseases*, but in global terms the handful of common conditions described in Chapters 15–17 are responsible for about half the present mortality due to inherited disease. Treatment is often impossible, and there is usually a 25–50 per cent risk of a recurrence in subsequent pregnancies. Families with affected members need a great deal of help in coping with chronic disability, inconvenient and boring management regimes, the danger of premature death, and fear of recurrence within the family. Prevention depends on identifying healthy carriers and informing them of their genetic risk, and the steps they can take to avoid it. In general, people are highly motivated to avoid having affected children. Carriers can often be detected through the family history (or, increasingly, by genetic screening offered to the whole population) and informed of their risk. A genetic disease is considered a suitable target for population screening if it is common and severe, and carriers can be easily detected.

In pragmatic terms it also proves helpful to distinguish between sporadic and inherited disorders. The *sporadic* group includes most malformations and chromosomal disorders: they occur randomly, rarely, and unpredictably during gamete formation or embryonic development, and can affect any pregnancy. There is usually relatively small risk for relatives. By contrast, *inherited* diseases affect a limited number of families. They are carried by many healthy people, on whom they confer a high and life-long genetic risk. When suitable tests are available, carriers can be identified and risks predicted.

Medical geneticists are involved in differential diagnosis and counselling for all these types of disorders, and the dividing line between the groups is far from clear. For example, some congenital malformations are due to chromosomal disorders and some are inherited; some chromosomal disorders are also inherited.

About 40 per 1000, i.e. half a million of the 13.6 million infants born annually in Europe, suffer from a severe congenital anomaly. Even under

Table 1.3 Established community genetics services

Primary prevention
Of Rhesus haemolytic disease, by post-partum use of anti-D globulin.
Of congenital Rubella, by immunization of girls.
Of congenital malformations: possible prevention of neural tube defects by folic acid and of some others by control of maternal diabetes.
Avoidance of mutagens and teratogens, such as radiation, alcohol, and certain drugs.

Neonatal screening for early treatment
Examination of the newborn for congenital malformations, e.g. congenital dislocation of the hip.
Biochemical tests for phenylketonuria, congenital hypothyroidism, and sickle cell disease

Prenatal screening for fetal abnormalities
For congenital malformations: ultrasound, fetal anomaly scan, maternal serum alphafetoprotein estimation
For chromosomal abnormalities: maternal age, maternal serum factors
For inherited diseases: family history; carrier screening for haemoglobinopathies and Tay-Sachs disease

the best conditions almost 30 per cent of them die perinatally or in child-hood. Over 40 per cent may be successfully treated (mostly by surgery) and almost 30 per cent (mostly with genetic disease) will suffer chronic severe disability, assuming they survive.

A great deal can be done during pregnancy and early infancy to avoid life-long chronic disease. Some serious disorders can be prevented by avoiding environmental risks to pregnancy, many can be diagnosed during pregnancy in time for the option of selective abortion, and some can be detected in early infancy in time for successful treatment. Table 1.3 summarizes currently available 'community genetics services' (Modell *et al.* 1990). Most depend on some form of population screening. The main emphasis in this book is on these services, which cannot be carried out effectively without the participation of primary care workers.

Key references

Baird, P. A., Anderson, T. W., Newcombe, H. B., and Lowry, R. B. (1988). Genetic disorders in children and young adults: a population study. *American Journal of Human Genetics*, **42**, 677–93.

Modell, B., Kuliev, A. K., and Wagner, M. (1991). *Community genetics services in Europe*. WHO Regional Office for Europe, Public Health In Europe Series. (In press.)

Royal College of Physicians (1989). *Report on prenatal diagnosis and genetic screening; community and service implications*. The Royal College of Physicians, London.

WHO (1985*b*). WHO Advisory Group on Hereditary Diseases. *Community approaches to the control of hereditary diseases*. Unpublished WHO document HMG/WG/85.4. May be obtained free of charge from: The Hereditary Diseases Programme, WHO, Geneva, Switzerland.

Part 1

Basic genetics

2. Genetics and DNA

Inheritance

Mendel's famous experiments with peas and other plants in the second half of the nineteenth century established the following basic facts:

1. Inheritance operates through the transmission of defined units that Mendel called elements and were later renamed genes.
2. Each individual has two sets of genes for every characteristic, one set inherited from each parent.
3. When the effects of the two genes affecting a given characteristic are different, one may be 'dominant', i.e. determine the characteristic, while the other may be recessive, i.e. have no evident effect. However, its effects may re-emerge in subsequent generations when two recessive genes for the same feature are inherited together.
4. The sets of genes handed on from the four grandparents are re-arranged in the germ cells of the parents, and so are handed on to their offspring in new combinations.

The term 'Mendelian inheritance' describes the transmission of characteristics according to these basic genetic principles. Figure 2.1 shows that a child inherits one copy of each gene from each parent, and one copy of half the genes of each grandparent.

A person's genetic constitution, i.e. the information contained in their DNA, is described as their 'genotype', while their physical characteristics constitute their 'phenotype'. The genotype sets the limits of what is phenotypically possible, while the phenotype is determined by the interaction of the genotype with the environment. For example, a person may be short because their genotype makes them short, or because, though they have the genotype to be tall, malnutrition or chronic infection in childhood has prevented them from reaching their full genetic potential. In the latter case, the next generation may be taller than their parents if they are reared in a healthier environment. However, when the genotype dictates shortness, children cannot become tall in any environment.

The significance of Mendel's observations was fully appreciated only after the development of microscopical techniques made it possible to examine the structure and behaviour of cells. It was then observed that at cell division, the nuclear material becomes separated into thread-like 'chromosomes' (p. 33), which divide in an orderly manner, so that a complete set is retained

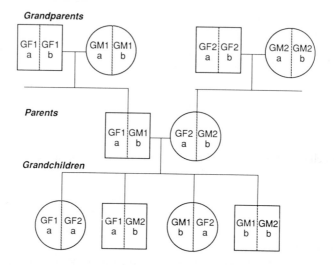

Fig. 2.1 Assortment of grandparents' genes in grandchildren. Three generations are shown. For each homologous gene pair, grandchildren have one gene from one maternal grandparent and one gene from one paternal grandparent. If the genes can be distinguished either directly or by associated 'markers', their inheritance can be tracked through the family. If the grandparents are related, it is possible for grandchildren to inherit two identical mutant genes (see Chapter 9).

in every cell. When gametes (sperm and ova) are formed, the chromosomes can be seen arranging themselves in pairs, interacting, and then separating into single sets so that each gamete contains half the number of chromosomes seen in other tissues. Clearly, the genes are contained in the chromosomes, which constitute the physical basis for Mendel's laws of inheritance, and form the nucleus which controls the cell's activities.

Mendel's general principles describe the inheritance of *genes*, but few human *characteristics* are controlled by only one gene. Hair colour is a rare example of a characteristic that is inherited in a simple Mendelian manner: dark hair is dominant, while fair hair is recessive. Two people with dark hair or one with dark and one with light hair can have light-haired children, but two light-haired people cannot have dark-haired children. Most characteristics such as height or skin colour are determined by many genes, which, though inherited according to Mendel's laws, interact to give quite complex patterns of inheritance. In order to trace inheritance clearly, it is necessary to identify specific characteristics controlled by a gene or family of genes. This is not always easy: biochemical characteristics are often easier to define and trace than physical ones. Inherited diseases have been particularly helpful in increasing our understanding of human genetics because many show clear-cut Mendelian inheritance. In such conditions, the changes observed in the disease must be due to disturbance of a single gene, which may subsequently be identified.

The cell

Human beings, like other organisms, are composed of cells and their products. The components of a typical basic cell and their main functions are shown in Fig. 2.2. There are over 200 different cell types in humans, the cells of different tissues being adapted for specific functions e.g. as nerve, muscle, or liver cells, by having specific structural proteins and modified organelles

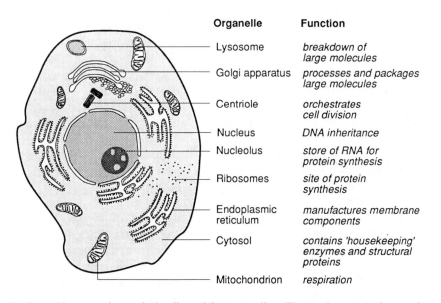

Organelle	Function
Lysosome	*breakdown of large molecules*
Golgi apparatus	*processes and packages large molecules*
Centriole	*orchestrates cell division*
Nucleus	*DNA inheritance*
Nucleolus	*store of RNA for protein synthesis*
Ribosomes	*site of protein synthesis*
Endoplasmic reticulum	*manufactures membrane components*
Cytosol	*contains 'housekeeping' enzymes and structural proteins*
Mitochondrion	*respiration*

Fig. 2.2 Diagram of a typical cell, and its organelles. The *nucleus* controls protein synthesis, and hence all the activities of the cell. The *mitochondria* produce energy by controlling oxidation of food substances. They are most abundant in cells with a high energy consumption such as neurons and muscle cells. *Lysosomes* are responsible for digesting biological materials that have been taken up by the cell. They are most abundant in cells which engulf and digest foreign organisms, such as macrophages and white blood cells, and in the liver. The *endoplasmic reticulum* is a network of surfaces within the cell; it is continuous with the nuclear membrane and cell membrane. The *cytosol* contains soluble proteins, and structural proteins such as the contractile fibres of the mitotic spindle, and actin and myosin in muscle. *Ribosomes* are the machinery for protein synthesis. They may lie loose in the cytosol when they are making soluble proteins, or may be lined up on the endoplasmic reticulum when they are making proteins to be secreted by the cell. The *cell membrane* is a highly organized, essentially fluid structure carrying specific proteins in particular ratios. It can be invaginated into, and nipped off within, the cell. New areas may be formed by portions of the endoplasmic reticulum opening out onto the cell surface. The *Golgi apparatus* accumulates proteins to be secreted by the cell.

(intracellular structures). Some cell types and specific proteins involved in the inherited diseases discussed later are listed in Table 2.1.

Cells are made up of the basic biological components — carbohydrates, fats, proteins, and nucleic acids — together with water and salts. These large molecules are built up by metabolic processes from simpler basic structural units, sugars, fatty acids, amino acids, and nucleotides (Table 2.2).

Nucleic acids and proteins

Nucleotides have a unique role as subunits of the deoxyribonucleic acid (DNA) and ribonucleic acid (RNA) which store genetic information. The DNA in the nucleus directs the synthesis of proteins, which control the structure and activity of each cell, and so of the entire organism. The structure, and hence the function, of each protein is determined by the sequence of nucleotides in a specific section or sections of DNA, defined as the gene(s) for that particular protein. Though each nucleus contains the complete genetic code for the whole organism, as a rule only the genes relevant for the function of the particular cell are active. When a gene is active, its genetic code is copied into 'messenger' RNA. This passes out of the nucleus into the cytoplasm, and provides a blueprint for the proteins produced there. DNA is replicated by enzymes, which are themselves proteins whose structure has been determined by DNA. Thus DNA, through its control of the cell machinery, has the fundamental property of living matter, the power of self-replication.

The amino acid sequence of each protein is determined by the DNA of its coding gene(s). Some genes and so some proteins are much longer than others. Proteins characteristically fold up (spontaneously) into a specific shape determined by their amino acid sequence. The fact that there are 20 different amino acids offers an almost unlimited range of variation in sequence and therefore in shape. Many proteins are made of subunits controlled by different genes, e.g. adult haemoglobin is composed of two α and two β chains. Proteins have many roles in the cell: some proteins such as collagen or elastin are primarily structural, but others, for example enzymes, are extremely active biochemically.

All cells share some basic biochemical and reproductive processes, so all contain basic structural proteins and a set of 'housekeeping' enzymes including respiratory enzymes for generating energy in the form of adenosine triphosphate (ATP), metabolic enzymes for handling sugars, fats, proteins, and nucleic acids, and enzymes for DNA replication and protein synthesis. In addition, most cells manufacture specific proteins concerned with their function in the organism (see Table 2.1). Every cellular process is carefully controlled by sophisticated mechanisms. It seems that very little happens randomly, and there are numerous backup or 'fail safe' mechanisms. Nevertheless, the processes of the living cell are so interdependent that a distur-

Table 2.1 Some cells and specific proteins involved in inherited diseases

Specific cell	Specific protein	Disease
Muscle cell (myoblast)	Dystrophin	Duchenne muscular dystrophy
Red blood cell	Haemoglobin (O_2 transport)	Thalassaemia
	G6PD (enzyme)	G6PD deficiency
Fibroblast ⎫ Osteoblast ⎭	Collagen (structural protein)	Osteogenesis imperfecta
Epithelial cell (sweat gland, lung, pancreas)	CFTR (ion transport)	Cystic fibrosis
Nerve cells	β hexosaminidase A (catabolic enzyme)	Tay–Sachs disease
Liver cells	Phenylalanine hydroxylase (enzyme)	Phenylketonuria
Liver and other cells	Factor VIII of the clotting cascade	Haemophilia A

Table 2.2 Basic building blocks of living organisms

Basic unit	Polymer	Main functions
Sugars (many types)	Carbohydrates	Energy storage and metabolism
Fatty acids (many types)	Fats	Cell structure (membranes etc) Energy storage
Amino acids (22 types)	Proteins	Intra- and extra-cellular structures (collagen, elastin) enzymes
Nucleotides (5 types)	DNA	Material of inheritance: controls structure and function of cell
	RNA	Mediates between DNA and proteins (also nucleoside triphosphates, ATP, GTP, and UTP are involved in energy supply and control)

bance of one step can have far-reaching consequences.

In many tissues, cells are continually dying or being worn away, and must be constantly replaced from less differentiated 'stem cells'. Often the daughter cells become 'terminally differentiated', i.e. they lose the power to divide and become completely dedicated to their specific function. For example, red blood cells lose their nucleus and intracellular organelles, and live only for four months; keratinized cells of the skin live for about 4 weeks, and the lining cells of the gut live for only 48 hours. Other terminally differentiated cells such as nerve cells retain all their organelles, and may live as long as the individual does.

Structure of DNA

The electron micrograph in Fig. 2.3 shows that DNA consists of two paired chains of nucleotides. Fig. 2.4 shows the structure of the DNA. Each nucleotide is composed of one of four possible nitrogenous bases (adenine (A), guanine (G), cytosine (C), or thymine (T)) plus a sugar (deoxyribose) and a phosphate ion. The phosphate ions form bridges between sugar molecules to make a 'backbone' of alternating sugar and phosphate molecules, from which the bases protrude sideways like the teeth of a comb. Two such strands are paired face to face along their whole length, by hydrogen bonds formed between their bases. Each nucleotide base can accept only one other base as a partner: C must pair with G (by three hydrogen bonds), while A must pair with T (by two hydrogen bonds). Therefore the one strand determines the sequence of the other. The pairing of the chains leads the double strand of DNA to adopt a spiral structure in which the two chains twist around each other to form the 'double helix'. Individual hydrogen bonds are weak, but the multitude involved in the double helix produces very firm bonding between the strands. However, in the laboratory, conditions can easily be adjusted to weaken the hydrogen bonds and separate the DNA strands.

The order of the nucleotide bases on those DNA sequences that represent genes determines the amino acid sequence of the corresponding protein. Three nucleotide bases (a 'triplet') code for one amino acid, the order of the

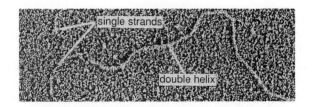

Fig. 2.3   An electron micrograph of a DNA strand. The thicker double strand can be seen to the right. To the left the two complementary strands have become separated. Reproduced from Alberts *et al.* (1989) with permission.

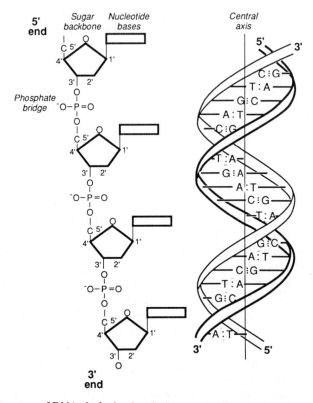

Fig. 2.4   Structure of DNA. *Left*: the chemical structure of DNA. Organic bases (purines or pyrimidines) are attached to sugar molecules, which are linked by phosphate bridges. The asymmetrical attachment of the phosphate to the sugars gives each strand of the DNA molecule a direction, running from a 3′ to a 5′ end. This is the direction in which DNA is synthesized. *Right*: the double helix. DNA strands will pair with each other only if they have a series of exactly 'complementary' matching mucleotide bases. The two paired strands run in opposite directions (one 3′ to 5′, the other 5′ to 3′). Awareness of the directional nature of DNA is important for understanding the diagnostic methods described in Chapter 10.

three bases determining which of the 20 amino acids available is to be used in protein synthesis. There are also specific triplets which code for 'start' and 'stop' signals. The 'genetic code' is shown in Fig. 2.5.

The *genes* are the sections of DNA that hold the code for the amino acid sequence of a protein or a protein subunit (though not all genes code for proteins: some code for RNA, see below). Since the code is in the bases which are on the inside of the double-stranded DNA molecule, the helix must to some extent be opened for the code to be read, either in the normal course of protein synthesis, or in the laboratory by molecular biologists. In nature

First position (5′ end)	Second position				Third position (3′ end)
	U	**C**	**A**	**G**	
U	UUU } Phe UUC UUA } Leu UUG	UCU UCC } Ser UCA UCG	UAU } Tyr UAC UAA } Stop UAG	UGU } Cys UGC UGA Stop UGG Trp	U C A G
C	CUU CUC } Leu CUA CUG	CCU CCC } Pro CCA CCG	CAU } His CAC CAA } Glu N CAG	CGU CGC } Arg CGA CGG	U C A G
A	AUU AUC } Ileu AUA AUG Met and start	ACU ACC } Thr ACA ACG	AAU } Asp N AAC AAA } Lys AAG	AGU } Ser AGC AGA } Arg AGG	U C A G
G	GUU GUC } Val GUA GUG	GCU GCC } Ala GCA GCG	GAU } Asp GAC GAA } Glu GAG	GGU GGC } Gly GGA GGG	U C A G

Fig. 2.5 The genetic code. Each triplet of three bases codes for one amino acid. There are more permutations of the four coding bases (64) than there are amino acids (20). Therefore each amino acid is coded for by more than one triplet. It is the last base in the triplet that tends to vary, the first two being more important in specifying the corresponding DNA base. This is known as 'degeneracy' of the genetic code. Abbreviations for amino acids are as follows (in order of appearance in the figure). Phe = phenylalanine, Leu = leucine, Ileu = isoleucine, Met = methionine, Val = valine, Ser = serine, Pro = Proline, Thr = Threonine, Ala = Alanine, Tyr = Tyrosine, His = Histidine, Glu N = Glutamine, Asp N = Asparagine, Lys = Lysine, Asp = Aspartic acid, Glu = Glutamic acid, Cys = Cysteine, Trp = Tryptophan, Arg = Arginine, and Gly = Glycine. UAA, UAG, and UGA are signals to stop. AUG codes for both methionine and START — all protein synthesis starts with a methionine which may later be removed.

this is done by enzymes that duplicate DNA (DNA polymerases) or copy it to RNA (RNA polymerases). Specific sequences in the non-coding DNA around the gene allow the RNA polymerase to bind, and may determine in which tissue a gene is expressed. One example is the expression of the insulin gene in pancreatic islet cells.

The sugar–phosphate links in DNA are not symmetrical: the phosphate bridge is attached to the 5th carbon atom of one sugar and the 3rd of the other. Therefore the sugar–phosphate backbone has a *direction* (5′ → 3′), and the complementary DNA strand runs in the opposite direction (3′ → 5′).

DNA can be synthesized or decoded only in the $5' \rightarrow 3'$ direction. Both strands act as templates for DNA replication, but only one strand functions as the coding strand for RNA and protein synthesis: the other ensures that the molecule remains stable and can be reproduced. The non-coding strand cannot be decoded because it is like mirror-writing and does not contain the specific base sequences that allow the attachment and operation of RNA polymerase. Even if the non-coding strand were decoded, it would produce a 'nonsense' protein — a grossly abnormal protein with no function in the cell. Abnormal proteins are usually rapidly degraded by cellular scavenging mechanisms.

Only a small fraction of total human DNA is occupied by sequences that code for protein. By far the greater part consists of non-coding *intergene segments*. Fig. 2.6 shows that genes themselves include both coding sequences (*exons*) and non-coding sequences (*introns*). Genes for different proteins differ greatly in size and in number of introns and exons. Table 2.3 summarizes present information on the genes involved in the inherited diseases in Table 2.1.

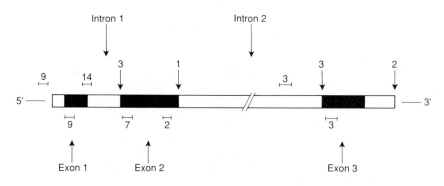

Fig. 2.6   Diagram of the β-globin gene. This relatively small gene (1.5 kilobases long) contains three exons (shaded boxes) and two introns (open boxes). Non-coding flanking DNA is shown as a single line. A short section of DNA at either end of the exons is included in the mRNA. The numbers and sites of the common mutations that can cause β-thalassaemia are indicated.

Protein synthesis

RNA is an essential intermediary for most protein synthesis. It is very similar to DNA except that the backbone sugar is ribose, not deoxyribose, the base thymine (T) is replaced by uracil (U), and it is not arranged in a double helix. It is able to pair with complementary sections of DNA and RNA. It directs the production of proteins according to the code laid down in DNA. RNA is coded for mainly by nuclear genes; some originates from mitochondrial DNA.

Table 2.3 Some genes for disorders discussed in this book

Disorder	Gene	Location on chromosome	Gene length (kb[1])	No. of exons	Length of mRNA (kb[1])	Length of protein (amino acids)	Molecular weight (Daltons)
α thalassaemia	α globin	16 p 13.1	1.3	3	.56	141	16 200
β thalassaemia Sickle cell disease	β globin	11 p 14	1.5	3	.58	146	15 800
G6PD deficiency	G6PD	X q 28	17.5	13	2.3	515	59 265
PKU	Phenylalanine hydroxylase	12 q 22	90	13	2.4	451 xx	51 872
Tay–Sachs disease	β N-acetylhexaminidase A	15 q 22	35	14	2.6	529	50 000
CF	CFTR	7 q 31	250	24	6.5	1480	168 138
Haemophilia A	Factor VIII	X q 28	186	26	9	2332	240 000
Duchenne muscular dystrophy	Dystrophin	X p 21	2000	60	14	4000	400 000

[1] 1 kb = 1 kilobase = 1 thousand base pairs of DNA

Protein synthesis is summarized in Fig. 2.7. When a gene starts to be expressed, RNA polymerase first separates the two DNA strands and then copies the sequence of the coding strand to form complementary single-stranded RNA within the nucleus. This process is called 'transcription'. The introns, and parts of the non-coding DNA at both ends of the gene, are also copied. The complementary RNA is then further processed. The introns are removed by 'splicing' (i.e. their ends are pinched together and the loop so formed is cut out), and a stabilizing 'cap' and a long chain of adenosine nucleotides (a poly-adenosine tail) is added to the two ends (Fig. 2.8). The resulting messenger RNA (mRNA) enters the cytoplasm through the pores in the nuclear membrane.

Though mRNA determines the type of protein synthesized by the cell, it makes up only a small proportion of cytoplasmic RNA (Fig. 2.9). Most is ribosomal RNA, which is produced by special regions of the chromosomes and stored in the nucleolus. In the cytoplasm it associates with numerous proteins to form ribosomes — relatively huge particles that contain the machinery for protein synthesis.

Once in the cytoplasm, mRNA associates with the ribosomes, which are the site of 'translation' of the genetic code. Small transfer RNAs (tRNAs) carry individual amino acids to the mRNA–ribosomes complex; one end binds a specific amino acid while the other has three bases complementary to the corresponding triplet of the mRNA. Ribosomes move along the mRNA one triplet at a time. They bind the tRNA that is complementary to

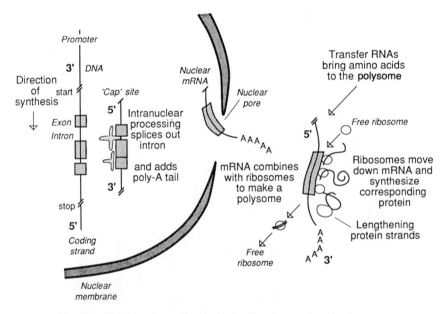

Fig. 2.7 Outline of protein synthesis. Details are given in the text.

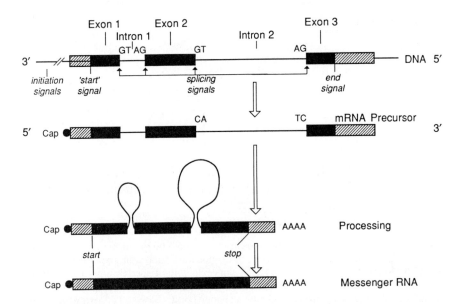

Fig. 2.8 Formation of messenger RNA. The full length of the gene, including 5′ and 3′ flanking sections, is transcribed to produce messenger RNA precursor. This is then processed extensively in the nucleus: a cap is added at the 5′ end, and introns are removed by 'splicing'. The beginning and the end of the introns have characteristic bases, GT and AG in the DNA, CA and TC in the RNA. Splicing enzymes recognize these sequences, bring them together, and pinch out the intervening RNA loop. Finally a tail of adenosine nucleotides up to 200 residues long (the poly-A tail) is added to the 3′ end of mRNA.

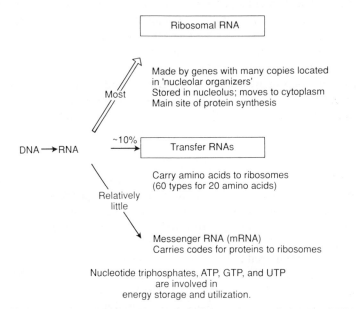

Fig. 2.9 Nucleic acids and nucleotides and their functions. All forms of RNA are involved in different aspects of protein synthesis.

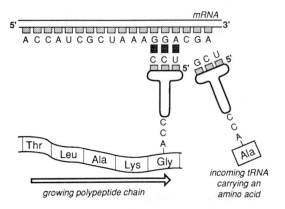

Fig. 2.10   Diagram showing how proteins are synthesized on the ribosomes. Nucleotides in mRNA are matched three at a time to complementary triplets on particular transport RNAs (tRNAs). The other end of a tRNA molecule holds a specific amino acid. When RNA matching occurs, this amino acid is brought into a position where it can be added to the end of the growing protein chain by a peptide bond. (Based on Alberts *et al.* 1989.)

that triplet, hold the specific amino acid it is carrying, release the tRNA, and join adjacent amino acids with peptide bonds to make proteins (Fig. 2.10). A procession of ribosomes moves along the mRNA, spinning out lengthening protein chains behind it. When a ribosome reaches a stop signal, it detaches and the protein chain is released. All protein synthesis occurs in the cytoplasm; even nuclear proteins are made in the cytoplasm and returned to the nucleus.

Since cells maintain numerous proteins, they contain many different mRNAs, mostly in a few copies only. However, when a protein is synthesized in bulk, there are many copies of the corresponding mRNA: for example, most mRNA in developing red blood cells is globin mRNA, and in mature B lymphocytes (p. 79) most of it is immunoglobulin mRNA.

Chromosomes

Germ cells are 'haploid' i.e. they contain only a single set of chromosomes, but the vast majority of plants and animals are diploid, i.e. contain two complete chromosome sets, one derived from the female and one from the male parent. Corresponding chromosomes from each parent are described as *homologous*.

Somatic human cells contain two haploid DNA complements of 3 000 000 000 base pairs each, which are estimated to include 50–100 000 genes. If stretched out fully, one DNA complement would be 9 metres long, and if the sequence of its base pairs were printed it would make a book with more than half a million pages. To make DNA controllable and accessible, it is packaged in a highly organized way in to 23 chromosomes, each containing a single long unbroken strand of DNA consisting on average of 130 million base pairs, and including 2–4000 genes. The DNA in each chromosome is associated with basic nuclear proteins at a number of different levels

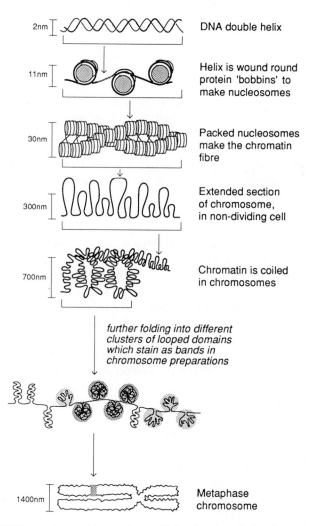

Fig. 2.11 DNA associates with proteins and is intricately coiled to form chromosomes. Steps in coiling are shown from above downwards, starting with the DNA double helix. This is first wound round disc-shaped proteins to make 'nucleosomes', which appear like beads along the string of DNA. The nucleosomes are then packed to form chromatin fibres. The chromatin fibres are then looped, and coiled to form a chromosome. Chromosomes may be further folded to form looped domains which are packed together to make the bands that can be seen in stained chromosome preparations. (Based on Alberts *et al.* 1989).

of complexity, as shown in Fig. 2.11. A great deal of biological energy is devoted to maintaining the structure of the chromosomes, and controlling their form and function.

A human diploid chromosome complement is shown in Fig. 2.12. Individual chromosomes can be recognized more easily if they are stained with dyes that give them a characteristic banded appearance. Fig. 2.13 shows a diagram of a single set of 23 banded human chromosomes, at the metaphase of mitosis. Though their lengths are different, each chromosome has a centromere, which is responsible for ordered movement during cell division, with an arm on either side. The longer is called the Q arm; the shorter (P) arm is sometimes insignificantly short. Both males and females have 22 pairs of homologous chromosomes (autosomes) but the 23rd pair differs between the sexes. Females have two homologous X chromosomes which carry many genes: males have one X chromosome paired with a very small Y chromosome which carries very few genes, but determines male sex. The X and Y chromosomes are called sex chromosomes for this reason.

The human genome and individual variation

The site a gene occupies on a chromosome is called its *locus* (i.e. place). The genes are (nearly) always arranged in the same order on a given chromosome in the whole population, and it is possible to work out this sequence and so to 'map' the chromosomes. Everyone has two loci for a given gene, one on each homologous chromosome. At about 60 per cent of loci there is little or no variation in protein product between individuals; and therefore we know that there is little or no variation in these parts of the DNA between individuals. The structure of the chromosomal DNA-binding proteins (histones) for example, is identical in most organisms. However, differences between genes at a minority of loci account for differences between individuals. If that were not so, we would all be identical.

At some loci, a number of 'alleles' occur within the population. Alleles are alternative possible genes that usually code for slightly different versions of the same protein. They differ by one or more nucleotides — for example, Hb S and Hb A are alleles that can occupy the β globin locus. These variant forms were originally produced by mutation. When an allele is common, like the different blood groups, it is called a polymorphism. When it is rare and unusual it is considered a mutant form. When the protein products of alleles differ in electric charge or some such property, they can be distinguished in the laboratory.

People can be homozygous (carry two identical alleles) or heterozygous (carry two different alleles) for a gene with two or more allelic forms, but no-one can have more than two of the alleles available in the population for one characteristic. For example, there are three possible alleles of an enzyme that controls the synthesis of the common blood group substances (A, B, and

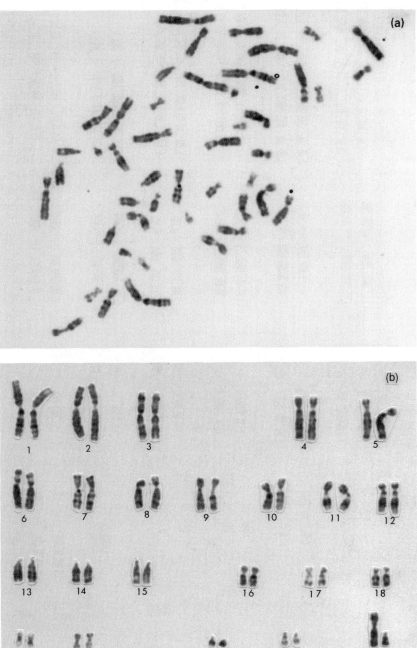

Fig. 2.12 (a) A typical preparation of human chromosomes. To analyse the preparation, individual chromosomes are often cut out from a microphotograph of the preparation and arranged as in (b). Photographs kindly provided by Dr Joy Delhanty

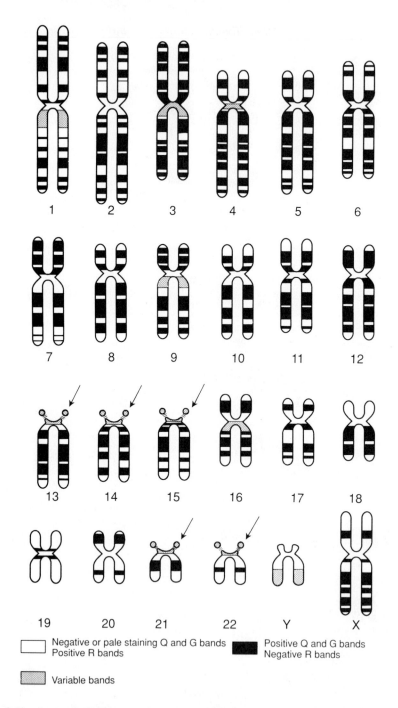

Fig. 2.13 A standard diagram of human chromosomes, to show their characteristic banding patterns. The arrows above chromosomes 13, 14, 15, 21, and 22 show the nucleolar organizers. (Modified from Paris Conference 1971.)

H (= O)), and everyone has a selection of any two of these (AA, AB, AO, BO, OO (p. 336)). At the locus for the plasma protein α-1 antitrypsin (p. 228) there are 30 common alleles, each differing from the others by one amino acid and separable by electrophoresis, and everyone has a selection of two of these. There is an immense number of possible alternatives at the four closely related loci of the tissue histocompatibility antigen complex (the HLA system, p. 84), making for a very wide range of individual variation at this site.

Changes (mutations) in DNA that alter a coding sequence and modify a protein are often harmful. If they reduce the ability of the affected person, or their offspring, to survive and reproduce, they are eliminated from the population by natural selection. As a result, coding sequences remain relatively constant, i.e. are 'conserved'. The non-coding DNA sequences in inter-gene segments tolerate more mutations without bad effects, so these regions are much more variable than the coding sequences. Such harmless individual variations have turned out to be very useful in genetic diagnosis (p. 137).

Many genes for producing ribosomal RNA are clustered together on the very short P arms of chromosomes 13, 14, 15, 21, and 22 (Figure 2.13). These regions are called 'nucleolar organizers' because they cluster together to form the nucleolus. The number of RNA genes in these clusters is extremely variable between individuals, probably changing continually with the generations due to unequal crossing over (p. 39). Even the microscopic appearance of these regions can differ considerably between individuals.

Chromosomes are most tightly coiled when they go through cell division; in non-dividing cells they become considerably unwound. However some portions of the chromosome, on either side of the centromere, remain tightly coiled even in non-dividing cells, and are called 'heterochromatin'. These regions contain many relatively simple serially repeated DNA sequences called 'minisatellite' DNAs because they resemble the repeated satellite sequences that code for ribosomal RNA, on the short arms of chromosomes mentioned above. They are not coding sequences but may have a structural function. In mammals, they are composed of multiples of sections of DNA a few hundred nucleotides long. The numbers of repeats and their arrangement varies so much between individuals that differences can often be seen in chromosome preparations. These sequences are important practically because their number and variability provides the basis for the 'genetic fingerprinting' that is now often used to identify individuals for forensic, legal, and medical purposes.

About 30 per cent of the human genome is made up of such highly repeated sequences, some located around the centromere, the rest dispersed throughout the genome. About 60 per cent is occupied by introns and inter-gene segments with a unique sequence (i.e. a sequence not duplicated elsewhere in the genome). Only 1–2 per cent consists of sequences coding for protein.

Though there are at least 5 million single base differences between individuals — more than 1 per 1000 base pairs — and thousands of differences in the number of units in the repeating sequences, the DNA from all human beings has a similar basic arrangement and organization. The concept of the 'human genome' refers to this characteristic pattern of the DNA of the human species.

Is all human DNA in the chromosomes, and is all chromosomal DNA human?

It seems that the borderline between organisms may be less distinct than we, with our emphasis on individuality, often imagine. Mitochondria have their own genome — a double-stranded molecule of DNA — which resembles the DNA of bacteria in being circular, in some of its sequences, and in not containing introns. It produces some mitochondrial proteins, while others are produced by chromosomal genes. This naturally suggests that some cellular organelles arose from symbiosis between more primitive organisms and bacteria, which finally became incorporated as a constitutional part of the cell. Rather similar arrangements can be seen today e.g. in giant clams in coral reefs that harbour symbiotic algae within their cells.

It also seems that the DNA of some viruses (retroviruses like the AIDS virus) can move in and out of animal chromosomes. Hitherto we have been aware of their presence only when they cause disease, but DNA methods for identifying characteristic sequences are revealing apparent integrated virus DNA that causes no symptoms. Up to 1 per cent of the normal mouse genome may be made up of incorporated viral fragments: the corresponding proportion of the human genome is unknown. We do not know whether this 'hitch-hiking' DNA is really useless, or does any harm, or even offers evolutionary possibilities.

Dosage effects, X chromosome inactivation, and sex determination

The normal expression of genes is very sensitive to what are known as 'dosage' effects. In most cases, two active genes are necessary for the full development of a characteristic. If one of the two copies of a gene is non-functional because of a mutation, the deficiency may cause disease (such a mutation is dominant) or may be compensated for by the remaining gene and other back-up mechanisms (such a mutation is recessive mutation). Similarly, if additional copies of a gene are present, unbalanced gene activity can sometimes cause disease. The importance of dosage effects is underlined by the biological arrangements that exist to ensure equal and balanced expression of the genes carried on the X chromosome in males and females.

Since males have only one X chromosome and females have two, a single dose of genes carried on the X chromosome is clearly sufficient to ensure normal functioning. Equal expression of these genes in males and females is ensured by inactivation of one X chromosome in females from a very early stage in development. The inactivated chromosome becomes tightly condensed and can be seen in the nucleus of female cells as the 'Barr body'; this makes it possible to establish genetic sex, for example by examining desquamated epithelial cells from buccal smears, even though they are not dividing. In each cell, the X chromosome either of maternal or of paternal origin is inactivated, and once inactivation has occurred it is permanent; all that cell's descendants have the same X chromosome inactivated. In the placenta and embryonic membranes, it is the paternal X chromosome that is inactivated, but in the embryo itself, there is no statistical difference between the numbers of cells with an inactivated maternal X chromosome and those with an inactivated paternal X chromosome. The phenomenon of X chromosome inactivation, which is called Lyonization after Mary Lyon who discovered it, has important implications for diagnosis of female carriers of X-linked genetic disorders (p. 245).

The Y chromosome is responsible for determining maleness. In the absence of a Y chromosome, the fetus will develop into a female (e.g. in XO or XXX syndromes); in its presence, the fetus develops into a male (e.g. in XXY syndromes). The gene determining maleness is on the short arm of the Y chromosome. In a few families, crossing over between the X and Y chromosomes has transferred the male-determining gene to the short arm of the X chromosome. XX individuals carrying such a chromosome are males, while XY individuals whose Y chromosome has lost the male-determining gene are female (McLaren 1990, 1991). It is thought that the male-determining gene exerts its effect by causing the supporting cells of the gonad to start developing into a testis, while in the absence of the gene product, the gonad will become female. The gonad then determines whether the germ cells that migrate into it develop as ova or sperm, and also controls the development of other sexual characteristics by secreting the appropriate sex hormones.

It appears that certain characteristics are expressed differently in an offspring depending on whether identical genes or segments of chromosomes were inherited from the mother or the father (genomic imprinting). An example is a severe form of early onset Huntington's disease which can occur if the Huntington gene originated from the father (Hall 1990).

Key references

Alberts, B., Bray, D., Lewis, J., Raff, M., Roberts, K., and Watson, J. D. (1989). *The molecular biology of the cell*, (2nd edn). Garland Publishing Inc., New York.

Gelehrter, T.D. and Collins, F.S. (1990). *Principles of medical genetics*. Williams and Wilkins, Baltimore.

Watson, J.D., Hopkins, N.H., Roberts, J.W., Steitz, J.A., and Weiner, A.M. (1988). *The molecular biology of the gene*, (4th edn). The Benjamin Cummings Publishing Company Inc., California.

3. Cell division: an opportunity for change

The growth, functioning, and division of most living cells falls into a common pattern described as the *cell cycle*, summarized in Fig. 3.1. The observed cell cycle time for different tissues ranges from 1–100 days or more, most of the variation being in the length of time the cell spends in the G_1 (gap) phase, when it is not involved in cell division, but is busy with its normal functions. An orderly resumption of the cycle is essential for normal growth: uncontrolled cell growth may result in cancer (p. 71). Various protein factors control cell growth and division.

Mitosis

Mitosis is the process by which the diploid cells of the body (somatic cells) divide and multiply. Before division can occur, the contents of both the nucleus and the cytoplasm must double in amount. In the G_1 phase of the cell cycle, the chromosomes are extended in the nuclear sap, attached at specific points to the nuclear membrane and the nucleolus, and RNA polymerases move along the chromosomes synthesizing the mRNAs needed for cell growth and function.

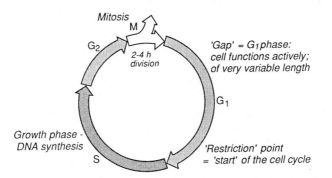

Fig. 3.1 The cell cycle. M represents mitosis. The G_1 ($=$ gap) phase is a period of protein synthesis and cell growth. Its duration varies depending on the type of cell. In rapidly dividing cells, G_1 is short: is slowly dividing cells like liver cells it can be extremely long. At a particular point towards the end of the G_1 phase there is a 'start' signal, and the cell then has to proceed through the rest of the cycle. In the S (synthesis) phase, DNA is replicated. The G_2 phase follows and leads to mitosis.

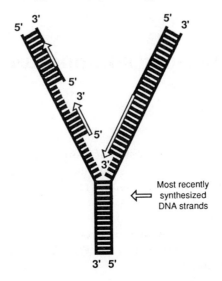

Fig. 3.2   DNA replication. DNA polymerase can synthesize DNA only in the 3′ to 5′ direction of the template, so the enzyme moves steadily down one strand synthesizing a continuous complementary strand in the 5′ to 3′ direction. As it does so, it opens up the double helix and exposes the opposite strand, on which synthesis occurs in the opposite direction in short lengths. The mechanism is known as 'back-stitching'.

When the signal for cell division is given, the G_1 phase ends and the S (synthesis) phase, in which the DNA replicates, begins. To replicate, the two complementary strands of DNA must first separate, so that DNA polymerase can use them both as templates for making new DNA strands (Fig. 3.2). The process cannot be simple, because of the coiled form and immense length of the DNA strand. The strand is worked on by a battery of enzymes that cut it at specific points and separate the strands so that new complementary DNA can be made on shorter, more manageable sections. It is thought that it is while DNA is in the single-stranded state that it is most vulnerable to the small mutations that are so important in evolution and genetic disease. Histones and other DNA-binding proteins are synthesized in the cytoplasm at the same time, and enter the nucleus and combine with the replicated DNA to make new chromosomal material. At the end of the S phase and during the G_2 phase, the nucleus is essentially tetraploid ($4n$ where n is the haploid number of chromosomes). It contains twice the diploid amount of chromosomal material, each chromosome consisting of two identical sister 'chromatids' united only at the centromere.

In mitosis (M phase), summarized in Fig. 3.3, the chromosomes first condense (i.e. become tightly coiled, as in Fig. 2.11) and then separate into two identical diploid sets; the cytoplasm is then divided, and two daughter nuclei

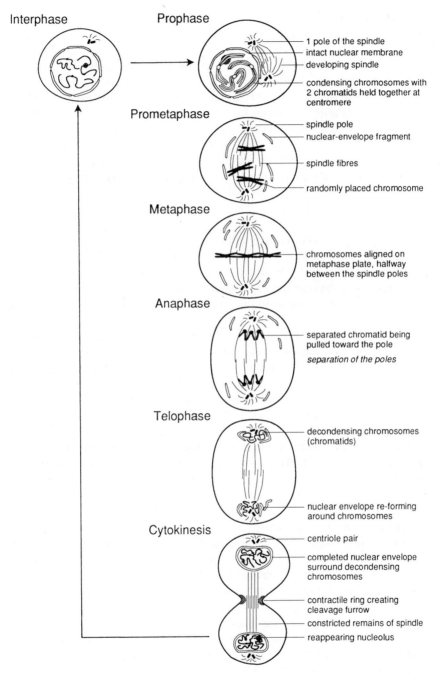

Fig. 3.3 Mitosis.

are reconstituted. Though two daughter cells are genetically identical, they do not necessarily have the same phenotype: one may be a stem cell dedicated to further reproduction which re-enters the cell cycle, and one may become terminally differentiated. Chromosomes can be studied only when cells are actively dividing, because only then are they sufficiently condensed to be seen clearly and distinguished from one another. This is why cells (e.g. from amniotic fluid) must be grown in culture before any chromosomal abnormality can be identified. It also explains why chromosomes are usually shown as two chromatids attached to a centromere, their form just prior to separation at the anaphase of mitosis (see Figs 2.11 and 2.12), rather than as single elongated strands, the form in which they usually exist.

Meiosis

As a rule, the germ cells (which will ultimately give rise to haploid gametes, either ova or sperm) appear very early in embryonic development, often a surprisingly long way from their final location. In humans they arise in the yolk sac outside the rear end of the embryo, and then migrate into the genital ridges at the back of the abdominal cavity, where gonads develop ready to receive them.

Once in the gonads, the germ cells multiply by mitosis like other somatic cells, until the time comes for the gametes to be formed. Since these contain only one haploid set of 23 chromosomes, the germ cells' chromosomes must be separated into single sets. This is achieved by another type of cell division called meiosis (meaning 'reduction' in Greek), in which one chromosome replication is followed by two cell divisions, so halving the number of chromosomes in each cell. Meiosis occurs only in germ cells and only in the gonads. In the male testis it continues throughout adult life, but in the female most ova are formed before birth, and remain in the ovary suspended half-way through meiosis until ovulation starts at the onset of puberty. Meiosis is compared with mitosis in Fig. 3.4.

Exactly as in mitosis, the chromosomes have already replicated before the start of meiosis: the cell contains a tetraploid ($4n$) amount of DNA, and each chromosome consists of two chromatids joined together only at the centromere. However, when prophase starts, each chromosome pairs up with its homologous chromosome and becomes attached to it, probably using specific recognition sites scattered along their length. This specific pairing during meiosis is one of the most important phenomena in genetics (see below). The centromeres of each chromosome pair become physically joined, each double centromere being attached by one spindle fibre to each centriole. The fused centromeres separate again at anaphase, and two single sets of whole chromosomes, each still consisting of two chromatids, are separated by the contracting spindle fibres, so that in the daughter cells the number of chromosomes is halved (to $2n$).

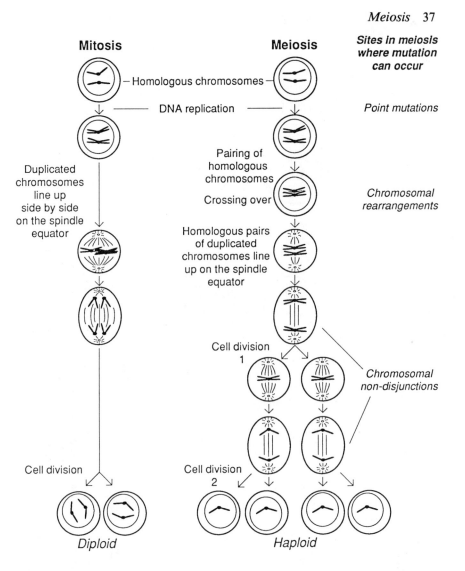

Fig. 3.4 Comparison of mitosis and meiosis. For simplicity only one pair of homologous chromosomes is shown. The main points where a mutation may occur are indicated.

The second meiotic division closely resembles mitosis, without any intervening G_1, S, or G_2 phase. The chromosomes simply divide at the centromere, and the two sets of chromatids are separated into two daughter cells that now contain only a single ($1n$) set of chromosomes.

In sperm, the second meiotic division immediately follows the first, but in ova it occurs only after fertilization. Figure 3.5a and b show the different

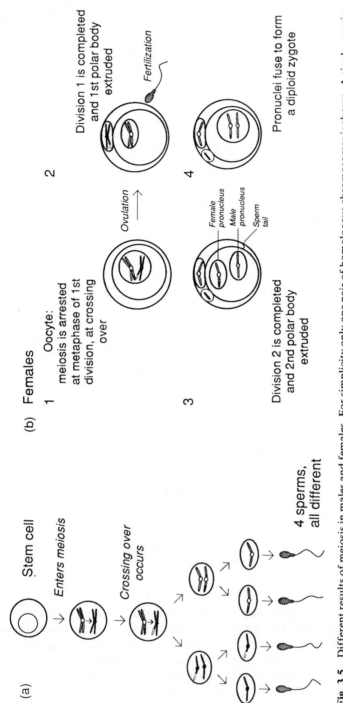

Fig. 3.5   Different results of meiosis in males and females. For simplicity only one pair of homologous chromosomes is shown. A single crossing over event has taken place. (a) In males, a single meiosis leads to four sperm, each genetically different from the others. (b) In females, the first meiotic division happens at ovulation (1 → 2), and one set of chromosomes is extruded in the first polar body. The second division is completed only after fertilization (3 → 4), when the second set of chromatids is extruded in the second polar body. Only one out of the four products of meiosis is a potentially viable germ cell.

results of meiosis in male and female germ cells. In the male, all four daughter cells become sperms, but in the female only one of the four becomes the mature ovum, the other three being discarded as the first and second polar bodies, at ovulation and fertilization respectively.

The main function of meiosis is to produce variation between individuals as a basis for natural selection. All four haploid sets of chromosomes produced during meiosis differ from each other and from the parent cell, because chromosomal material of maternal and paternal origin is reassorted (recombined). In order to explain this, it is necessary to discuss more fully the events at metaphase and anaphase of the first meiotic division.

The first possibility for re-assortment arises because, although one set of chromosomes was originally derived from the mother and one from the father, chromosomes of maternal and paternal origin are mixed randomly in the sets that go to each pole in the first meiotic division. If one parent has contributed two different specific characteristics controlled by genes situated on different chromosomes, there is a 50 per cent chance that they will be separated during meiosis. With 23 pairs or chromosomes, many different combinations can be generated simply by re-assorting whole parental chromosomes. However, more variation is needed to allow species the flexibility to survive competitively with others.

Much greater possibilities for recombination are provided by 'crossing over' between chromatids of homologous chromosomes while they are paired during the first meiotic division. The physical union of homologous chromosomes occurs not only at the centromere but also at intervals along the chromatids, where special homologous regions of corresponding chromatids also become physically fused. The paired chromatids then break at equivalent sites and reunite with each other, so exchanging homologous sections and creating new combinations of genes on the same chromatid. Crossing over at meiosis is the rule rather than the exception. It occurs at several sites on each chromatid, so all four chromatids produced during a given meiosis are different from each other, and from those produced by other meiotic divisions in the same individual.

When two genes (for different characteristics) are far apart on the same chromosome, crossing over in the intervening length of chromosome is so likely that there is a random (50 per cent) chance that they will be separated during meiosis, and end up in different germ cells. However, if they are close together on the same chromosome, there is a greater chance that they will be passed on together, the chance increasing the closer together they are. Conversely, genes for characteristics that are consistently inherited together must be close together on a chromosome. For instance, the gene for ankylosing spondylitis is very closely associated with the gene for HLA type B27, and the two are usually transmitted together. Genes for characteristics that are usually inherited together are described as 'linked' (Fig. 3.6).

Observation of linkage is a basic genetic tool for mapping the arrangement

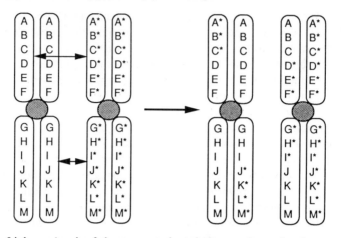

Fig. 3.6 Linkage. A pair of chromosomes in meiotic metaphase, showing two crossing over events (indicated by double headed arrows) (usually there are more). Genes with identifiable differences on the two chromosomes are marked A to M on one chromosome and A* to M* on the other. Crossing over has occurred between C and D on two chromatids, and between I and J on two chromatids. The result is four different chromatids. Note that genes which are close together on the chromosome (B and C, or G and H) are unlikely to be separated by crossing over, and are likely to be inherited together. Genes that are further apart (A, M or G, M) are more likely to be separated.

of genes in the genome, and in genetic research and diagnosis. On the (not strictly correct) assumption that the probability of crossing over in meiotic metaphase is the same all along the chromosome, the frequency of recombination (i.e. separation of inherited characteristics that were previously together) can be used to measure how far apart two genes are on a chromosome. This is how 'maps' of the chromosomes are constructed. Figure 3.7 shows some results of using modern linkage methods for the human X chromosome. Similar maps for all the human chromosomes are updated annually (McKusick 1988).

In view of the very large number of genes on each chromosome, the differences between alleles, the rearrangements produced by crossing over at meiosis, tandem repeats, and the many differences in non-coding regions, two chromosomes of a given type are never identical. That is why they are described as *homologous chromosomes*. Each person has two sets of homologous chromosomes, each one of which is slightly different from the corresponding chromosome of anyone else, including their parents, though all variations (except those due to new mutation) must be inherited from, and traceable to, the parents. This makes it possible to trace the parental origin of particular chromosomes, e.g. when an additional chromosome 21 is present in a child (or fetus) with Down syndrome.

Genetic recombination during meiosis ensures that the grandparents'

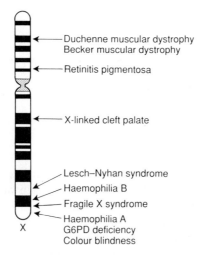

Fig. 3.7   Map of the human X chromosome, showing the loci of some genes involved in selected inherited disorders (Based on McKusick 1988).

genetic contributions are re-assorted in so many ways that each gamete is unique. An adult human male may produce up to 1.5 thousand million sperms in his lifetime, and each one is different from the others—i.e. he could have fathered a quarter of the present human population, and no two offspring would be identical. Females produce far fewer gametes, but each is also unique.

Key reference

Alberts, B., Bray, D., Lewis, J., Raff, M., Roberts, K., and Watson, J. D. (1989). *The molecular biology of the cell*, (2nd edn). Garland Publishing Inc., New York.

4. Fertilization and embryonic development

Germ cells

Female germ cells multiply rapidly within the developing ovary at around 10 weeks of embryonic life, enter meiosis, and remain arrested in the first meiotic division (with the chromosomes already paired and apparently in the process of crossing over) until the onset of puberty. Most of the 3 million or so primary oocytes present in the ovary at birth subsequently degenerate. Only about a quarter of a million survive until puberty, and of these, only a few hundred are discharged during a woman's reproductive life. In humans then, individual oocytes remain arrested in the first meiotic division for 13–45 years. This long pause may account for the fact that meiosis does not always continue normally when it is resumed.

Before each ovulation, one or two oocytes grow rapidly. At ovulation they complete their first meiotic division. One (diploid) set of chromosomes is extruded in the first polar body and the other is retained as the 'female pronucleus' (see Figure 3.5b). The ovum halts in prophase of the second meiotic division unless fertilization occurs.

Sperm are produced in the testes continually from the onset of puberty, one spermatogonium dividing into four spermatids, each with a haploid chromosome complement (Figure 3.5a).

Fertilization

Fertilization usually takes place in the fallopian tube (Fig. 4.1). Many sperm may penetrate the transparent zona pellucida (the thick, non-adhesive capsule that surrounds the ovum), but the first to dock on to the cytoplasmic membrane of the ovum starts a rapid surface change which blocks the entry of further sperm. As soon as the sperm nucleus (the male pronucleus) enters the ovum, the female pronucleus returns to the surface near the first polar body to complete its second meiotic division, and extrudes a single (haploid) set of chromatids in the second polar body. The haploid female pronucleus then sinks back and fuses with the haploid male pronucleus to form a diploid zygote. All its cytoplasm and organelles and half its chromosomes are derived from the ovum; the sperm contributes only chromosomes and a centriole.

The chromosome complements of egg and sperm appear almost identical,

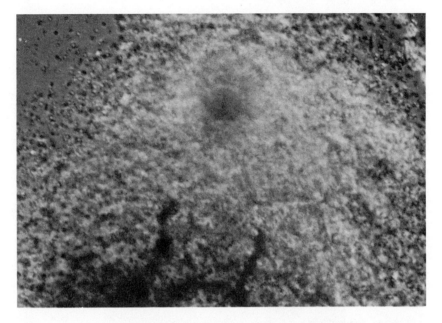

Fig. 4.1 Fertilization, showing that the egg is surrounded by a cloud of cells (the cumulus), discharged with it from the ovary. The egg is surrounded by sperm, which must penetrate the cumulus as well as the zona pellucida to reach the egg. (Photograph kindly supplied by Dr Paul Serhal.)

but several large regions are 'imprinted', i.e. their expression is permanently modified as a result of passing through the male or female germ line. For example, several sets of genes in the chromosome set inherited from the father are activated to promote development of the placenta, while other sets of genes from the mother are activated to promote development of the embryo. The effects of paternal and maternal imprinting are balanced in normal embryos: mutations leading to imbalance contribute to genetic disease (p. 64).

Development of the embryo

The first steps in embryonic differentiation occur at a very early stage. The fertilized ovum, still inside the zona pellucida, has functioning 'house-keeping' systems but most other genes seem to be switched off. In the first four days after fertilization, it travels down the fallopian tube, dividing as it does so to form a hollow ball of cells, the blastocyst (Fig. 4.2). This has a slight thickening in one side, the inner cell mass. A signal arising at a very early stage determines that this will become the embryo, while the

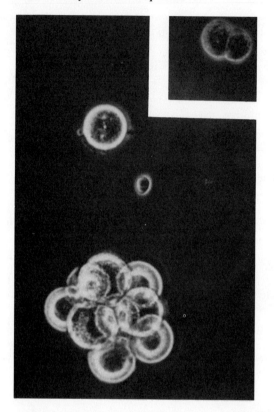

Fig. 4.2   A human morula. The fertilized egg has divided to produce a cluster of cells. The zona pellucida has been removed in the laboratory. At this stage if one cell is removed, the embryo may continue normal development. If a diagnosis can be reliably performed on a single cell, pre-implementation genetic diagnosis may be possible. (Photograph kindly supplied by Dr Peter Braude.)

more numerous cells that form the surface of the blastocyst will become trophoblast and ultimately chorion and placenta. These cells begin secreting human chorionic gonadotrophin (hCG) in sufficient quantities to be detectable in the maternal serum: the function of hCG is to maintain the corpus luteum until the embryo is well implanted and the placenta takes over secretion of hormones that maintain the pregnancy. This early secretion of hCG allows the diagnosis of fertilization, even if no pregnancy ensues.

By the fifth day post-fertilization, the ovum enters the uterus and hatches from the zona pellucida. The trophoblastic cells are highly active. They multiply extremely quickly and invade and digest the endometrium of the uterus, so that the pre-embryo physically burrows beneath it to become implanted. Their external surface lacks the usual cell surface (HLA) antigens (p. 84) that could provoke rejection by the mother's immune system. The trophoblast is responsible for providing nutrients to the developing embryo by digesting maternal tissue, and secretes fluids into the early embryonic sac so that it expands. It develops finger-like chorionic villi with an inner core that produces red blood cells and blood spaces which link up to form a vascular system connected to the developing embryo.

As the chorionic cells originate from the zygote, they reflect the genetic make-up of the embryo, and can be used for diagnostic sampling and genetic studies of the embryo. The chromosomes of chorionic villus material can be studied relatively quickly, because the cells are dividing very rapidly (see p. 150).

By about 5 weeks of embryonic life (7 weeks from the last menstrual period), the developing embryo becomes surrounded with a second membranous sac, the amnion, into which it secretes amniotic fluid. The two membranes—amnion and chorion—become closely apposed and almost inseparable by about 10 weeks of embryonic life. Amniotic fluid contains cells originating from the embryo, and is often sampled to obtain cells for prenatal diagnosis (p. 149).

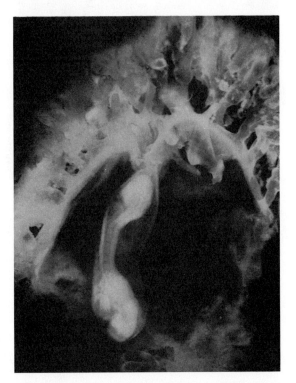

Fig. 4.3 A human embryo at 5 weeks from the onset of the last menstrual period, i.e. at 3 weeks of embryonic life. At this stage the fetal heart beat can already be seen on ultrasound examination. (Reproduced from Boyd and Hamilton, 1970, with permission)

Table 4.1 Timing of development of main organ systems in human embryos

Gestation (weeks)	Length of embryo (mm)	Digestive system	Urogenital system	Vascular system	Nervous system	Body form
4.5	1.5	Gut not distinct from yolk sac	Allantois present	Blood islands on chorion and yolk sac	Neural groove indicated	Flat embryonic plate
By 8	12.0	Identifiable tubes of the liver, stomach, and intestines. Yolk sac stalked and extra-amniotic	Cloaca subdivided, sexless gonad and genital tubercle prominent. Mesonephros reaches caudal limit	Heart has general final form. Haemopoiesis in the liver	Three primary flexures of the brain present. Bulging cerebral hemispheres	Head dominant in size. Limbs recognizable. Upper jaw components separate. Lower jaw halves fused.
By 12	40.0	Intestine has withdrawn from umbilical cord and assumed internal position. Anal canal formed	Kidney formed and can excrete. Bladder present. Testes and ovary distinguishable. Vaginal sac forming	Main blood vessels in correct place. Lymphatic system developed. Red cells without nuclei predominant	Cerebral cortex has typical cells. Visible olfactory lobes. Final internal structure of spinal cord	Erect head and well formed limbs
By 18	112.0	Liver secretes bile, pancreatic ducts present, as are gastric and intestinal glands. Meconium present	Final shape of kidneys, uterus, and vagina; testes still undescended. Prostate and seminal vesicles present	Haemopoiesis in bone marrow and spleen. Condensed heart muscle	Brain has developed final morphology. Cervical and lumbar enlargement of spinal cord	Body has outgrown head which has 'human' face with typical eyes, nose, and ears and developing hair. Sex of fetus obvious. Spontaneous active muscle

The timing and stages in the differentiation of organ systems are summarized in Table 4.1 and the early embryo is illustrated in Figs 4.3 and 4.4. Important points to note are:

1. Most differentiation occurs early; much of later intra-uterine life is dedicated to growth.
2. The effects of disturbance on the embryo depend on the stage of development at which the disturbance occurs.
3. The membranes and amniotic fluid are very important for prenatal diagnosis.

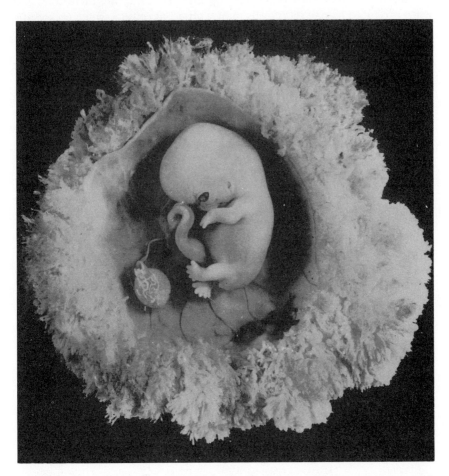

Fig. 4.4 A human fetus at 8 weeks from the LMP (last menstrual period), i.e. at 6 weeks of embryonic life. Chorionic villus sampling can first be done at this stage. The chorion has been opened, the amnion removed, and the yolk sac, often conspicuous on the ultrasound picture, can be seen in the bottom left-hand part of the space. (Reproduced from Boyd and Hamilton (1970), with permission.)

Embryonic differentiation

Individuals start from a single cell but end as a fully differentiated adult. This means that despite the fact that each cell contains the full programme for embryonic development within its nucleus, as cell division proceeds cells in different regions become different from each other, according to a predetermined plan whose pattern is laid down in the DNA. It seems that a limited number of switching genes are involved, which switch on or switch off whole sets of genes in different parts of the organism at each stage of development. Some factors influencing differentiation are diffusible, and when produced by one set of cells induce adjacent cells to switch on specific sets of genes. When groups of genes are switched on other groups are irreversibly switched off, so that as differentiation progresses the number of alternative options available to each cell or tissue type is steadily reduced (Fig. 4.5).

A mutation in a switching gene can have marked effects on a developing embryo, e.g. in *Drosophila* it can lead to the replacement of an entire antenna by an entire leg. The DNA sequence of some such *Drosophila* switching genes (homeotic genes) has been defined (Ingham 1988). Human

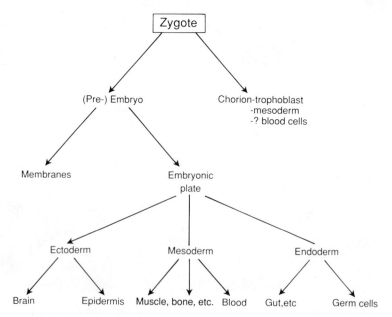

Fig. 4.5 The 'cascade' of embryonic development. Each step is irreversible. Switching genes are involved in the selection of each step. The same switching gene may have different effects in different tissues and at different stages of development. Only a few main steps are shown.

and other animal genomes include almost identical homeotic genes, also concerned with switching embryonic cells from one developmental path to another. Mutations in homeotic genes may be responsible for some congenital malformations such as Hirschprung's disease, which is due to developmental absence of ganglion cells from the lower bowel wall. It is the commonest cause of neonatal intestinal obstruction, occurring in 0.2/1000 babies.

Present evidence suggests that about 100 genes are involved in the control of growth during embryonic development. Some produce growth-promoting factors that cause localized proliferation of cells for a given period of time, and so, for instance, may cause a limb bud or the lower jaw to form. Others suppress or activate these growth-promoting genes. Disturbance of a gene that produces a growth factor, or of one that promotes or suppresses it, can lead to cancer (p. 75). These genes were first identified through research on cancer and so have been called 'proto-oncogenes' i.e. genes that can develop into cancer-generating genes. We prefer to call them growth-controlling genes.

In the embryo, many proteins, growth and differentiation factors, and even structures such as the primitive kidneys or the notochord appear at a particular stage, perform a transient function in organizing the next step in development, and then disappear. In the adult no evidence may remain of the presence of genes determining such transient characteristics, but sometimes a trace of the gene product may be detectable, as in the case of the embryonic plasma protein alphafetoprotein (AFP). Though AFP is very useful in prenatal diagnosis of congenital malformations, its function in the embryo is still unknown. It is manufactured in the yolk sac in early embryos and later in the liver, reaching a maximum concentration of about 10 per cent of plasma proteins at about 14 weeks' gestation. The level then decreases again. It has practically disappeared by the time of birth, and is virtually absent in adults. During pregnancy, a very small amount crosses the placenta and enters the maternal circulation. An increased maternal serum AFP level can indicate a break in the continuity of the fetal skin caused by a neural tube defect or a malformation of the gastrointestinal tract (p. 331).

Fetuses with Down syndrome show a slight lag in the pace of normal development. They tend to be smaller than average for the gestational age, and factors produced by the fetus or placenta that can be detected in the mother's serum may be higher than usual when they are declining in normal pregnancies (human chorionic gonadotrophin), or lower when they are rising in other pregnancies (AFP, unconjugated oestriol). Such differences can be very useful for screening for Down syndrome in pregnancy (p. 334).

Many other proteins besides AFP have a developmental sequence. For example, human haemoglobins are composed of four protein chains, two α-globin chains and two β-globin chains. Several different genes in the α-globin gene cluster on chromosome 16, and the β-globin gene cluster on chromosome 11 produce different globin chains. These genes turn on and off sequentially during embryonic developmental to produce a sequence of five haemoglobin types, with appropriate oxygen-carrying capacity for the embryonic stage at which they appear (Table 4.2). Adults have about 97 per cent Hb A and about 2.5 per cent Hb A$_2$; less than 1 per cent of Hb F remains. Many other proteins go through similar developmental sequences, so it seems possible that as much of the human genome is concerned with embryonic development as with maintaining extra-uterine life.

Congenital malformations

Disturbances in the sequence of development are sometimes the result of a chromosomal or genetic abnormality (e.g. Down syndrome is a result of trisomy 21) or of an environmental insult such as a viral infection, but many occur apparently spontaneously. The commonest serious abnormalities leading to death or chronic disease occur in the development of the central nervous system, of the heart, and of the kidneys. Internal malformations are often undetected at birth, and may have insidious effects leading to late-presenting chronic disease. For example, coarctation of the aorta can lead to hypertension and left ventricular failure. The following are among the commonest types of congenital malformations.

Abnormalities of the central nervous system

(1.5–6 per 1000 births in Europe)
About 3 weeks after fertilization, the neural groove develops in the dorsal ectoderm of the embryo; it deepens and the margins fuse to form the neural tube. Failure of the anterior end to close at this stage causes anencephaly and encephalocele; a defect of the posterior end leads to spina bifida and meningomyelocele (p. 257). The incidence of these severe disorders differs considerably in different populations, possibly because of differences in maternal diet and genetic susceptibility.

Abnormalities of the heart

(4–8 per 1000 births in Europe)
The four-chambered heart develops from a single tubular organ. Most cardiac abnormalities in infants at birth are due to some degree of failure of development of the divisions of the heart. Major abnormalities include a one-chambered ventricle, or transposition of the great vessels so that deoxygenated blood is pumped into the peripheral circulation and oxygenated to the lungs. The commonest defects of all arise from incomplete closure of the ventricular or atrial septum, or the ductus arteriosus. Atrial septal defects and patent ductus arteriosus are universal at birth, because both are

Table 4.2 Developmental sequence of haemoglobin

Embryonic age	Type of haemoglobin				
	Embryonic		Fetal	Adult	
	Gower 1_2 $\zeta_2\epsilon_2$	Gower 2 $\alpha_2\epsilon_2$	HbF $\alpha_2\gamma_2$	HbA $\alpha_2\beta_2$	HbA$_2$ $\alpha_2\delta_2$
4–6 weeks	++	+	−	−	−
5–9 weeks	+	++	+	−	−
8 weeks–birth	−	−	++	<10%	−
>6 months of age	−	−	<1%	*ca.* 97%	<3%

++ : the dominant Hb at each stage; + : some of this Hb present; − : none of this Hb present characteristic of the fetal circulation. After birth the ductus usually closes as a result of the stress stimulus of birth.

Abnormalities in the urinary tract

(9–16 per 1000 in Europe)
These abnormalities include failure of the kidney to develop at all, fusion of the kidneys, and abnormalities of the ureters. Many cause obstruction of the urinary tract and lead to hydronephrosis. As with other congenital malformations, the milder abnormalities are the commonest. Many, particularly when they are unilateral, are not diagnosed at birth, but may cause the child to present with a urinary tract infection at some time during infancy or childhood. One of the commonest causes of urinary tract infections in children is a disturbance of the angle at which the ureter enters the bladder, so that urine refluxes up the ureter when the bladder contracts.

Similar developmental problems can occur in any other organ system, and cause abnormalities ranging from the lethal to the mildly cosmetic. Most families include one or two individuals with at least a mild disturbance of morphogenesis.

Twins

There is evidence from ultrasound examination in early pregnancy that about 5 per cent of pregnancies start as twins (Robinson and Caines 1977). However, in the majority the second sac regresses early in the first trimester, so only about 1 per cent of pregnancies end with live-born twins.

Identical (monozygotic) twins develop from a single egg fertilized by a single sperm and so are genetically identical. They occur in about 1 per 200 pregnancies, irrespective of the mother's ethnic group. Twinning may arise from the very beginning by independent development of the first two blastomeres, or relatively late, by the formation of two embryos on a single embryonic disc. Congenital abnormalities including neural tube defects are more common in identical twins than in singletons or non-identical twins. This may be because identical twins have a common placenta and sometimes a common circulation, which may allow them to interfere with each other's development; e.g. fetus-to-fetus transfusion can lead to complications of polycythaemia in one twin and anaemia in the other.

Non-identical (dizygotic) twins develop from two ova fertilized at the same time, and are genetically no more closely related than normal siblings. The natural frequency of non-identical twinning varies with ethnic group, from about 2 per cent of pregnancies among Africans, through about 1 per cent of pregnancies for Europeans, to about 1 in 400 among orientals. The frequency of non-identical twinning is now rising in Europe as a side-effect

Table 4.3 Some conditions in which twin studies have helped to establish the importance of genetic factors in disease

Condition	Concordance[a] in monozygotic twins / Concordance in dizygotic twins
Club foot	10
Congenital dislocation of the hip	14.8
Cleft lip and palate	6.4
Cancer	1.6
Coronary heart disease	2.4
Diabetes mellitus	4.9
Atopic disease (asthma, eczema)	11
Hyperthyroidism	15.1
Psoriasis	4.7
Gall stones	4.1
Tuberculosis	2.3
Sarcoidosis	5.9

(Based on Vogel and Motulsky 1986)
[a] When one of a twin pair has the condition, concordance is the percentage frequency with which the other twin develops the same condition

of the treatment of infertility with drugs that induce multiple ovulation, and *in vitro* fertilization programmes.

Twins have long been of particular interest for geneticists. For example, information on the relative contribution of genes and environment to disease can be obtained by comparing the extent to which identical and non-identical twin pairs suffer from similar disorders (Table 4.3). Such comparisons have been very useful in defining the hereditary component in many normal characteristics, as well as in disease.

Key references

Alberts, B., Bray, D., Lewis, J., Raff, M., Roberts, K., and Watson, J.D. (1989). *The molecular biology of the cell*, (2nd edn). Garland Publishing Inc., New York.
Walbot, V. and Holder, N. (1987). *Developmental biology*. Random House, New York.

5. Mutation, evolution, and sex

Differences in DNA sequence between individuals have originated at some time in the past through spontaneous changes in the DNA called *mutations* (derived from the Latin, 'to change'). Mutations are essential for evolution: they occur in all living organisms, and generate variation so that the species as a whole can respond flexibly to environmental changes. Genetic disease is a by-product of these evolutionary mechanisms.

Most mutations are accidents of DNA replication that occur during the S phase of the cell cycle preceding cell division. Most occur in association with mitosis in somatic cells. **Somatic mutations** can affect only the person in whom they occur. Fortunately, most have little or no effect or kill the cell in which they arise, but some mutations in growth-controlling genes lead to benign tumours or to cancer (see Chapter 6). Somatic mutations arising during development of the embryo may also be responsible for some congenital malformations.

This chapter is concerned with mutations that arise in germ cells. **Germinal mutations** affect all the cells of any offspring who inherit them, and may be passed on to their descendants. Many germinal mutations have no observable effect on the offspring. Some are neutral and may become polymorphisms (p. 26) or die out because the people carrying them do not reproduce for some unrelated reason. Others are harmful. A mutation that leads to disability or early death is likely to prevent reproduction of the individual and will therefore not be passed on to the next generation. However, harmful *recessive* mutations and those causing disease late in life may be passed on from parent to offspring. Yet other mutations are beneficial (e.g. confer protection against an infection) and may become established in whole populations.

Table 5.1 indicates the stages at which mutations are known to occur during the development and maturation of germ cells (see also Figure 3.4). The broad types of mutations and some of their effects are summarized in Table 5.2. Those encountered in clinical practice include the following types:

1. Changes in the sequence of DNA (point mutations). These are more common in the germ cells of males than in those of females.
2. Chromosomal rearrangements, such as deletions, inversions, and translocations (p. 65). These often arise from errors during crossing over at the first meiotic division.
3. Chromosomal non-disjunctions (p. 61). These arise from errors in chromosome separation during either the first or the second meiotic divi-

Table 5.1 Points in germ cell development where mutation can occur

Process	Result	Comment
DNA replication	Point mutation	Point mutations are more common in males. In females, only about 20 mitoses occur in the germ line, all in the fetus. In males, >700 occur in a lifetime
	Rearrangements[a]	Possibly more common in male germ cells
Meiosis 1	Aneuploidies	More common in female germ cells
Meiosis 2	Aneuploidies	More common in female germ cells

[a] Rearrangements include deletions, insertions, translocations, inversions, etc. They may be small (involving 1 or 2 base pairs) or large enough to be visible in the light microscope.

Table 5.2 Types of mutations and their possible effects

1. **Point mutation** — Many neutral, some pathological

2. **Chromosomal rearrangements**
   Small duplication or deletion — Resemble point mutations, but probably more are pathological
   Major visible in microcope — Affect many genes; rarely neutral

3. **Non-disjunction**
   Of whole chromosome set

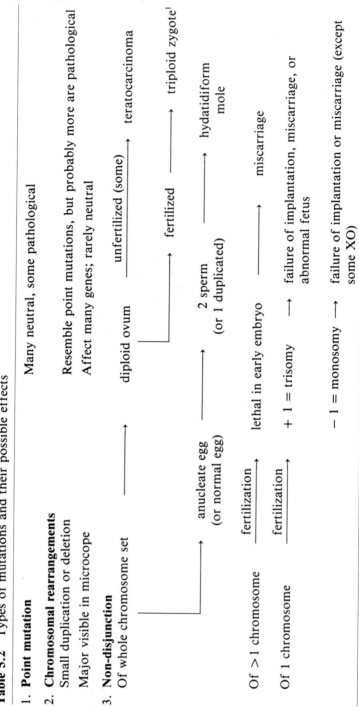

anucleate egg (or normal egg) $\xrightarrow{\text{fertilization}}$ diploid ovum

diploid ovum $\xrightarrow{\text{unfertilized (some)}}$ teratocarcinoma

diploid ovum $\xrightarrow{\text{fertilized}}$ (2 sperm (or 1 duplicated))

$\xrightarrow{}$ hydatidiform mole

$\xrightarrow{}$ triploid zygote[1]

Of >1 chromosome $\xrightarrow{\text{fertilization}}$ +1 = trisomy $\longrightarrow$ lethal in early embryo

Of 1 chromosome $\xrightarrow{\text{fertilization}}$ +1 = trisomy $\longrightarrow$ failure of implantation, miscarriage, or abnormal fetus $\longrightarrow$ miscarriage

−1 = monosomy $\longrightarrow$ failure of implantation or miscarriage (except some XO)

[1] Leads to miscarriage, still-birth, or neonatal death.

sion. They lead to aneuploidy or polyploidy and are more common in the germ cells of females than in those of males.

Point mutations

The faithfulness with which DNA is copied is quite remarkable. Its double-stranded structure is largely responsible for maintaining its stability. Even in resting DNA, random molecular movements often cause one base to be replaced by another, but the altered base causes a distortion in the double helix because it cannot pair correctly with its partner on the opposite strand. Mismatches are usually detected and corrected by special 'proof-reading' and repair enzymes that recognize distortions in the DNA, and restore the original sequence by cutting out the mismatched base and replacing it with the only one that will fit correctly. However, some changes that happen during DNA replication (possibly while DNA strands are briefly un-paired) are copied into the new complementary strand: such changes will be perpetuated.

The known types of point mutations are summarized in Table 5.3. A point mutation may involve replacement of one base pair by another, or deletion or insertion of one or a few base pairs. Since the coding sequences of genes occupy a relatively small proportion of the entire length of human DNA, most mutations occur in non-coding sequences, though there do seem to be certain 'hot spots' where they are particularly likely to occur.

A point mutation in a coding section of DNA may change a nucleotide triplet, but does not always change the amino acid it codes for, because several triplets code for each amino acid (see Figure 2.5). Even when an amino acid is changed, if the properties of the new amino acid resemble those of the old there may be no change in protein function. However, changes that alter the charge or shape of a protein usually affect its function. An insertion or deletion of anything more or less than three base pairs (or a multiple of three base pairs) causes a 'frame-shift' that throws the entire code of the mRNA out of phase, and results in a 'nonsense' protein. Some point mutations change a triplet coding for an amino acid to a 'stop' signal, so that ribosomes detach themselves prematurely from mRNA, leaving short, non-functioning protein fragments. Such inappropriate proteins and fragments are quickly digested within the cell.

Probably all possible point mutations have occurred more than once in the history of the human race. Certainly most inherited diseases have arisen independently several times, as a result of several different mutations.

Why are some mutations dominant while others are recessive? In view of the flexibility of biological systems, it is understandable that halving the quantity of particular protein often causes few problems for carriers. It is less easy to understand why a single dose of an abnormal gene can cause serious problems, but the example of osteogenesis imperfecta is illuminating.

Table 5.3 Types of point mutation

Change in the DNA		Result
Substitution		
Of 1 base pair in coding sequence		Altered protein
In stop or start sequence		No protein or shortened or lengthened protein
In non-coding sequence	not controlling gene function	No change
	controlling gene function	Change in rate of protein synthesis
Deletion in coding sequence		
Of 1 or 2 base pairs		Frame-shift, usually producing a 'nonsense' protein
Of 3 base pairs		One amino acid absent (e.g. cystic fibrosis)
Larger	involving coding sequence	Often no protein
	involving non-coding sequence	Effect variable
Unequal crossing over		Compound gene product

This is a dominantly inherited condition in which the bones are fragile and liable to spontaneous fractures from birth. It is caused by a mutation in one of the collagen genes. The collagen molecule is a triple helix containing two α-1 chains and one α-2 chain. A mutation in the gene encoding the α-2 chain would affect 50 per cent of collagen molecules, and might be tolerated. In contrast, if there was a mutation in gene encoding the α-1 chain, at least 75 per cent of collagen molecules would contain either one or two abnormal chains, and this is highly likely to cause disease (Sykes 1987).

Mutations in coding sections of the DNA often disturb normal functioning and so are gradually eliminated by natural selection because individuals with the mutation die before they have had as many children as others. Therefore most coding sequences of DNA are 'highly conserved', i.e. show little variation between individuals. When every detail of a protein's structure is vital for its function, its coding DNA is highly conserved. The sequences of growth-controlling and homeotic genes (p. 48) in species as far apart as mice and men are almost identical, reflecting their fundamental importance in embryonic growth and differentiation. Some non-coding sections of DNA that control the switching on and off of genes are also highly conserved. When a highly conserved region of DNA is found, it must have an important function that depends on its precise sequence, even if there is no clue to what this function might be. By contrast, the 20 amino acid long fibrinopeptides that are discarded from fibrinogen when clotting begins will tolerate almost any change in their amino acid sequence without change in function.

Since mutations in many non-coding regions of DNA have little effect, such regions can be relatively variable. Many neutral mutations are present in non-coding DNA and are simply incidental polymorphisms that mark the descendants of the first person in whom they appeared. This reservoir of polymorphisms in non-coding sequences is very helpful in locating important genes, and achieving carrier and prenatal diagnosis (p. 137).

The human mutation rate

Information on the human mutation rate is very limited. What evidence we have is largely based on the observed birth incidence of people with dominant or X-linked genetic conditions that have arisen as new mutations, i.e. in a previously unaffected family. Most such studies show that mutations arise more often in the father's germ cells than in those of the mother, and that the mutation rate increases with increasing paternal age (Fig. 5.1) (Vogel and Rathenberg 1975). Though there are some exceptions (in Duchenne muscular dystrophy for example, new mutations seem to arise equally often in sperm and ova (Winter and Pembrey 1982) the human mutation rate seems to depend more on paternal age than maternal age. This is probably because the number of mutations accumulated in a given cell line should reflect the

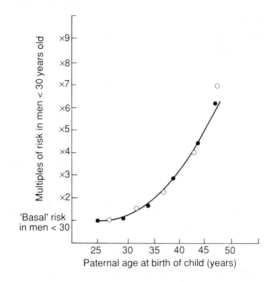

Fig. 5.1   Curve showing increase in risk of a new mutation at birth, in relation to the age of the father. It is only possible to give relative figures for the rise in mutation rate. The rate that would prevail if all fathers were less than 30 years old is taken as the base-line ($=1$). The rate in subsequent 5 year age groups is calculated as a multiple of the base-line, to give a 'relative paternal-age-related mutation rate' (Modell and Kuliev 1990).

number of cell divisions in its ancestry. In females each oogonium (stem cells giving rise to ova) undergoes about 21 mitotic divisions during embryonic life. There are no more divisions, so there should be little possibility of further point mutations thereafter. By contrast, the possibility of new mutation is always open to male germ cells since males produce sperm throughout adult life. The chance of a new mutation appearing in offspring would be expected to rise in a linear fashion with paternal age, but the observed rate rises faster than this (Figure 5.1). Some additional factor must be involved: it has been suggested for example that DNA repair mechanisms become less efficient with increasing age (Vogel and Rathenberg 1975).

Some conditions caused by new dominant mutations, like achondroplasia (p. 237) (Stevenson 1957) and osteogenesis imperfecta (p. 57) (Carothers *et al.* 1986) are highly visible and can be detected relatively reliably at birth. They are known as 'sentinel phenotypes' because observed changes in their incidence might, in principle, be used to monitor changes in the human mutation rate associated with nuclear accidents or other environmental hazards (Czeizel and Kis-Varga 1987).

Chromosomal aneuploidies

Errors during meiosis that change the number of chromosomes are very common. Chromosomal abnormalities include aneuploidies, in which one or more chromosomes are present in too many, or too few copies, and poly-ploidy, in which whole extra sets of chromosomes are present (e.g. 3n = 69 or 4n = 92 chromosomes in humans). Aneuploidies can arise at the first or second meiotic division when paired homologous chromosomes fail to sepa-rate from each other and move to the same pole of the meiotic spindle instead of to opposite poles (non-disjunction) (Fig. 5.2). Table 5.4 sum-marizes information on the frequency of chromosomal abnormalities in gametes, and after conception (Plachot *et al.* 1987).

In early pregnancy, several chromosomes may be involved, but severe abnormalities cause failure of fertilization, implantation or embryonic development. A small minority of abnormal fetuses with less severe abnor-malities, e.g. an additional copy (= trisomy) of a small autosome (21, 18, or 13) or an additional or missing sex chromosome, or with mosaicism (see below) survive the first trimester of pregnancy. Trisomy involving a larger autosome containing hundreds of extra copies of genes inevitably causes severe disturbance of development and function, but abnormalities of the sex chromosomes in general have much less severe consequences.

More than 50 per cent of spontaneous abortions are due to a chromosomal abnormality that has not interfered with implantation, but does not allow normal development (Alberman and Creasey 1977). Both the incidence of spontaneous abortion and the birth incidence of infants with chromosomal aneuploidies increase with maternal age (Fig. 5.3); in 80 per cent of cases the extra chromosome is of maternal origin (Mikkelsen, 1988). The reason for the increase in chromosomal non-disjunction with maternal age is not

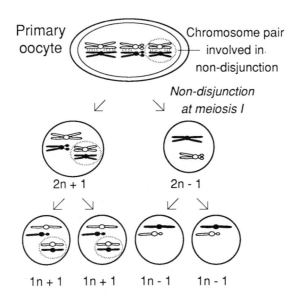

Fig. 5.2 Formation of aneuploid gametes by non-disjunction.

Table 5.4   Natural selection against chromosomal abnormalities

Stage of development		% chromosomally abnormal	
Gametes	Oocytes	32	
	Sperm	8	
Fertilized ova		38	24% of chromosomally abnormal ($=9\%$ of total) fail to cleave
Pre-implantation embryos		29	48% of chromosomally abnormal ($=14\%$ of total) fail to implant
Implanted embryos		15	>90% of chromosomally abnormal ($=14\%$ of total) abort spontaneously
Newborn babies	Total	0.6	
	Clinically significant	0.2–0.3	

(Based on Plachot *et al.* 1987)

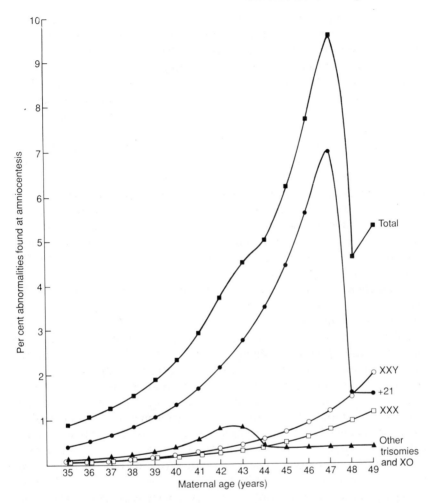

Fig. 5.3 Maternal age is strongly related to the incidence of chromosomal aneuploidies. The figure shows data for fetuses at amniocentesis. The incidence at birth is about 30% lower, due to late miscarriage of chromosomally abnormal fetuses. Separate curves are shown for the different classes of aneuploidy. XO is one of the commonest abnormalities at conception, but most affected fetuses abort early, so it is relatively uncommon at amniocentesis. This is particularly so after the age of 40, accounting for the decrease in incidence in 'other trisomies and XO' shown above. The apparent fall in incidence of fetuses with trisomy 21 after maternal age of 46 may be an artefact due to small sample size, or the result of an increased rate of early miscarriage of chromosomally abnormal fetuses. (Based on Ferguson-Smith and Yates 1984)

understood, but is commonly thought to be due to changes that take place at the centromeres of the chromosomes during the many years that the oocytes remain suspended in metaphase of the first meiotic division.

The relationship does not seem to vary with place or ethnic group. Maternal age provides a first step towards screening for women at increased risk of having chromosomally abnormal children. A woman who has already had one child with a chromosomal aneuploidy is at increased risk of having a second affected child.

Chromosomal aneuploidies are very rarely transmitted to offspring. Though the sex chromosome aneuploidies often have relatively mild phenotypic effects, XXY males and XO females are infertile, and there is no increased incidence of chromosomal abnormalities in the children of XXX females or XYY males. Children with the severe autosomal aneuploidies + 13 or + 18 almost always die in infancy. People, especially females, with Down syndrome may be fertile and can transmit their abnormality but rarely reproduce because of their psychological characteristics and their medical and social situation (p. 289).

Non-disjunction can sometimes involve the entire chromosome set, leading to one gamete with two sets of chromosomes and one with none. Occasionally, while still in the ovary, an ovum duplicates its chromosomes and develops without fertilization. As there are no paternally imprinted chromosomes to promote development of a placenta the result is an ovarian teratocarcinoma — a malignant tumour arising from embryonic cells, with a chaotic organization of differentiated tissues. If an ovum with two chromosome sets is fertilized, the result is triploidy (3n). An excess of maternally imprinted chromosomes leads to an underdeveloped placenta, and usually to spontaneous abortion. Some triploid zygotes contain a double set of paternal chromosomes as a result of fertilization by two sperms, or by one sperm which doubles its chromosomes. Here an excess of paternally imprinted chromosomes usually leads to a hydatidiform mole, in which there is uncontrolled proliferation of chorionic tissue (Vejerslev *et al.* 1987). This causes vaginal bleeding and sometimes leads to a chorion carcinoma.

Mosaicism

The term mosaicism describes a person or tissue whose cells are partly of one genetic make-up and partly of another. About 3 per cent of people with autosomal and 10 per cent with sex chromosomal aneuploidies are in fact mosaics, with a mixture of normal and aneuploid cells (e.g. a mixture of normal cells and cells with three copies of chromosome 21). In two-thirds of cases of mosaicism, the zygote was originally trisomic, but the extra chromosome was lost from one cell at an early cell division, leading to the embryo becoming a mosaic of trisomic and normal cells. Mosaicism for an autosomal aneuploidy usually causes a severe disorder similar to that in fully trisomic individuals, but mosaicism for a sex chromosome aneuploidy causes very little problem in many cases (Chang *et al.* 1989). Varying degrees of mosaicism may contribute to the wide clinical spectrum of chromosomal disorders.

Chromosomal rearrangement

Chromosomal rearrangements caused by errors in crossing over during meiosis include translocations, inversions, deletions, and duplications. They are diagnosed by chromosome studies and are fairly common. Primary care workers are likely to come across people in whom one has been diagnosed because of a family history of a heritable chromosome abnormality, or because of recurrent abortion or infertility. A chromosome rearrangement may be found incidentally in a fetus undergoing prenatal diagnosis for some other reason.

Some rearrangements between homologous chromosomes are neutral. For example, unequal pairing and crossing over between minisatellite sequences on homologous chromosomes is common and harmless, but leads to inherited differences between individual chromosomes that provide the basis for 'genetic fingerprinting' (see p. 29).

However, if non-homologous chromosomes cross over accidentally at meiosis, a translocation results. Translocations are of two broad types (Fig. 5.4).

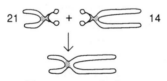

Chromosome fusion

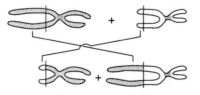

Reciprocal chromosome
translocation

Fig. 5.4 Types of translocations. Two types of chromosomal rearrangements are common. In a chromosomal fusion (Robertsonian translocation), two chromosomes join to become one. In a reciprocal translocation, two non-homologous chromosomes exchange parts, to produce two rearranged chromosomes.

1. *Robertsonian translocation* (chromosome fusion). This usually occurs within the group of chromosomes whose short arms carry nucleolar organizers which contain genes for ribosomal RNA (chromosomes 13, 14, 15, 21, and 22). Two such chromosomes may fuse at the centromere to form a single large chromosome, losing their nucleolar organizers in the process. The commonest chromosome fusion involves chromosomes 14 and 21: this is usually harmless.
2. *Reciprocal translocations.* These arise when crossing over occurs accidentally between chromatids of two non-homologous chromosomes during meiosis. They may involve large or small parts of chromosomes, and can be seen clearly by chromosome banding techniques (see p. 26).

Once a translocation has occurred in a germ cell, three possible types of gametes may be formed. 50 per cent will have an unbalanced chromosome constitution, 25 per cent will have a balanced chromosome constitution (i.e. will have the correct number of copies of genes but in an unusual arrangement), and 25 per cent will be normal.

An unbalanced chromosome constitution usually leads to severe physical and mental disability. People with a balanced rearrangement are normal, but make the classes of gametes mentioned above and so risk having abnormal children. However, in practice this risk is much lower than might appear because the 50 per cent of gametes with an unbalanced chromosome constitution are less viable than the others. This is particularly so for sperm, so fewer unbalanced rearrangements are inherited from fathers than from mothers with a balanced rearrangement. Females may have repeated miscarriages, and it seems that on average, translocation carriers have rather fewer children than the rest of the population.

Chromosomal rearrangements are relatively common. About 2 per 1000 of the healthy population carry a balanced translocation, about 0.4 per 1000 of these being new mutations. About 0.6 newborns per 1000 have an unbalanced rearrangement, about 0.2 per 1000 of these being new mutations. These figures show that new chromosomal rearrangements appear relatively frequently, but are also lost from the population quite rapidly by selection against unbalanced gametes, embryos, and individuals.

Evolution

Evolution operates on species rather than on individuals. In evolutionary terms the only criterion of individual 'fitness' is success in surviving to reproductive age and leaving as many copies of one's genes to the next generation as others do. To evolve, a species must continually generate enough individual variation to allow natural selection of the individuals that are best adapted to prevailing environmental conditions. Diploidy and sex are mechanisms that allow for speed and flexibility of response to environmental change.

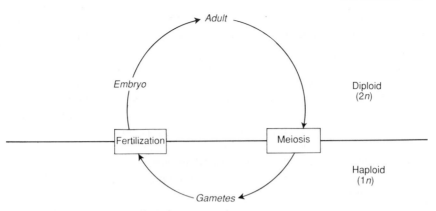

Fig. 5.5 The haploid/diploid cycle.

The life cycle of most species involves an alternation between haploid ($1n$) and diploid ($2n$) states, with sexual reproduction in between (Fig. 5.5). Human beings are diploid for most of their life cycle: only the gametes are haploid, until they fuse to form the diploid zygote. Even simple organisms such as bacteria that are haploid for most of their life cycle usually have a brief diploid phase that allows sexual reproduction. Diploidy and sexual reproduction require so much energy and organization that at first sight they seem inefficient; however, they are so widespread because they are vital for flexible evolution. Diploidy allows new mutations to persist long enough within a species to be tried out in different genetic combinations and environmental conditions. Sexual reproduction allows a species to act as a single entity from the evolutionary point of view by allowing new genes to spread rapidly and enter new combinations. Once diploidy and sex were introduced by one set of organisms, others had to follow suit or lose the evolutionary race.

Sexually reproducing organisms have an evolutionary advantage over asexually reproducing ones. For example, suppose two highly advantageous mutations, a and b, occur at about the same time in different members of an asexual population that reproduces only by budding (producing clones). Both improved clones will expand until the whole population are members of either clone a or clone b. However, the advantages of a and b can be combined only if one occurs as a second independent mutation within the other clone, and this can take a long time. When it does occur, the combined advantage will probably make the third clone predominant. Evolution can occur much faster in sexually reproducing populations, because two advantageous mutations that arise in different individuals can be relatively rapidly combined in their offspring.

Diploidy is necessary for sexual reproduction to occur, but also has other advantages. A diploid organism has two copies of every gene and is likely to be more robust than a haploid one, because it can tolerate more mutated genes within the genome. This also allows the species a reserve of genetic variation. Many mutations that are neutral or deleterious in some circumstances are advantageous in others. Recessive apparently deleterious mutations can persist in the genome long enough to be tested thoroughly, and sex allows them to be tried out in different genetic combinations. Sex also provides the necessary mechanism for getting rid of recessive disadvantageous genes, by leading to the birth of homozygotes who, if 'unfit', die without reproducing and take two disadvantageous genes with them.

The haemoglobin disorders, thalassaemia, and sickle cell disease illustrate the above concepts and demonstrate human evolution in practice. They are lethal in homozygotes but protect carriers (heterozygotes) against death from falciparum malaria. Falciparum malaria causes high mortality in infancy. If, for example, a sickle cell mutation occurs in a region where falciparum malaria is common, mortality among offspring who are carriers falls below that of normal infants, and more carriers survive to reproductive age – i.e. carriers are 'fitter' than normal people. Such a reproductive advantage can cause carrier frequency to rise quite rapidly, though other forces oppose this tendency. As carrier frequency rises, the chance of two carriers mating, which depends on the square of their frequency, rises even faster (Fig. 5.6). Every time such couples reproduce (in natural conditions), a quarter of their offspring die in early infancy from homozygous sickle cell disease, and two genes leave the population. Finally, loss of sickle genes through the deaths of homozygotes statistically balances the advantage to carriers, and the frequency of the gene stabilizes. Its final prevalence in a population represents the maximum protection that this lethal recessive gene can provide to that particular population under those circumstances. This situation is known as a 'balanced polymorphism'. Other lethal recessive mutations that occur in 1 per cent or more of a population, such as phenylketonuria, cystic fibrosis, congenital adrenal hyperplasia, and α-1 antitrypsin deficiency probably also represent balanced polymorphisms and are likely to confer some advantage on carriers.

A balanced polymorphism is not ultimately a satisfactory state. However, it has the advantage that it makes the abnormal advantageous gene common, so that new mutations are likely to occur in or near the gene. Several evolutionary mechanisms may then improve the situation. For example, a second mutation occurring near the first may protect homozygotes from lethal effects: the mild form of sickle cell disease common in Saudi Arabia seems to have this advantage. Alternatively, if a less lethal mutation arises that protects carriers but is not lethal to homozygotes, it will replace the first: haemoglobin E is replacing β-thalassaemia in South-East Asia for this reason. More radical solutions are possible. For instance, a duplicated normal gene may become paired with the advantageous mutant by unequal

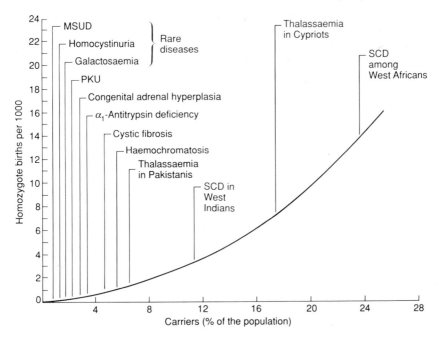

Fig. 5.6 Relationship of the percentage of the population who are carriers of a recessively inherited disorder, to the number of homozygotes born/1000 births. Various disorders are indicated. MSUD = maple syrup urine disease; PKU = phenylketonuria; SCD = sickle cell disease. (Reproduced from Royal College of Physicians (1989), with permission.)

crossing over between misaligned regions of homologous chromosomes during meiosis. Such a combination — which simultaneously enriches the genome, preserves the advantages and reduces the disadvantages of a mutated gene — may spread rapidly. The existence of sequences of closely related genes side by side on the same chromosome, like the globin or immunoglobulin genes (p. 80) shows that such unequal crossing over leading to gene duplication and diversification has been a powerful evolutionary mechanism in the past.

Key references

Alberts, B., Bray, D., Lewis, J., Raff, M., Roberts, K., and Watson, J. D. (1989). *The molecular biology of the cell*, (2nd edn). Garland Publishing Inc., New York.
Bodmer, W. F. and Cavalli-Sforza, L. L. (1976). *Genetics, evolution, and man*. W. H. Freeman and Co., San Francisco.
Edwards, J. H., Lyon, M. F, and Southern, E. M. (eds) (1988). *The prevention and avoidance of genetic disease*. The Royal Society, London.

Vogel, F. and Motulsky, A. G. (1986). *Human genetics: problems and approaches*, (2nd edn). Springer-Verlag, Berlin.

WHO (1986). Prevention of avoidable mutational disease: memorandum from a WHO meeting. *Bulletin of the World Health Organization*, **64**; 205-16.

6.   Cancer

Cancer is an acquired genetic disease, as it is usually the result of somatic mutations in the DNA of body cells. Most environmental carcinogens exert their effects by causing mutations (WHO 1986). In industrialized countries, about a quarter of the population will eventually die of a malignancy, and it seems likely that one of the most important future applications of new genetic knowledge will be in the cancer field.

The control of growth in the adult body must be very finely balanced, so that organs and tissues retain their structure and function. Even a slightly increased growth rate starting with one cell, e.g. in the basal layer of the epidermis, will in time cause a localized lump (for example a benign wart). Benign tumours can occur in almost any tissue. For a tumour to be malignant, in addition to increased growth, the cells need to be 'transformed'. This means that they lose both some of their distinguishing features and their ordered response to the usual environmental constraints that keep cells fixed, so that they can invade other tissues – i.e. malignant cells can metastasize. In tissue culture, cancer cells can be recognized because they pile up on each other, in contrast with normal cells which form an orderly single layer; they need fewer growth factors than most other cells, probably because they make some of their own; and they can go on dividing indefinitely instead of dying after from 20–50 divisions, as most normal cells do. They retain these characteristics in culture or when re-introduced into experimental animals, indicating that they are due to inherited changes in the DNA.

Cancers are classified according to the type of tissue from which they originate. Table 6.1 summarizes data on their incidence in the United States in 1986.

Origin of cancers

Most cancers are thought to originate from a single somatic cell whose DNA has undergone one or several specific mutations. Somatic mutations that cause cancer must affect growth-controlling genes and must arise in stem cells. Mutations in terminally differentiated cells such as epidermal or intestinal epithelial cells that are lost from the body would not be perpetuated.

An important piece of evidence that a DNA mutation is a fundamental cause of malignancy comes from the study of the rare recessively inherited disorder xeroderma pigmentosum, in which the enzyme system that repairs

Table 6.1 Relative incidence of different types of cancer in the USA

Site of cancer	Approximate percentage of all new cancers per year
Digestive organs	23
Lungs	16
Breast	13
Reproductive tract	18 (including cancer of the cervix, 2%)
Urinary tract	7
Muscles and connective tissue	1
Lymphomas and leukaemias	8
CNS and eye (gliomas and retinoblastomas)	2

(Based on Alberts *et al.* 1989)

DNA damage caused by ultraviolet light is defective. As a result, exposure of skin to sunlight leads to somatic mutations, and so to multiple skin cancers.

There is also good evidence for the single cell origin of cancers. For example, in patients with chronic myeloid leukaemia, the leukaemic cells differ visibly from normal cells, since they contain the so-called Philadelphia chromosome. This chromosome results from a somatic translocation between the long arms of chromosomes 9 and 22. Different patients have slightly different sites of translocation, but in any given person all the malignant cells have the identical translocation, strongly suggesting that they represent a clone descended from a single mutated ancestral cell.

Some of the DNA changes involved in causing cancer may be inherited, some may arise by chance, and others may be caused by environmental carcinogens. These include ionizing radiation (such as X-rays) which typically causes chromosomal translocations and breaks, virus infections which may result in foreign DNA being incorporated into the host genome, and chemical carcinogens such as those found in tobacco smoke. Some substances are so carcinogenic that if they are in contact with tissues for long enough, cancer is bound to develop. For example, workers who distilled 2-naphthylamine, a compound used in the chemical industry in the early part of the century, almost invariably developed cancer of the bladder. Most recognized carcinogens increase the mutation rate in bacterial cultures, and this has become a standard method for testing substances for carcinogenicity.

In a human life, about 10^{16} somatic cell divisions occur, and there is an approximately one in a million chance of a mutation arising in each gene at each cell division. In a lifetime, therefore, probably every cell accumulates at least one somatic mutation, and about 10 000 million mutations will have occurred in cells scattered throughout the body. Probably 100 times more mutations will have happened in non-coding than in coding sequences of the DNA. Thus the really perplexing question is not why does cancer occur, but why does it not happen more frequently?

The relative rarity of malignant transformation suggests that usually between three and seven independent mutational events must happen within the same cell before it can become cancerous, fewest for leukaemias, more for most carcinomas. If one mutation were enough, then the chance of cancer occurring should be the same in each year of life, but in fact this is not so. The incidence of most cancers rises with age. Thus in the United States there are less than 50 deaths from cancer of the large bowel per million people under 40 years of age per year, but the figure rises to about 400 by the age of 80.

Early cancer studies led to the suggestion that the first event in the natural history of a developing cancer is 'tumour initiation', i.e. a predisposing mutation in a cell, arising either spontaneously or as a result of exposure to a mutagenic agent such as radiation, a virus, or a carcinogen such as tobacco. Tumour initiation will lead to disordered cell division only if the cells are subsequently exposed to a 'tumour promoter', which stimulates cell division, perhaps leading to several small benign growths. The longer the cells are stimulated, the more small tumours with an increased mitotic rate are produced, and the greater the chance that further somatic mutation leading to malignancy will occur in one of them.

Some such theory of cancer production is necessary to explain the long incubation period between exposure to a potential carcinogen and the development of a malignancy, and the slow and unpredictable evolution of cancer within areas of disturbed growth. The fact that malignant change is not usually detectable until the original mutated cell has produced millions of descendants is not enough to account for this delay. For example, the peak incidence of leukaemia in Hiroshima and Nagasaki occurred about eight years after the atomic bombs were dropped, in about 1953. By analogy, the peak effect of the nuclear accident at Chernobyl is to be expected in about 1995.

There is now good evidence that cancer evolves stepwise (Fig. 6.1). For example, in chronic myeloid leukaemia there is a pre-malignant phase in which circulating white cells with the characteristic Philadelphia translocation are present but do not proliferate uncontrollably. This is followed by a rapidly progressive malignant phase in which many additional chromosomal changes are present in the malignant cells. In some women, the epithelium of the uterine cervix contains patches of 'dysplasia' (somewhat disorganized differentiation). These often regress, but in some cases gradually become more marked and in a few, if the lesion is not removed, cancer of the cervix develops.

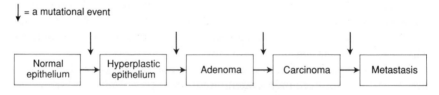

Fig. 6.1   Steps in the development of a carcinoma.

Progression of a malignancy, once it has started, apparently depends on a process of 'natural selection' for the most rapidly growing cells. The more rapid the growth, the greater the likelihood that other mutations that alter growth characteristics will occur. Naturally, the fastest growing cells overgrow the slower, so as time passes the cells typical for the cancer may change, may grow faster, and may become increasingly invasive, as described above for chronic myeloid leukaemia.

The incidence of many cancers differs in different countries. Populations that move from one country to another tend to take on the cancer pattern of the host population, confirming that environmental factors are very important in causing malignancy. Some environmental factors may act as tumour initiators, others may stimulate initiated cells to grow.

Oncogenes

Modern theories of the fundamental causes of cancer revolve around the concept of oncogenes. Oncogenes are modified versions of normal growth-controlling genes (proto-oncogenes, p. 49). The modification may involve either mutation of the growth-controlling gene itself, or a change adjacent to the gene, or both. Understanding of oncogenes started with studies of tumour viruses, and of the chromosomal abnormalities associated with some malignancies.

Viruses that cause cancer are quite common in laboratory animals but uncommon in humans (Table 6.2). Most tumour-producing viruses are retroviruses whose genome consists of a single strand of RNA; the virus reproduces itself by using the ribosomes of the host cell to make an enzyme, reverse transcriptase, which copies its RNA into DNA, which can then multiply. Sometimes this DNA copy is inserted into host cell DNA and reproduces there.

Tumour viruses may stimulate tumour formation in one of two ways. Some contain a 'viral oncogene', a growth-promoting sequence that causes malignant transformation of the host cell. Others include 'promoter' sequences that stimulate the host cell to reproduce the viral DNA. If by chance a viral promoter is inserted into host DNA next to a growth-controlling gene, it may induce over-expression of the gene, or may switch it on again if it was switched off. In these cases the host gene becomes an oncogene.

Once viral oncogenes had been identified and their DNA sequenced, it was soon found that the normal human genome contains almost identical genes, which were named 'proto-oncogenes'. They are involved in growth regulation and embryonic development. At least half of these genes code for a protein kinase—a type of enzyme known to be closely involved in control of cell growth. In fact, viral oncogenes are normal mammalian growth-controlling genes that have been picked up in the past from various hosts and are 'hitch-hiking' with the virus. They have presumably become incorporated into viral genomes because they create favourable conditions for viral replication, by causing host cells to proliferate.

Over 80 per cent of human cancers arise without any obvious relationship to a virus, but proto-oncogenes are still involved. Point mutations in growth controlling genes are certainly involved in most cancers (Harrris 1991). In addition, chromosomal rearrangements can occur in somatic cells, and some

Table 6.2 Viruses known to be involved in carcinogenesis in humans

Virus	Associated malignancies
Hepatitis B	Liver cancer
HTLV1	Adult T cell lymphomas
Some papilloma (=wart) viruses	Cervical cancer
Epstein–Barr virus	Nasopharyngeal carcinoma
	Burkitt's lymphoma (in children in Africa)
	Can also cause cancer in immunosuppressed patients

are associated with specific tumours in humans (e.g. chronic myeloid leukaemia). These rearrangements usually involve a chromosome break very close to the site of a proto-oncogene, and the rearrangement either stimulates increased activity of the proto-oncogene or stops it being switched off normally.

There is now good evidence that oncogenes are involved in most cancers. This is encouraging, because despite the incredible variety of cancers, if a limited number of oncogenes are responsible, a limited number of approaches, focussing on these genes, may be needed for cancer prevention and control.

Inheritance and cancer

A few rare types of cancer are inherited in an almost Mendelian fashion. In other types of cancer, e.g. cancer of the breast or colon, there may be a familial tendency, while in many others there is no evidence at all of a hereditary element. The reason for this spectrum of heritability is now beginning to be understood. It seems that in some cases, both of the alleles for a particular growth factor in a somatic cell must mutate, in order to allow uncontrolled production or activity of growth factor.

Familial cancers

Amongst the familial cancers are retinoblastomas, fibroblastomas, and osteosarcomas. In these families, a recessive mutation of one growth-controlling gene is transmitted within the family, so that all that is needed for cancer to develop is a somatic mutation of the allele on the corresponding chromosome. For example, retinoblastoma depends on deletion of a gene that suppresses a growth factor. This deletion may be inherited in the heterozygous form and causes no problems as long as it remains heterozygous. However, when by chance the identical mutation occurs on the remaining normal gene in a retinal cell, a malignant retinoblastoma may develop. People who have inherited the genetic predisposition usually develop bilateral retinoblastomas, confirming the high frequency with which spontaneous somatic mutations occur.

In many 'familial' cancers, e.g. Wilms' tumour or retinoblastoma, non-familial cases also occur, i.e. these cancers can be either inherited or sporadic. Sporadic cases usually occur in people with a normal genotype. They are usually unilateral, because they require the coincidence of two random mutations, one in each copy of a gene in the same cell, a very rare event, to produce malignancy.

The tissue-specific nature of the inherited malignancies illustrates the complexity of the factors causing cancer. For example, in familial retinoblastoma, the inherited heterozygous mutation is present in every body cell, and there is a high chance of a somatic mutation converting it to homozygosity in any cell, but in these families malignancy arises only in the eye. Other families have a mutation in exactly the same growth factor, but a second somatic mutation usually causes cancer only when it occurs in bone, leading to familial osteosarcoma.

Some familial tumours are 'dominant', e.g. familial adenomatous polyposis coli and neurofibromatosis. In these cases, a single (heterozygous) mutation allows local growth at certain points, so that numerous polyps of the gut, or neurofibromas, are formed. The increased rate of cell multiplication in these benign tumours allows the possibility of further somatic mutation, and a cancer may arise. Within these families, offspring have a 50 per cent chance of inheriting the genetic predisposition to cancer.

Members of families with a known genetic predisposition to cancer should be actively sought out and referred for specialist advice. Some need regular examination, e.g. of the eye or colon, in order to avoid malignancy or detect it early. In many families at risk for malignant polyposis coli, at present the most effective way of preventing cancer is to remove the colon. With DNA technology it is now becoming possible to offer definitive carrier and prenatal diagnosis to such families.

A small proportion of common cancers, such as cancer of the colon or the breast, also appear to be familial. There is now good evidence that these families are also transmitting a recessive mutation in or near a specific growth-controlling gene, which increases susceptibility to specific cancers. It only requires a somatic mutation at the same locus in a single cell in the appropriate tissue to allow cancer to develop.

Common, generally non-familial, malignancies

Studies of familial cancers have helped to show that particular growth-controlling genes are involved in many of the common non-familial forms of cancer. Recent evidence suggests that a number of different changes in different growth suppressor genes and oncogenes may be necessary for the development of many malignancies. These findings are likely to find clinical applications in the near future. For example, it may prove possible to screen families or the population to identify people with a genetic predisposition to particular cancers, and offer them surveillance. It also seems likely that improved understanding of the molecular basis of particular forms of cancer will help in selecting appropriate forms of treatment. It may even become possible, ultimately, to develop tailor-made treatment for particular patients by identifying the growth factors involved, and either switching off the gene or blocking its activity in specific tissues.

Key reference

Alberts, B., Bray, D., Lewis, J., Raff, M., Roberts, K., and Watson, J. D. (1989). *The molecular biology of the cell*, (2nd edn). Garland Publishing Inc., New York.

7. The immune system

Defence against infections is an absolute priority for living organisms. All vertebrates have an immune system which destroys invading organisms and foreign toxic proteins such as those produced by bacteria. Normal immune reactions are responsible for some congenital and genetic disease such as blood group incompatibility between mother and fetus, the vulnerability of the fetus to congenital infections, autoimmunity, and the limitations of tissue grafting. There are also some rare inherited disorders of the immune system.

The immune system, though spread throughout the body, is estimated to weigh as much as the brain or liver. It includes two main groups of lymphocytes. B cells develop in the fetal liver and the adult bone marrow and produce circulating antibodies. T cells develop in the thymus and control cell-mediated reactions. Both types originate from haemopoietic stem cells, which are also precursors of red and white blood cells. T cells are involved in all immune responses, but B cells are easier to study because the soluble antibodies they produce can be collected so easily from blood samples. The cells of the immune system interact with each other continuously, through direct contact or through chemical messages carried in the blood stream.

The main tasks of the immune system are:

1. To recognize the multitude of foreign antigens, including viruses, bacteria, and toxins.
2. To distinguish self from non-self antigens.
3. To mount the appropriate destructive response to foreign antigens.

Any substance that can stimulate an immune response is called an antigen. Most large molecules can stimulate a specific response, so there is an enormous number of potential antigens. Remarkably, mammals have a set of lymphocyte clones ready to identify and react specifically to most of them, including many that the body will never encounter. The cells of a lymphocyte clone are derived from a single ancestral cell that has become differentiated to react to a specific antigen: all its descendants inherit its specific characteristics.

The characteristics that make a molecule, or a part of it, recognizable to the immune system are its shape and properties such as solubility in water or in lipids, rather than exact chemical structure. Different parts of one large molecule may be recognized, with different degrees of intensity, by several different clones of cells of the immune system, and one clone may react to a range of different antigens with some shape characteristics in common. This partly accounts for the ability of the immune system to react to almost all foreign proteins, and for cross-reactions of antibodies to a range of antigens.

Lymphocyte clones react to an encounter with their specific antigen either by producing a circulating antibody (humoral immunity) or by binding directly to the antigen (cell-mediated immunity). Antibodies are immunoglobulins, and account for almost 20 per cent of the plasma proteins.

Genetics of antibody formation

All the antibody molecules produced by a given clone of B cells are identical. They are Y-shaped, with two identical heavy and two identical light polypeptide chains (Fig. 7.1). There is an antigen-binding site on each arm of the Y: in any given molecule these are identical. Most antigens have more than one antigenic site and an antibody molecule often binds two separate antigens, bending at a 'hinge' region where the Y forks. In this way antigens and antibodies can form large complexes that are recognized and removed by, for example, phagocytic cells.

The differences between the antibodies produced by the cells of different B cell clones depend on specific, different amino acid sequences at the antigen-binding site of the antibody molecule. These differences are generated by a carefully regulated system for producing diversity from a limited number of basic genes, rather than by millions of different genes coding for different antibody molecules.

The heavy and light chains are determined by three allelic pairs of gene 'pools' on three different chromosomes. One pool on chromosome 14 codes for the heavy chains, and two pools on chromosomes 2 and 22 code for two alternative types of light chains (κ and λ chains). That is, there are six gene pools per B cell. In a B cell that is actively synthesizing antibody, only one of the two gene pools for a heavy chain and one of the four gene pools for a light chain is functional. The remainder are permanently inactivated, perhaps in the same way that one X chromosome is inactivated in females (p. 31).

In the two active gene pools, the characteristic light and heavy chains that will make up the specific antibody to be produced are composed by random selection of three out of a very large number of 'proto-gene segments'. These three DNA segments come together to make a single mRNA that codes for the entire heavy or light chain. The intervening proto-gene segments are physically cut out of the chromosomal DNA, so a cell's choice is irreversible: and all its descendants must make exactly the same antibody. The descendants are therefore a clone.

More than 10 million different clones of B cells are generated by this mechanism alone. The number of possible combinations is probably further

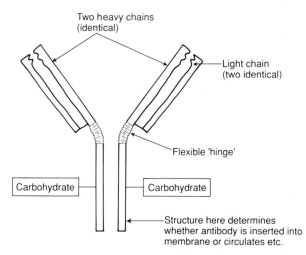

Fig. 7.1 Diagram of an antibody. Two heavy chains and two light chains are paired to give a Y-shaped molecule. The region of each heavy chain at the fork of the Y is flexible creating a mobile 'hinge'. The antigen-binding sites at the distal end of each arm of the Y include the variable sections of both the heavy and light chains. They have the same configuration in all antibody molecules produced by a single clone of cells. This identity is essential for effective antibody action, which often depends on a single antibody molecule binding two different antigen molecules. The characteristics of the regions of the heavy chains at the tail of the Y determine whether the antibody molecules are inserted into a cell's membrane or released into the plasma.

increased by somatic mutation within the genes. Apparently most possible clones are generated during embryonic development and remain ready to respond to stimulation by an appropriate antigen.

Antibodies are further modified to fall into one of the five classes shown in Table 7.1, by being attached to other proteins or polysaccharides. The largest antibody classes are IgM and IgG.

The first light and heavy chains produced in an unstimulated B cell are combined with a protein characteristic for a macroglobulin (IgM), which ensures that they are inserted into the B cell's membrane rather than being secreted into the plasma. In mature organisms, if a B cell encounters antigen that binds to this membrane-bound antibody, it is activated (see below) and starts to divide (Fig. 7.2). Some offspring develop into plasma cells that produce a large quantity of *soluble* IgM (macroglobulin) antibody, which binds the same antigen but is secreted into the blood stream. This 'primary response' takes several days. The plasma cells involved are terminally differentiated i.e. will die without offspring. The remaining offspring of the stimulated clone become 'memory cells' which circulate freely, constantly ready to react to their specific antigen, so immunity, once acquired, usually lasts lifelong. When a memory cell encounters its antigen, it divides to produce more

Table 7.1   Classes of antibody made by B cells

Class	Function and site
IgA	Main antibodies in secretions such as milk, saliva, tears, and intestinal and lung secretions
IgD	Produced by B cells that have not yet encountered their antigen
IgE	Binds on mast cells and basophils causing release of histamine and serotonin. The basis of acute allergic reactions
IgG	Most important type of antibody in the secondary response. Crosses from mother to fetus to provide passive immunity
IgM	First antibodies produced in the primary immune response. Does not cross placenta

cells capable of making antibody. Their DNA undergoes further irreversible changes to code for a further class of antibody, usually IgG (γ globulin), which is secreted into the circulation in very large amounts. This is the 'secondary response'.

The fact that the characteristic antibody produced in the primary response is IgM and in the secondary is IgG, and the greater speed of the secondary response, is useful in predicting risks to the fetus when a woman is exposed to certain infections during pregnancy. The presence of IgM in maternal plasma is associated with a risk of infection for the fetus, while the presence of IgG, which crosses the placenta, indicates that the mother was already immune, and the fetus will have been protected from infection.

Antibodies can inactivate harmful molecules such as toxins, and block the 'docking' antigens on the surface of invading organisms that allow them to enter cells. On their own they cannot dispose of invaders, but they include components that attract white cells and various T cells, and activate phagocytes, which can. They also often activate the complement system, a complex set of plasma proteins that lyses foreign cells, literally by creating leaky pores and inserting them into the invader's cell membrane.

Cell-mediated immunity

Fundamentally similar genetic mechanisms are thought to be involved in the more complex processes of cell-mediated immunity, which are controlled by many kinds of T cells, distinguishable only by their surface characteristics. T cells can bind antigens only when they are on the surface of other cells and

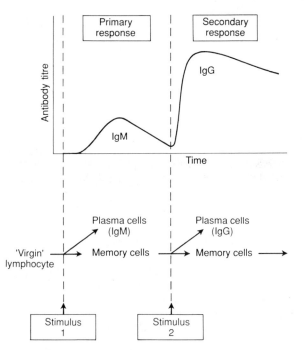

Fig. 7.2 The immune response to the first, and to subsequent challenge with a given antigen (see text). In the 'primary response' 10–14 days pass before the plasma titre of IgM reaches its maximum. The IgM level falls rapidly during the next 4 weeks. After a second exposure to the same antigen, the 'secondary response' happens far more quickly, the antibody produced is IgG, the antibody titre is higher, and the response lasts longer. IgG antibody production often continues indefinitely, and from now on the individual will respond very briskly to this particular antigenic stimulus.

associated with one of the HLA glycoproteins (see below). T cell 'antigen receptors' are the equivalent of antibodies and somewhat similar to them in structure (Fig. 7.3).

There are three main classes of T cells. *Cytotoxic T cells* kill virus-infected cells. They can identify infected cells because some of the viral protein produced within the cell is broken down and fragments are held out in the 'jaws' of class 1 HLA antigens on the cell surface. Exposure to antigen stimulates cytotoxic T cells to multiply, like B cells. Vaccines containing attenuated viruses, such as polio, measles, mumps, and rubella vaccines, evoke this type of immune response.

Helper T cells are necessary for both B and T cells to destroy foreign antigens. The helper cells 'recognize' antigens, usually fragments of protein, which become associated with class II HLA proteins on the surface of antigen-presenting cells (some B cells, thymus epithelial cells, dendritic cells

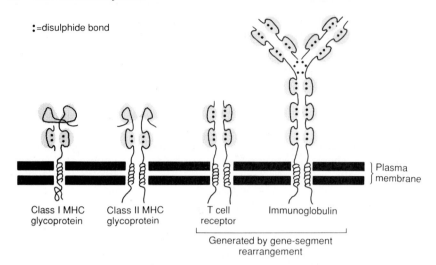

Fig. 7.3 Antibodies, T cell receptors, and the HLA class I and class II glycoproteins are fundamentally similar in structure. All consist of a number of 'domains' with a similar basic structure, arranged in different patterns in the different types of immune protein. It is highly likely that they developed during evolution by multiplication and diversification from a common protein ancestor. (Modified from Alberts *et al.* (1989), with permission.)

of lymphoid organs, Langerhans cells in the skin, and some macrophages). Helper cells stimulate the appropriate B or T cell clones to respond, by secreting chemical mediators called interleukins. The central role of helper T cells is evident in patients with Acquired Immunodeficiency Syndrome (AIDS). The HIV retrovirus, which causes AIDS, leads to the death of helper T cells and gradually reduces the effectiveness of the immune system, so patients eventually succumb to infections that healthy people could easily overcome. *Suppressor T cells* dampen the response of other T or B cells to antigens and help to regulate the immune response and avoid autoimmune disease.

The human leucocyte antigen (HLA) proteins

One set of proteins that plays a very important part in immunity comprises the polymorphic proteins of the human major histocompatibility complex (MHC). In clinical practice they are usually called the HLA (human leucocyte antigen) complex, because they were first identified in white cells. One set, *class I*, occurs on all nucleated cells and forms up to 1 per cent of their membrane. *Class II* HLA proteins are found only in the membranes of lymphocytes and macrophages (and other antigen-presenting cells).

Class I HLA proteins are produced by three adjacent genes, derived by duplication of a single ancestral gene at some time in the distant past. As there is a very large number of possible alleles at each HLA locus, most people have six different class I HLA proteins on the surfaces of their cells. Since the HLA genes are close together on the short arm of chromosome 6, they are usually inherited together (linked), and there is a one in four chance that two children of the same parents will carry the same set of HLA genes, i.e. will be 'HLA identical'.

The main role of the class I HLA proteins on most cell surfaces is to 'present' fragments of both normal and abnormal proteins from inside the cell to T cells. The configuration of the HLA proteins is highly attractive to cytotoxic T cells, which are constantly 'inspecting' them and their bound protein fragments. They do not react to HLA-bound fragments of self proteins. However, if for instance viral fragments are bound to the HLA proteins, T cells protect the organism by destroying the virus-infected cell. Class II HLA proteins have a similar role in presenting fragments of non-viral antigens (which have been broken down inside the cell) to helper T cells.

It is thought that their role in initiating immune reactions accounts for the extraordinary degree of polymorphism of the HLA proteins. Viruses and bacteria are under strong selection pressure to adapt their surface proteins genetically to resemble host self proteins, so that they can evade detection when bound by HLA antigens, and avoid the host's immune defences. To cope with their adaptations it is vital for host species to develop a wide range of variation in antigen-binding HLA proteins: and this in turn leads to variations in susceptibility to specific infections between individuals and populations.

If by chance a pathogen mutates to a form that can evade recognition when bound to common HLA proteins, it may sweep as a pandemic through the population. If the infection is lethal, only individuals with HLA combinations that can still pick up the altered antigen and present it to white cells will survive. Episodes such as the bubonic plague (the black death) in fourteenth century Europe, or the repeated smallpox epidemics prior to Jenner's introduction of vaccination, may have been partly responsible for determining the present pattern of HLA alleles prevalent in Europe. One clear example is that of Dutch settlers in Surinam. In the 19th century, most of the migrants died in a typhoid epidemic, and today, the distribution of HLA genes in this Dutch sub-population is quite different from that in the Netherlands. Similar selection pressures may explain the very different frequencies of different HLA combinations found in different populations, and the vulnerability of some populations, such as Eskimos, to infections such as measles or influenza, that they had not encountered prior to contact with Europeans.

Some diseases are commonly associated with certain HLA types (Table 7.2). The link between HLA B27 and ankylosing spondylitis is probably the best-known example at present. This association means that the diseases in

Table 7.2 Some diseases are associated with particular HLA types

HLA allele	Associated disease	Relative risk[1]
A3	Haemochromatosis	4
B27	Ankylosing spondylitis	90
Bw47	Congenital adrenal hyperplasia	15
Cw6	Psoriasis	4
DR4 only	Diabetes mellitus	6
DR3 only		3
Dr3 + DR4		33
	Chronic active hepatitis	2
DR3, DQw2	Myasthenia gravis	3
	Coeliac disease	17
DR2, DQwl	Multiple sclerosis	4
DR4	Rheumatoid arthritis	4–6
DR2	Narcolepsy	34
DR2, DR3 C4A XQ0	Systemic lupus erythematosis	?

(Based on Scriver *et al.* 1989)
[1] Relative risk is the risk that carriers of these HLA haplotypes run of developing the associated disorder, relative to the rest of the population (= 1).

question must have a genetic component: either the relevant gene or genes are on chromosome 6 next to the HLA loci, or the patient's particular HLA pattern predisposes to the disease by its effects on the immune system.

Self, non-self, and tissue tolerance

Many of the original B cells that develop in the embryo form antibodies which can react with the individual's own antigens ('self' antigens). Embryos are usually insulated from foreign antigens, so to avoid inappropriate immune reactions to self, B cells are programmed to die if they encounter their corresponding antigen during early fetal life. Consequently, only B cells with 'non-self' antibodies on their surface survive, and the immune system becomes incapable of reacting to 'self' antigens. It starts to function and produce antibodies only after 20–24 weeks' gestation.

The ABO blood group substances (p. 336) provide an example of the distinction between self and non-self. Adults, but not fetuses, possess naturally occurring antibodies to groups other than their own. For example, a person of blood group A has circulating anti-B, but not anti-A antibodies. These antibodies are formed during early life in response to A and B antigens on the surface of the *Escherichia coli* bacteria that colonize the gut. Babies with A antigen are 'tolerant' to substance A, their lymphocyte clones capable of making anti-A antibodies having been eliminated, but they can and do make anti-B antibodies.

Tolerance limited to self leads to intolerance of others. In an incompatible blood transfusion the recipient's naturally occurring antibody destroys donor blood cells, and may incidentally kill the recipient as well. Tissue grafts are rejected unless donor and recipient are HLA identical. This is because HLA proteins are designed to attract the attention of T cells when their conformation is changed to one which the T cells do not recognize as 'self'. T cells therefore react to foreign HLA antigens on the surface of grafted cells with the same hostility as to host HLA antigens binding a foreign protein.

Tolerance to some foreign proteins can sometimes be induced by, for example, injecting antigen repeatedly (e.g. in wasp sting desensitization), or in association with an immunosuppressive drug. When bone marrow is grafted from a group A donor to a group B recipient, the recipient stops producing anti-A antibodies, i.e. becomes tolerant to the A substance. However, A and B blood group substances are not strong antigens. It is impossible to induce tolerance to really strong antigens like the HLA proteins, and tissue grafts must be HLA compatible.

It is extremely rare for two unrelated individuals to be HLA identical: identical twins are always compatible, but siblings have only a one in four chance of compatibility and parents are rarely HLA compatible with their children. The chance that two randomly selected individuals will be HLA compatible is less than one in a million. Because tissue transplantation is so important, tissue banks are being set up to store a list of potential donors for people needing a transplant, even though such matches tend to be less successful than matches within the family.

Intra-uterine infections with organisms that are easily controlled by adults can seriously damage a fetus. If a fetus is infected with a virus before 20 weeks' gestation, it cannot mount an immune response and so cannot destroy the virus. It may even become tolerant. Newborns infected *in utero* e.g. with rubella virus, cytomegalovirus, or human immunodeficiency virus (HIV) may be seriously affected, or may become lifelong virus carriers.

In mature organisms, the mechanism of tolerance can become disturbed in ways that are not yet understood, but are strongly influenced by the individual's genetic make-up. Autoimmune diseases, such as rheumatoid arthritis, disseminated lupus erythematosus, myasthenia gravis, thyroid disease etc. show a strong familial tendency.

Key reference

Alberts, B., Bray, D., Lewis, J., Raff, M., Roberts, K., and Watson, J. D. (1989). *The molecular biology of the cell*, (2nd edn). Garland Publishing Inc., New York.

Part 2

Genetic diagnosis and counselling

8. Patterns of inheritance: the family history

Genetic disease affects not only individuals but also their families. Study of the family history is an important first step in diagnosing genetic disease, but the absence of a history of the disease in the family rarely excludes a genetic basis for the disease. When a disease is suspected to be of genetic origin, the family needs a diagnosis, prognosis, and recommendations for treatment for the affected member. In addition they need advice about the risk of recurrence in further children, reproductive risks to other family members, and ways of reducing such risks. All this depends on precise diagnosis.

Figure 8.1 summarizes the steps in genetic diagnosis. Depending on the current state of knowledge of the condition, progress towards a precise genetic diagnosis may become stuck at one of these stages. Some inherited forms of mental retardation, blindness, or deafness, can as yet be diagnosed only at the clinical level. In these cases risk prediction for relatives must be based on the inheritance pattern alone. When a disease follows a Mendelian pattern of inheritance, this suggests that it is due to a defect in a single gene. When interactions of more than one gene, or interactions between genetic make-up and the environment, are involved, the pattern of inheritance will be less clear (p. 106).

A biochemical or chromosomal diagnosis can be made in an increasing proportion of cases, and in about 10 per cent of conditions whose biochemical basis is known, DNA studies have now led to precise identification of the gene responsible at the DNA level. In these cases it is often possible to decide which healthy relatives are carriers and which are not, and to offer prenatal diagnosis to those at risk. The further possibility of detecting genetic traits directly in people without a family history of the disease introduces new possibilities and problems, discussed in Chapter 13 on population screening.

The family tree in primary care

In the UK, a person registering with a general practitioner is encouraged to arrange an introductory consultation, during which brief details of the personal, past, and family history are entered on a summary card. This is placed in the front of the person's notes and acts as an aide-memoire for future consultations. The card includes space for details of first-degree

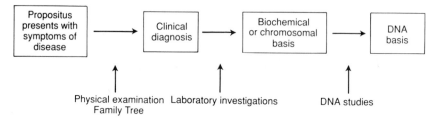

Fig. 8.1   Stages in genetic diagnosis.

relatives (Fig. 8.2), and in fact a simple family tree, which should be updated every few years, can be constructed very quickly during a ten-minute consultation.

The primary care family tree highlights both events with emotional significance (psychogenic chest pain in a patient of 55 may be related to the death of his father from a heart attack at the same age) and genetically significant events such as premature death of a parent with ischaemic heart disease, or congenital dislocation of the hip in a sibling. At present, family doctors use the family tree most frequently to obtain a short-hand view of the dynamics of a family: whilst cataloguing significant family events, recent crises such as divorce, or death or severe illness of a parent or sibling often become evident.

To meet the increasing role of genetics in medicine, the primary care family tree can be developed into a basic genetic family tree with very little extra effort. The aims should be:

1. To document the frequency of common conditions with familial implications, including common conditions of later life.
2. To detect families and individuals who may benefit from seeing a clinical geneticist or other specialist.
3. When a patient has a disease with genetic implications, to identify relatives who may benefit from information and surveillance.
4. To collaborate with specialists once a genetic diagnosis has been made.

Two types of disease pattern may emerge in a family tree, suggesting either multifactorial or Mendelian inheritance. A multifactorial pattern is more characteristic of the common diseases of adult life (see Table 8.5). Some families have more than their fair share of cardiovascular disease, auto-immune conditions, or cancer, while other luckier ones are marked by longevity. A plasma lipoprotein profile may be suggested for the siblings and children of patients with premature ischaemic heart disease, and measurement of intra-ocular pressure for relatives of someone who has just been found to have glaucoma.

Clear-cut inherited diseases tend to manifest themselves early, so

NAME			D.O.B.		Practice No.	
Cultural group		sex	Reg.		Family No.	
Civil State	date	Occupations		Year	Housing	

PARTNER	CHILDREN

FAMILY HISTORY	Key workers/relatives/services etc
F	
M	
siblings	PRIORITY FACTORS

PERSONAL HISTORY AND RISK FACTORS	date		date
Allergies		Depressive illness	
Atopy		Hypertension	
Cigarettes		Diabetes	
Alcohol		Obesity	
Contraception		Carcinoma	

SERIOUS PAST ILLNESS/OPERATIONS	date	BP	WGT	Cx smear	Breasts	Urine	date

Fig. 8.2 An example of a standard patient data sheet used in primary health care, with space for the family history. Though e.g. M and F are used for mother and father, accepted genetic symbols could be used to make a basic genetic family tree (with thanks to the James Wigg Practice).

Mendelian inheritance is in general more characteristic of diseases that present in childhood. Clinical findings that might suggest the possibility of genetic disease in babies and children include the following:

1. Intra-uterine death, still-birth, or neonatal death. These are sometimes due to chromosomal abnormalities, or inherited metabolic diseases. All infants that are still-born or die in the neonatal period should be examined by a paediatric pathologist. Their parents may require genetic as well as bereavement counselling.
2. 'Birth defects'. Most congenital disorders that are evident at birth are not inherited (p. 254), but parents of affected children often believe they are, and many need genetic counselling. The result is usually quite reassuring as far as risks for the rest of the family are concerned.
3. Unusually deep or prolonged neonatal jaundice (see p. 108).
4. Death in infancy. This can be caused by numerous genetic conditions of slightly later onset, such as sickle cell disease or Blackfan Diamond anaemia (pure red cell aplasia). There have been several suggestions that inherited metabolic disorders can contribute to sudden infant death syndrome (cot death, or SIDS).
5. Failure to thrive, repeated respiratory infections, or specific symptoms such as a bleeding tendency.
7. Mental retardation, deafness, blindness, and autism.

Drawing a family tree

In theory, there is no limit to the potential size of a family tree, since according to the Old Testament we are all descendants of Adam and Eve. However,

Table 8.1 Types of relatives

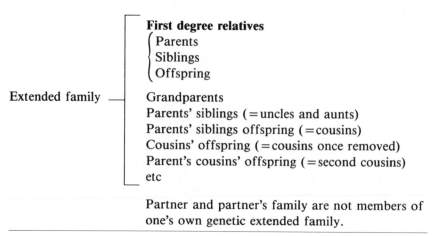

First degree relatives
(Parents
{ Siblings
(Offspring

Extended family

Grandparents
Parents' siblings (=uncles and aunts)
Parents' siblings offspring (=cousins)
Cousins' offspring (=cousins once removed)
Parent's cousins' offspring (=second cousins)
etc

Partner and partner's family are not members of one's own genetic extended family.

the practical possibilities are limited because most people have only about two or three generations' worth of information about their family.

A pedigree can be drawn at two levels. A simple pedigree limited to first-degree relatives (Table 8.1) is usually sufficient in primary care. It is only necessary to progress further if a genetic disease is suspected in a family member (the first family member in whom the condition is suspected is called

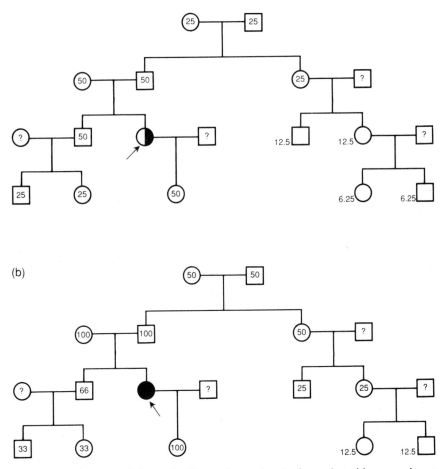

Fig. 8.3 Importance of the family history for tracing carriers, when either a carrier or an affected person is identified. (a) A heterozygote (carrying a single disease gene) is identified (half-shaded, indicated by arrow). The % risk to different relatives of carrying the same gene is shown inside or beside each symbol. This applies equally for recessively inherited conditions (e.g. cystic fibrosis, haemoglobin disorders, Tay–Sachs disease) and for dominantly inherited conditions (e.g. Huntington's chorea, adult polycystic disease of the kidney). (b) A homozygote for a recessively inherited disorder is identified (dark shading, indicated by arrow). The % risk to different relatives of carrying the same gene is shown inside each symbol. The chance that a healthy sibling is heterozygous is 66%, as two-thirds are carriers and one-third are normal.

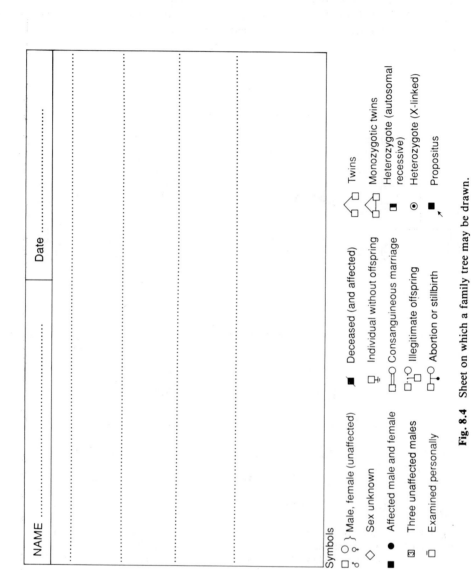

Fig. 8.4 Sheet on which a family tree may be drawn.

the propositus). Clinical geneticists usually draw an extended family tree, including as many members as possible: the rules are clearly presented by Harper (1988). If a family member carries an inherited trait, many members of the extended family may carry the same trait (Fig. 8.3), and it may be advisable to offer them carrier testing if it is available (see Mouzouras *et al.* 1980).

Important points to note in drawing a family tree are:

1. Names.
2. Deaths, age at death, and cause of death when known.
3. Consanguineous marriages.
4. All pregnancies of the couple under discussion (and with previous partners) and their outcome, including miscarriages, still-births, neonatal and childhood deaths, and their causes when known.
5. Any possibility of mistaken paternity, illegitimacy, artificial insemination by donor (AID) etc., although this type of information may be too sensitive to enter on a family tree.

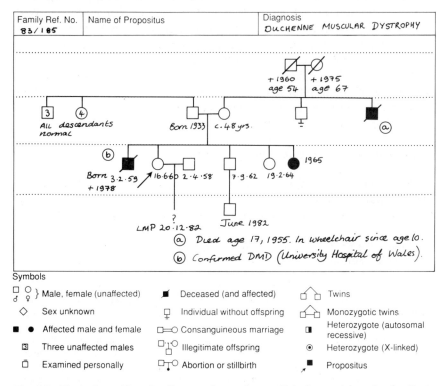

Fig. 8.5 Typical working family tree drawn by a clinical geneticist; the family is transmitting a gene for Duchenne muscular dystrophy. The arrow indicates the propositus, in this case a pregnant married woman who wanted to know if she was at risk for having an affected child. (Reproduced from Harper (1983), with permission).

Figure 8.4 is a typical blank sheet with the recognized symbols, that can be photocopied and used for drawing a family tree for patients when needed. Figure 8.5 shows an example.

We suggest that our readers draw their own family tree at this point. Though relatively few families have members with clear-cut inherited diseases, neutral inherited features such as freckles, and disorders with a genetic component, such as thyroid disease, cancer, heart disease, diabetes, hypertension, or psychotic illness, may be found in every family. It is interesting to note positive traits such as those associated with intelligence (p. 362) as well as negative traits, and instructive to assess one's family's overall genetic fitness (total reproductive performance of each generation, relative to the norm for the population at the time).

Mendelian patterns of inheritance

Dominant, X-linked, and recessive modes of inheritance are summarized in Fig. 8.6. A history of relatives in the extended family with the same or a similar condition may suggest dominant or X-linked inheritance. With recessively inherited disorders there is very rarely a family history: they are usually identified either because the diagnosis in the presenting child is clear-cut, or the condition recurs in a sibling.

Diseases with Mendelian patterns of inheritance make up only about 10 per cent of all congenital disorders (Table 8.2), but they are particularly important because they are often severe (Table 1.2) and involve a high, often avoidable genetic risk for family members.

Autosomal dominant inheritance (Fig. 8.6a)

In dominantly inherited conditions, disease becomes apparent when only one of a pair of chromosomes carries the mutant gene. There are, broadly speaking two groups of dominant disorders, sporadic and familial.

Sporadic disorders: Some disorders, such as severe forms of osteogenesis imperfecta, cannot be inherited because they cause death in infancy or pre-

---

in all carriers of a single abnormal gene. (b) X-linked inheritance. The disorder manifests itself in males who carry a single abnormal gene on their one X chromosome. Female carriers are usually protected by their normal X chromosome. In males, the X chromosome is always inherited from the mother, so if she is a carrier her sons have a 50% chance of being affected. Daughters inherit one maternal and one paternal X chromosome: if their mother is a carrier, they have a 50% chance of being carriers, while if their father is affected they are all carriers. The diagnosis of an affected male suggests that his mother is a carrier, and most female relatives are at risk for being carriers. X-linked disorders may be transmitted undetected in the female line for several generations (see lower part of figure). (c) Recessive inheritance. There is rarely a family history. (Reproduced with permission from the Royal College of Physicians 1989).

(a) Dominant inheritance

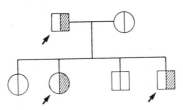

	Male
	Female
	Abnormal gene
	Affected individuals

(b) X-linked inheritance

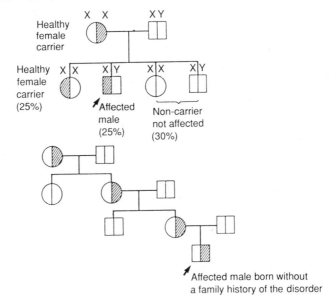

Affected male born without
a family history of the disorder

(c) Recessive inheritance

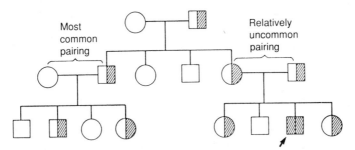

Fig. 8.6 Classical patterns of Mendelian inheritance. Carrier (heterozygote) status is shown by shaded half-symbols throughout. The arrows indicate individuals who have, or may develop, the disease. (a) Dominant inheritance. The disorder usually manifests itself

Table 8.2 Disorders with Mendelian inheritance account for only a small proportion of congenital disorders

Category of disorder	Incidence/1000 births	Inherited/1000 births
Congenital malformations	17–30	ca. 1–2[1]
Chromosomal (pathological)		
Aneuploidies	ca. 2	0
Inherited	ca. 0.6	<0.6
Inherited (single gene)		
Dominant	3.9–4.6	ca. 3–4[1]
X-linked	0.8–2.0	ca. 0.8–1.5[1]
Recessive	0.9–2.0	0.9–2.0
All disorders	25–41	6–10 (=24–30% of total)
Probable recessives, including those involved in malformations		3–4 (=10–12% of total)

[1]Excludes cases due to new mutations, which are quite common among disorders with dominant or X-linked inheritance, and also cause perhaps 1% of congenital malformations.
The Table excludes haemoglobin disorders and G6PD deficiency, which cause no pathology in some populations and a mortality of up to 2% of all births in others. They are discussed separately in Chapters 15 and 17.

vent reproduction. New cases are usually due to new mutations (p. 59), and therefore the recurrence risk within the family is low.

Familial disorders: Conditions like Huntington's disease, adult polycystic disease of the kidney, and familial polyposis coli have a relatively late onset in the reproductive or post-reproductive years. This allows them to be handed on within the family. The children of a person who develops a dominantly inherited disorder have a 50 per cent chance of carrying the same gene and suffering from the same disorder. Unless the disease is due to a new mutation, siblings of people with the disorder also have a 50 per cent chance of carrying the same gene. However, the statistical risk for relatives is quite strongly affected by the passage of time (Harper 1988). The likelihood that a son or daughter of a person with Huntington's chorea will develop the disease falls below 50 per cent as he or she passes the age when the disease presented in the parent. Thus a healthy adult of 20 whose father died of the disease at 50 is statistically more likely to develop it, than a fit person of 35 whose parent was dead by 40 (the average life expectancy of a patient being 16 years after diagnosis).

Rarely, both parents have the same dominantly inherited disease, (for example, achondroplasic dwarfs quite often marry each other, p. 239). On average, 25 per cent of their offspring will be normal, 50 per cent will be heterozygous (affected) and 25 per cent will be homozygous. Homozygotes for achondroplasia have severe skeletal abnormalities and usually die *in utero*, but homozygotes with Huntington's chorea are not clinically very different from heterozygotes.

Carriers of dominant disorders are not very common. Most heterozygotes for dominantly inherited disorders either have the disease or are likely to develop it, so heterozygote diagnosis usually amounts to presymptomatic diagnosis.

Inheritance of X-linked disorders (Fig. 8.6b)

A mutant gene carried on the X chromosome has a characteristic pattern of inheritance. As males have only one X chromosome, any mutant gene is necessarily 'dominant'. Males are described as *hemizygous* for genes on the X chromosome.

In females, mutant genes carried on the X chromosome usually have a recessive mode of inheritance, but the extent to which they are expressed varies unpredictably from one person to another. Normal females are 'mosaics', with the paternally derived X chromosome inactivated in approximately 50 per cent of their cells and the maternally derived X chromosome inactivated in the remainder. The net result is usually balanced expression of maternally and paternally derived genes (p. 31). However, individual variation in X chromosome inactivation can lead to wide clinical variations; e.g.

some female haemophilia carriers have a bleeding tendency, while others have a normal factor VIII level.

Carriers of X-linked disorders are relatively uncommon. Many are related to a known affected male, but some carry new mutations and some have inherited the abnormal gene in the female line, without a known affected male.

Autosomal recessive inheritance (Fig. 8.6c)

Genes for recessively inherited conditions are very common, but only become clinically apparent when an individual inherits the mutant gene from both parents. Since this depends on two carriers meeting by chance, there is rarely a family history. When both parents carry the same recessive gene, on average 25 per cent of their offspring inherit it from them both, 50 per cent are carriers, and 25 per cent inherit two normal genes.

When a first affected child is born with a rare recessively inherited disorder, it may be impossible to say whether the condition is inherited or not. Even when a second child of the same parents is affected, the inherited nature of the disease can be missed if the first died in infancy before a diagnosis was made. For rare recessively inherited diseases then, a 'family history' usually means that two children with the same disease have been born to the same parents. In countries where final family size is small this may never happen, so some recessively inherited disorders may never be diagnosed.

The birth incidence of homozygotes depends on couples of carriers meeting by chance, and so is proportional to the square of the carrier frequency (see Figure 5.6). Therefore, when the homozygote birth incidence is known, the carrier frequency can be calculated, and vice versa. The method shown in Table 8.3, using cystic fibrosis (CF) as an example, has been used extensively to obtain estimates of carrier frequency for many other conditions such as phenylketonuria, congenital adrenal hyperplasia, etc. This calculation can be done only where medical services are good enough for (almost) all affected individuals in a population to be reliably diagnosed.

Conversely, when carriers can be easily identified (as for the haemoglobin disorders or α-1 antitrypsin deficiency), the birth incidence of homozygotes can be worked out by reversing the calculation. This method has been used to estimate the birth incidence of infants with thalassaemia and sickle cell disease in developing countries without facilities for diagnosing homozygotes (WHO 1985). The curve in Fig. 5.6 allows either calculation to be made if either piece of information is to hand. It applies only for lethal, recessively inherited conditions.

Diagnosis and counselling of carrier couples after the birth of their first affected child ('retrospectively') permits them a range of choices if they wish to avoid the birth of a second affected child (p. 187). However, retrospec-

Table 8.3 Calculation of the birth rate of children with a lethal, recessively inherited disease when the carrier frequency is known, and vice versa

---

When the carrier frequency is known

In Britain 4.5% of the population, i.e. 1 in 22 people, carry the CF gene

The chance of one carrier marrying another is 1 in 22 × 1 in 22, so 1 in 484 couples are at risk of having affected children.

On average 1 in 4 of their children will have cystic fibrosis = 1 in 4 × 1 in 484 = 1 in 2000 of all children born

---

When the birth rate of homozygotes is known

In Britain 1 in 2000 children born suffer from cystic fibrosis (CF).

Since only 1 in 4 of the children of couples at risk are affected, 1 in 500 couples must be at risk (i.e. both carriers).

The carrier frequency is the square root of 500 = 1 in 22.4

---

tive diagnosis can have very little effect on the total birth rate of affected children. In practice, it is only possible to make a serious impact on the incidence of recessively inherited diseases if couples at risk can be detected and counselled 'prospectively' i.e. before they start their family. This requires population screening using unambiguous carrier tests that are independent of the family history. Population screening is already possible for Tay–Sachs disease and the haemoglobin disorders, and has led to a very significant decrease in their birth incidence. Pilot studies of carrier screening for cystic fibrosis are now under way.

Chromosomal conditions

Chromosomal rearrangements can show apparently non-Mendelian inheritance. Reasons for this have already been covered (p. 66).

Genetic heterogeneity

Most genetic disorders cover a range of severity. There are variations both in expression (the severity of the disorder in those who are affected) and in penetrance (the proportion of people carrying the gene who are clinically affected). For example, in most dominantly inherited conditions, there is considerable variation in age at onset and severity of the disorder. There may even be 'skipped' generations in which a known carrier of the gene shows no clinical evidence of the disease.

Even apparently clear-cut conditions like achondroplasia may be inherited in several different ways, indicating that mutations of several widely separated genes may cause a very similar clinical picture. Most inherited diseases due to mutation of a particular gene are caused by a number of different mutations, and slightly different mutations can lead to considerable differences in the clinical picture (see also p. 197). Differences in 'genetic background' may also cause marked variation. For example, rhesus haemolytic disease of the newborn can occur when the mother is Rh negative and the fetus is Rh positive, but not when there is also ABO blood group incompatibility between mother and fetus (p. 337).

Pitfalls and problems in making a genetic family tree

There are many difficulties in attempting to trace carriers, particularly of dominant or X-linked disorders:

1. Psychological factors are particularly important in dominantly inherited conditions, because carrier testing is usually carried out before symptoms are evident. In the absence of effective treatment, the personal implications are so serious that carrier testing for Huntington's disease, for example, has been taken up very cautiously (p. 232).
2. The inheritance pattern can be confused if family members who would have suffered from a late onset condition died earlier of some other cause, and when X-linked disorders are transmitted in the female line for several generations.
3. Many disorders (especially dominant ones) are very variable in expression.
4. Dominant and X-linked disorders are often due to new mutations. For example, only about two-thirds of boys with Duchenne muscular dystrophy have carrier mothers.
5. The diagnosis of a child with a dominantly inherited disorder due to a new mutation can lead to misinformation for the family if it is assumed that the recurrence risk in subsequent pregnancies is high. In reality it is usually negligible.

Such problems and the wide range of genetic heterogeneity means that, except in the simplest cases, it takes a clinical geneticist to determine the inheritance patterns of diseases.

The role of the primary care worker

Primary care workers have a key role (1) in suspecting genetic disease and referring families for diagnosis and counselling, and (2) in identifying possible carriers and ensuring that they receive any information and counselling they may need.

To be able to provide carriers of inherited diseases with information about risk and how to avoid it, one must be able to identify them. Many potential carriers of dominant or X-linked disorders can be identified through the family history. The next step, to distinguish clearly between carriers and non-carriers within the family, is possible only if specific biochemical or other tests are available. More such tests are becoming feasible, and carriers who could not be identified yesterday may be identifiable today.

It is therefore appropriate for a family doctor both to refer new cases, and also to review practice patients on whom a genetic diagnosis has already been made, to see if further help for the family may now be available. In case of doubt, it is appropriate to refer the family back to a clinical geneticist. A referral should include as much information as possible, including data on the extended family and photocopies of original reports.

Primary care workers are likely to be particularly heavily involved with carrier screening and counselling for the common recessively inherited conditions discussed in Chapters 15 and 16 (Table 16.1). A substantial proportion of the patients on a family doctor's list carry one of these

Table 8.4 Genetic risk for the family, when either a child, or one parent, is found to be heterozygous for a common recessively inherited condition

Example	Prevalence of heterozygotes	% chance that *both* parents are carriers	% chance that next child will be homozygous
Phenylketonuria	1–2	1–2	0.25–0.5
α1-antitrypsin deficiency	1–4	1–4	0.25–1
Thalassaemia in N. Indians	3	3	0.75
Cystic fibrosis in Europeans	4–5	4–5	1–1.25
Tay–Sachs disease in European Jews			
Thalassaemia in Pakistanis, plus consanguinity	6.5	18	4.5
Sickle cell disease in Afro-Caribbeans	12	12	3.0
Thalassaemia in Cypriots	17	17	4.3
Sickle cell disease in West Africans	>20	>20	>5

conditions. In ethnic groups at risk for haemoglobin disorders, there is a relatively high risk that a known carrier will choose another carrier as a partner, and have a homozygous child (Table 8.4).

Multifactorial disorders

Family and twin studies show a strong genetic element in many of the common chronic diseases of middle life and old age (Tables 4.3 and 8.5). These disorders occur in families more frequently than would be expected by chance, but less frequently than disorders caused by single gene defects with Mendelian patterns of inheritance. Many normal traits such as height, blood pressure, and intelligence are inherited in a similar fashion. It is thought that these characteristics are influenced by the interaction of several genes (polygenic effects) or of genetic and environmental factors (multifactorial effects). Congenital malformations are the main group of multifactorial disorders discussed in this book. A multifactorial disorder may be the endpoint of many different processes: some cases may be almost entirely genetic in origin, others almost entirely environmental.

Both the prenatal and postnatal environments may contribute to the development of certain multifactorial diseases. Barker and co-workers (1989, 1990) found that adult hypertension occurred more frequently in people who, at birth, had a combination of a low birth weight and large placenta. Also the death rate from ischaemic heart disease and obstructive airways disease was inversely related to weight at birth and at one year.

The empirically observed risk to relatives is an indicator of the genetic component in a condition i.e. its 'heritability'. (The nearer the heritability of a condition is to 100, the greater the importance of genetic factors (Vogel and Motulsky 1986)). The heritability of a condition can also be estimated from studies of its concordance in non-identical and identical twins (see Table 4.3). Such information is useful in answering patients' questions about the probability that their siblings will develop the same problem, or that they may pass it on to their children.

The following generalizations about multifactorial disorders are based on such empirical studies. Most of these conditions are very heterogeneous, and the generalizations reflect this.

1. There is an empirical 2–10 per cent risk of recurrence in siblings and offspring of an affected person.
2. For some types of conditions, the more severe the disease, the greater the risk of recurrence in other family members, and the more likelihood there is of a positive family history (i.e. the greater its genetic component is likely to be).
3. The risk of a further child with the disorder is increased if more than one sibling is affected.

4. The risk of recurrence in other family members diminishes rapidly in more distant relatives.
5. Some diseases are distributed unequally between the sexes. For example, pyloric stenosis, is six times more common in boys than in girls. A family history is much more likely if the affected individual is of the rarely affected sex.

Possibilities for sorting out the genetic and environmental components in common diseases are steadily improving, allowing recurrence risks to be worked out more exactly than by empirical data alone. For example, most family doctors are aware of the strong association between some conditions and certain histocompatibility antigens (Table 7.2). First degree relatives of a propositus with ankylosing spondylitis who carry the HLA B27 antigen are much more likely to develop the disease than those without the antigen.

In reality, describing a condition as 'multifactorial' simply means that we know there is a genetic element there, but we do not know much about it. Our ability to unravel the genetic and environmental components in a given disorder depends on the current level of knowledge. The limits of our knowledge are illustrated by the fact that we know so little about the genetic contribution to the differences in mortality and morbidity between the sexes (Silman 1987). In developed countries, a woman can expect to live about 6 years longer than a man, but the genetic, social, and environmental components of this difference remain uncertain. Infant mortality tends to be higher among boys than girls. Young women are less likely to be killed in an accident or by violence, alcohol, or drugs. Cigarette smoking is much commoner in European men, but the difference in smoking rate between the sexes is declining. In the 35–44 age group in the UK, the incidence of coronary heart disease (CHD) in men is more than 6 times that in women, and even in men aged 65–74, morbidity from CHD is almost twice that of women (Lerner and Kannel 1986). Between the ages of 55 and 64, the death rate of men from all causes is almost twice that of women. However, once a woman gets seriously ill, her chances of recovering seem little different from those of a man with the same illness, and the two groups have similar five-year survival rates for the common cancers (Silman 1987).

There is obviously a strong genetic input into these differences, sex being genetically determined. There may be differences in some of the effects of the X and Y chromosomes, but most differences seem likely to be secondary to the genetically programmed differentiation of the gonads in embryonic life, leading on to different effects of the hormones they produce in adult life.

Neonatal jaundice, diabetes mellitus, and coronary heart disease all follow the rules for multifactorial causation listed above, and may be used to illustrate the concepts of multifactorial inheritance.

Table 8.5 Genetic influences in common conditions

Condition	Risk to first degree relatives	Comments
Affective disorder	9% for relatives of patient with unipolar illness	Risk is approximately doubled if illness is bipolar (Owen and Murray 1988)
Alzheimer's Disease	3% for relatives of an isolated case	Much higher risk if there is a strong family history
Asthma	13% risk to relatives of extrinsic atopic patients	Affects up to 10% of the population
	5% for relatives of intrinsic cases	(Sibbald and Turner-Warwick 1979)
Baldness (severe early onset, male)		Probably autosomal dominant with expression only in males
Colour blindness	50% risk to both male and female children if an affected male marries a carrier female	X-linked occuring in about 8% of males and 0.4% of females

Diabetes (type I)	For children of a diabetic parent, and siblings of a diabetic child, 3% risk of diabetes in first 20 years of life	If both parents have type I diabetes, offspring have 20% risk of diabetes in first 20 years, and 50% at some stage
Diabetes (type II)	10% risk for 1st degree relatives	
Eczema	50% likelihood of some allergic problem if one parent is affected	Often associated with other evidence of allergy. May be autosomal dominant disease of variable penetrance
Idiopathic epilepsy (excluding febrile convulsions)	Up to the age of 20 years, 4% if one parent is affected	Increases to 15% if both parents have epilepsy, and 10% if one parent and a sibling have the disease
Migraine	Approximately 45%, increasing to 70% in offspring of two affected parents	
Multiple sclerosis	1%, increases to 5% if a parent and sibling are both affected	
Psoriasis	10% if one relative affected. 20% if two first degree relatives affected	Associated with HLA-CW6
Schizophrenia	10% if sibling affected, rises to 15% if parent affected	3% risk for second degree relatives

(Partly based on Harper 1988)

Neonatal jaundice

Factors contributing to neonatal jaundice do so by virtue of their effects on bilirubin metabolism (Fig. 8.7). Most bilirubin produced in the fetus crosses the placenta and is metabolized by the mother. Separation from the mother leads to the 'physiological jaundice' that develops in most infants by the second or third day of life. Synthesis of the liver enzyme responsible for conjugating bilirubin, UDP glucuronosyl conjugase, is stimulated by a rising bilirubin level, so physiological jaundice usually disappears by the tenth day of life. Neonatal jaundice is especially pronounced in premature infants because this enzyme is less readily induced. Jaundice appearing in the first day of life, or after the third, or deepening rapidly, or persisting unduly long, can be due to any of the genetic or environmental causes indicated in Fig. 8.7, and is often due to several combined. When severe, it can have serious consequences (p. 338). The Rhesus d gene and G6PD deficiency (p. 249) are well-known genetic causes of neonatal jaundice that can lead to a perinatal mortality of from 1–3 per 1000 in different populations, especially when combined with any of the other factors shown in the figure.

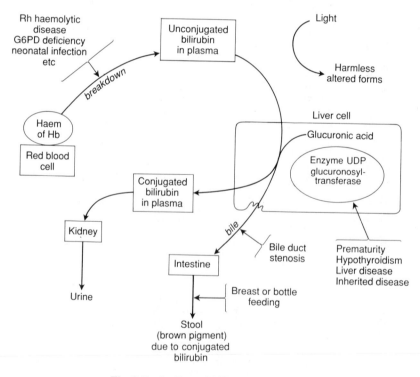

Fig. 8.7 Outline of bilirubin metabolism.

With time our understanding of other multifactorial conditions will equal the level of our understanding of neonatal jaundice.

Diabetes mellitus

The prevalence of diabetes mellitus is around 2 per cent, so one should expect about 200 patients in a group practice looking after 10 000 people: usually half will be undiagnosed. It occurs if the pancreas produces insufficient insulin in response to a rise in blood glucose, or if insulin is not used appropriately. It appears to be due to the interaction of several genes with environmental factors such as viral infections and obesity. The genetic element is rapidly becoming clearer. There is no evidence of any disturbance in or near the insulin gene itself (on the short arm of chromosome 11) in most forms of diabetes. However, there is evidence that genetic variants of the insulin receptor gene (on chromosome 19) are involved in some forms of the disease.

Type I diabetes classically presents in a child or young adult, with polyuria, polydipsia, malaise, loss of weight, and infections. Most patients need insulin and are at risk of developing ketoacidosis. Prevalence of type I diabetes in young adults is about 2 per 1000 in the UK, and may be increasing. Susceptibility is increased by genes linked with HLA types such as B8, B15, DW3, and DW4. It seems that some viral infections may evoke antibodies that cross-react with these HLA antigens on the surface of the pancreatic β cells, leading to their autoimmune destruction. Risks to first degree relatives of patients are shown in Table 8.5 (Harper 1988): risks to the fetus are discussed on p. 318.

People with **type II diabetes** develop resistance to the insulin they produce. Many are overweight, and the disease usually develops in middle or old age. The onset is usually less acute than in type I and the blood sugar can usually be controlled either by diet alone, or by a combination of diet and oral hypoglycaemic drugs. There are about three times as many patients with type II as with type I diabetes, and the genetic contribution is greater. Some ethnic groups, e.g. people originating from the Indian subcontinent, are much more susceptible than Europeans, and about 10 per cent of first degree relatives of type II diabetics will eventually develop the condition. If one of a pair of monozygotic twins develops type II diabetes, it is likely that the other will also develop it at some future time. A history of diabetes in a first degree relative is a good reason to avoid obesity.

Coronary Heart Disease (CHD)

Cardiovascular disease is responsible for about a third of deaths of men aged 25–64. About two-thirds of these deaths are due to coronary heart disease (CHD), occlusion of the coronary arteries as a result of atherosclerosis, which is a result of disturbed lipid metabolism. The transport and metabolism of lipids involves a large number of different types of molecules, with many alternative pathways and regulating mechanisms; disturbed lipid metabolism interacts with other factors leading to plaque disruption and clotting within the coronary arteries. Therefore it is not surprising that the prevalence of CHD varies with country, time, and a number of risk factors, some genetically and some environmentally determined.

A recent Finnish study of 1520 healthy men aged 40–59 years followed for 25 years, confirmed that a high serum cholesterol (> 8.4 mmol l, raised systolic blood pressure (> 150–160 mm Hg) and smoking 10 or more cigarettes a day, were each associated with about twice the chance of death from CHD compared with men without the relevant risk factor (Pekkanen *et al.* 1989). Other risk factors include diabetes mellitus, obesity, and a high alcohol intake. Many of these factors themselves have a genetic component.

Genetically determined susceptibility is particularly important when onset is before or during middle age. One of the most important single risk factors is a family history of premature CHD. Slack and Evans (1966) showed a marked increase in death from ischaemic heart disease in first degree relatives of an index case (Table 8.6). About 10 per cent of early onset CHD is due to single gene disorders of the lipid transport pathway (p. 00), but the rest are multifactorial in origin. As with other multifactorial conditions, when the affected person is of the sex less likely to get the disease (female in this case), relatives are at higher risk.

Table 8.6   Risk of death from ischaemic heart disease between the ages of 35 and 55 in first degree relatives of index patients with ischaemic heart disease

	Male index case (35–54 years)	Female index case (35–64 years)
Male first degree relative	1 in 12 (×5)	1 in 10 (×6.5)
Female first degree relative	1 in 36 (×2.5)	1 in 12 (×7)

(Based on Slack and Evans 1966)

The risk of coronary heart disease can be reduced by a cholesterol-lowering diet, stopping smoking, controlling weight and blood pressure, and taking regular exercise. In a study group, this approach decreased the amount of CHD by about 10 per cent over a 6 year period (WHO European Collaborative Group 1986). In the United States, between 1968 and 1979, the death rate from CHD (ages 35–74) fell from 8 to 5 per 1000 in men, and from 3.3 to 1.8 per 1000 in women (Marmot 1985), reversing the increase of previous decades. The decline has been far less in Europe.

From the primary care worker's point of view, the important question is, will increased understanding of the genetic aspects of common multifactorial disorders lead to screening tests that can be used to identify susceptible people? It seems likely, for instance, that screening for genetic predisposition to some cancers could become feasible in the future. This might be very useful when early diagnosis and treatment can improve the prognosis, and for conditions that can be prevented or ameliorated by appropriate changes in lifestyle. However, this entire area is clouded with uncertainties. In order to use such information, we need to know a great deal more both about the inheritance of these conditions, and how people react to knowledge of risk. The desirability of screening for risk factors for common disorders such as coronary heart disease is discussed in Chapter 13.

Key references

Harper, P.S. (1988). *Practical genetic counselling*, (2nd edn). Wright, Bristol.
Vogel, F. and Motulsky, A.G. (1986). *Human genetics: problems and approaches*. (2nd edn). Springer-Verlag, Berlin.

9. Consanguineous marriage

People who wish to marry a relative, usually a first cousin, often ask to see a clinical geneticist. The proposal may have legal, religious, social, and genetic dimensions.

Most societies consider marriage or sexual relationships with first degree relatives (Table 8.1) as incestuous, and forbid them by law, for genetic among other reasons (Anonymous 1981). Some societies (including most of Europe until the recent past) extend the prohibition to first, or even to second cousins. The underlying reasons for such wider prohibitions seem mainly social and anthropological (Fox 1966), but the emotional associations of incest often colour perceptions of consanguineous marriage. The couple may therefore have to deal with social prejudice, as well as with legal and genetic implications.

The frequency of cousin marriage has recently increased in the UK and other European countries due to migration from parts of the world where consanguineous marriage is positively favoured, such as Pakistan, the Middle East, North Africa, and Turkey (WHO 1986). It is therefore becoming important for primary care workers to be able to give basic helpful advice to young people who plan to marry a relative, and to their families.

Genetic implications of consanguineous marriage

A consanguineous marriage is one in which the partners have a common ancestor such as one or both grandparents, or a more distant relative. The three commonest types of consanguineous marriage are between first cousins, second cousins, or first cousins once removed. *Close* consanguineous marriage increases the likelihood that recessively inherited characteristics will emerge in the offspring, because the partners inherit a significant number of identical genes from their common ancestor (Table 9.1). Recessive genes that both partners inherit may then be passed on to their children in the homozygous state. Consanguineous marriage does not significantly affect characteristics with dominant or X-linked inheritance, since the way they are handed on is independent of how carriers select their partners. The table shows that the chance of a child being homozygous for a given recessive gene is one quarter of the chance that both partners carry the gene. These chances fall rapidly as relationship becomes more remote.

Figure 9.1 compares the relationship between heterozygote frequency and the birth rate of homozygous children with random mating (lower curve)

Table 9.1 Proportion of all genes identical in different types of relatives

Relative	Chance of any gene being identical (% of genes identical)	Chance for any gene that a child of this mating will inherit both copies from a common ancestor (%)
Parent, brother, sister	$\frac{1}{2}$ (50)	25
Grandparent, aunt or uncle; double first cousin[2]	$\frac{1}{4}$ (25)	12.5
First cousin[1]	$\frac{1}{8}$ (12.5)	6.25
First cousin once removed[1]	$\frac{1}{16}$ (6.25)	3.13
Second cousin[1]	$\frac{1}{32}$ (3.06)	1.5

[1] Common types of consanguineous marriages.
[2] Uncle–niece marriage is common in South India.

Uncle–niece   First cousins   First cousins once removed   Second cousins

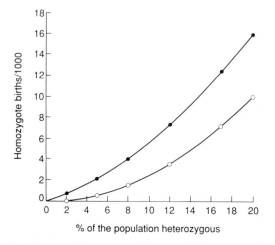

Fig. 9.1   Relationship between the per cent of the population who are heterozygotes and the birth rate of homozygotes for recessively inherited conditions, with 100% cousin marriage (upper curve) and random mating (see also Fig. 5.5). If the proportion of first cousin marriage in a population is known, a rough estimate of the population effects may be made using this chart (reproduced from WHO 1985*b*, with permission.)

with a situation in which 100 per cent of the population married their first cousin (upper curve). The convention increases the birth rate of homozygotes for recessively inherited traits in relation to the frequency of carriers. Most such traits are not pathological, so in one way the custom brings out the latent genetic diversity in a population. For example, in Afghanistan and Pakistan there are more blue-eyed children for a given frequency of this recessively inherited gene, than in 'randomly mating' populations.

Common consanguineous marriage also increases the birth frequency of children with recessively inherited diseases relative to the carrier frequency in the population. When a recessive gene is common, as β-thalassaemia is in Cyprus (17 per cent) or Sardinia (12 per cent), there is a relatively high chance that one carrier will marry another however they choose their partner, and the risk is approximately doubled in a first cousin marriage. However, genes for most recessively inherited disorders are carried by less than 1 in 200 of the population, and it is rare for a carrier to choose another carrier as partner, unless they are related. Thus consanguineous marriage particularly favours the manifestation of *rare* recessively inherited disorders, including metabolic diseases, deafness, some types of congenital malformation, and possibly mental retardation (Schull and Neel 1956). An increased birth rate of diseased infants is therefore to be expected in populations favouring consanguineous marriage, but existing evidence indicates that the effect may be relatively small (Bittles 1980). This may be because recessively inherited diseases cause a relatively small proportion of congenital disorders (Table 8.2), but there may also be other reasons.

Common consanguineous marriage increases the frequency with which the lethal recessive genes present in a population are *expressed*, but with the passage of time it may *decrease* the frequency of these genes because the homozygotes do not reproduce. When they die, two abnormal genes disappear from the population's gene pool, that would not have disappeared if they had been transmitted in the heterozygous state. The number of abnormal genes available to be passed on therefore falls with each successive generation, and the birth incidence of affected children would be expected to fall slowly back towards the level in populations without a consanguineous marriage pattern (Sanghvi 1966). Few studies conducted so far have been really satisfactory (Bittles *et al.* 1991), but there is evidence that this may be the case in Southern India, where monogamous close consanguineous marriage has been traditional, it is said, for several thousand years. However, it is not possible to generalize from these findings. Considerable differences between populations are likely, depending on the nature, frequency, and duration of the practice of consanguineous marriage.

In advising relatives who wish to marry, the geneticist first draws up a detailed family tree. If there is evidence that any family member may have, or may have died from, a genetic disease, investigations will be carried out to try to establish the true diagnosis. If the condition proves not to be inherited (e.g. Down syndrome or many congenital malformations), or not to be recessively inherited (e.g. a disorder due to a new mutation), the couple can be reassured. If the condition proves to be inherited, it may be possible to offer them carrier testing and prenatal diagnosis. In any case, they can be informed of their statistical risk of having children with the same disorder (Table 9.2).

However, in most cases there is no evidence of a genetic disorder in the family. The couple may then be informed of the empirically observed risk of consanguineous marriages in general. Where a couple are first cousins, both empirical studies and theoretical calculations indicate an additional total risk of still-birth, neonatal or childhood death, or congenital malformation of about 3 per cent (Harper 1988). When added to the general risk of about 2.5 per cent, this gives a total risk for such couples of 5–6 per cent. Put the other way round, they have a 94–95 per cent chance of a healthy baby in each pregnancy, as against the general population chance of about 97 per cent. It seems that most related couples find the discussion reassuring, and proceed to have a family.

Whatever its net effect on the incidence of congenital disorders, the convention of consanguineous marriage can, in principle, simplify the geneticist's work, because it alters the *distribution* of births of children with recessively inherited diseases. They tend to cluster within extended family groups instead of occurring sporadically throughout the population,

Table 9.2   The family history is particularly important in identifying risk of recessively-inherited disease where there is consanguineous marriage

Person under consideration   Per cent chance of an at-risk marriage

Status	Risk of being a carrier (%)	If partner unrelated	Additional risk if partner is:		
			1st cousin	1st cousin once removed	2nd cousin
Known carrier	100	n	12.5	6.25	3.13
Sibling has homozygous disease	66	0.66 × n	16.5	8.25	4.13
Sibling is a known carrier	50	0.5 × n	6.25	3.13	1.57

n = per cent carrier frequency in the population
Examples
The chance of an at-risk marriage for the sibling of a British Pakistani patient with β thalassaemia major who marries a first cousin is 0.66 × 6 (prevalence of β thalassaemia trait) + 16.5 = 20.5%.
The risk for the sibling of an English patient with CF who marries a first cousin is 0.66 × 4.5 + 16.5 = 19.5%.
These and many other risks can be investigated and clear information given.

and recessive genes that are present are likely to manifest themselves in a homozygote somewhere within the family. However, to exploit this advantage, appropriate genetic counselling facilities are necessary.

The convention of consanguineous marriage

Figure 9.2 (WHO 1985*b*) shows that nearly 20 per cent of the world population live in societies where consanguineous marriage is favoured. The convention has a long and respected history (Fig. 9.3) and often, but by no means always, coincides with Muslim religion. A recent study of the social implications of thalassaemia among Muslims of Pakistani origin in England (Darr 1990) casts light on its important social functions.

The most striking features of the families in this study grappling with the issue of prenatal diagnosis was that there was not a single case in which a husband overruled his wife's decision on whether to terminate a pregnancy or not; that none of the women were blamed for the disease; that both parents, and in some cases other members of the family, discussed the issue openly; that little stigma was attached to having an inherited disease in the family, illustrated by their open discussions in the support group meetings and subsequent visits to each other's homes; that women discussed amongst themselves their experience of prenatal diagnosis and termination at the first support group meeting; and that no interest has been shown in fetal sexing.

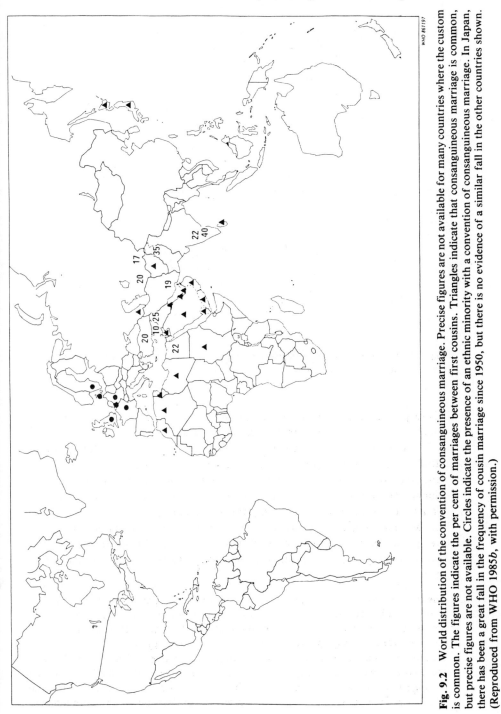

Fig. 9.2 World distribution of the convention of consanguineous marriage. Precise figures are not available for many countries where the custom is common. The figures indicate the per cent of marriages between first cousins. Triangles indicate that consanguineous marriage is common, but precise figures are not available. Circles indicate the presence of an ethnic minority with a convention of consanguineous marriage. In Japan, there has been a great fall in the frequency of cousin marriage since 1950, but there is no evidence of a similar fall in the other countries shown. (Reproduced from WHO 1985*b*, with permission.)

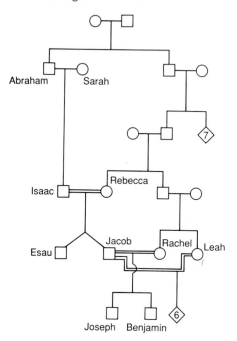

Fig. 9.3 Family tree of the patriarchs Abraham, Isaac, and Jacob. Both Isaac and Jacob made consanguineous marriages.

These points 'illustrate the power of the women over their reproduction, which serves as an indicator of their position in the family'.

This experience contrasted strongly with observations in Indian (Hindu) families with thalassaemic children, in groups that customarily avoid consanguineous marriage, both in London and in India (Sangani *et al.* 1990). Marked stigma tends to be attached to inherited disease; women may be blamed for bringing a sick child into the husband's family; some women hide the fact that their child is unwell even from their own parents or sisters, and are consequently very isolated and depressed. In addition, Indian Hindus sometimes request fetal sexing in order to avoid the birth of a female child (p. 187).

The researcher concluded that if the majority of the marriages within a society are in the same biraderi (loosely, = extended family) then it follows that the children belong to the extended family of their mother as well as their father. Hence females have as strong and supportive a network of relations as the males, including both the maternal and paternal sides of their family. It is because of this multiplicity of relationships that the majority of women are not isolated and blamed for bringing a disease into the husband's family, for to blame one's wife's family is akin to blaming one's own. Support structures for women as well as men and children are inherent in the kinship pattern of societies that practice consanguineous marriage. These structures are in effect created by that marriage pattern.

A Lebanese study also showed that many women consider consanguineous marriage protective (Khlat *et al.* 1986).

Genetic problems are not yet recognized as important in countries where consanguineous marriage is common because they begin to emerge only when the infant mortality is low (WHO 1985*b*). However, infant mortality falls suddenly from over 10 per cent to less than 2 per cent in groups originating from such areas who settle in Europe. At the same time, the frequency of consanguineous marriage may increase. For example, the proportion of first cousin marriages among British Pakistanis appears to have risen from about 32 per cent (in Pakistan) to about 55 per cent in the UK (Darr and Modell 1988), because of the social changes associated with migration. Genetic disease is beginning to be perceived as an important cause of childhood mortality and morbidity in these groups, and the relation of traditional consanguineous marriage to the pattern of genetic disease is beginning to be questioned.

In some countries, efforts are now being made to reduce the incidence of genetic disease through media campaigns against consanguineous marriage (Modell 1990). Taking into account what has been said above, it is clear that attempts to disrupt this traditional marriage pattern could do more harm than good. There is no evidence that such campaigns affect the frequency of consanguineous marriage, but they certainly make people feel bad about it.

When prejudice is widespread, as it is with respect to consanguineous marriage in Europe, similar destructive pressures may be applied unintentionally. In the study of British Pakistani families with thalassaemic children mentioned above, it was found that all the families had been told at least once by various health workers that the child was sick because the parents were related (Darr 1990). The real cause of the child's illness is that it has two genes for a specific disorder. It is true that the chance that this will happen is increased when the parents are related, but there can be no guarantee of avoiding an inherited disease by choosing an unrelated partner. The parents know that such 'explanations' are false because they live among related couples without sick children. The net result is that their confidence in health workers is undermined at a time when they particularly need their help.

'Random inbreeding' and genetic founder effects

Perceptions of the genetic implications of consanguineous marriage tend to be confused with the genetic effects of inbreeding in small isolated populations (see Eriksson *et al.* 1980).

When an island or a remote or newly-colonized area is settled by a *small number* of founding members, the gene pool available for their descendants is limited. For example, genes for many recessively inherited disorders are carried by less than 1 in 200 of the population. Genes for dominant and sex-linked disorders are far rarer. Therefore a population derived from less than 200 founding members will have the following genetic characteristics.

(1) The range of genetic variation represents only a limited sample of that present in the larger population of origin.

(2) Genes for inherited diseases introduced by founder members are likely to occur in more than 1 in 200 of subsequent generations.

(3) Some individuals contribute disproportionately to the gene pool for non-genetic reasons, ranging from different levels of sexual activity to random effects of natural catastrophes. In large populations such random effects tend to balance out statistically.

(4) As long as the population remains isolated, people can choose partners only within the group. This last is the only feature that is related to the issue of consanguineous marriage.

In such populations a pathological gene carried by only one of the founding members can be passed on to a disproportionate number of descendants, and a high frequency of a particular condition may arise. Such genetic 'founder effects' are common in South America and isolated parts of Europe such as northern Scandinavia and Russia. They are most conspicuous when they involve dominantly inherited disorders, as in the Venezuelan community in which the locus of the gene for Huntington's chorea was first mapped (Gusella *et al.* 1983), but the birth incidence of infants with a particular recessively inherited disorder may also be increased, as for hereditary nephrosis in part of Finland (Norio 1966, 1980). Conversely, many small isolated populations may end up with less than the average incidence of pathological genes, but they do not come to medical attention because they do not present specific genetic problems.

Founder effects are often mistakenly thought to illustrate the deleterious effects of inbreeding, but in fact they are genetic consequences of small population size, and the conditions most often involved are dominant or sex-linked disorders, whose frequency is almost unaffected by consanguineous marriage.

Role of the primary care worker

Populations with a tradition of consanguineous marriage have a special need for an informed approach in primary care, for an expert genetics service, and for counselling that is sensitive to social and cultural background. The genetic implications need to be seen in its social context, and misinformation and criticisms avoided.

For effective genetic counselling, it is essential to communicate with the *woman*. With British Pakistanis, language (usually Punjabi), Muslim religion, limited literacy in English, and the practices of purdah (in a modified form) and consanguineous marriage, make for special genetic counselling needs, at least for the present generation. The ideal counsellor for a British Pakistani Muslim family is a Punjabi-speaking woman, preferably Muslim, who is prepared to visit at home, several times if necessary (Royal College of Physicians 1989). It may be difficult to arrange such counselling in primary care, but ethnic counsellors are increasingly available at clinical genetics or sickle cell and thalassaemia centres.

Medical attention can be most productively focussed on the minority of families with positive evidence of genetic disease, and hence a potentially high recurrence risk.

1. When cousins from a large consanguineous family marry, or plan to marry, a careful family history should be taken. If there is any question of a genetic disease in the family, the couple should be referred for expert genetic advice.
2. If there is no indication of the presence of a genetic disorder in the family (the most likely situation), it is reasonable to inform the couple of the somewhat increased risk of a child with an inherited disorder, and to offer all available help. This should include advice on avoiding non-genetic risks to the pregnancy (dietary deficiencies, anaemia, rubella immunization, etc.), and the opportunity to consult a clinical geneticist if they wish.
3. Screening should be offered for common detectable inherited conditions such as thalassaemia, sickle cell disease or (in the future) cystic fibrosis.
4. Screening in pregnancy is particularly important (Chapter 21) and should include expert fetal anomaly scanning.
5. When a recessively inherited disorder is diagnosed in such a family, investigation of the extended family may permit many carriers and couples at risk to be identified and counselled before they have affected children. This opportunity can best be seized if an appropriate genetic counsellor is available (Royal College of Physicians 1989).

Key references

Bittles, A. H. (1980). Inbreeding in human populations. *Biochemical Reviews*, **50**, 108–17.
Bittles, A. H. *et al.* (1991). Reproductive behaviour and health in consanguineous marriages. *Science*. (In press.)
Fox, R. (1966). *Kinship and marriage*. Pelican Books, London.
Harper, P. S. (1988). *Practical genetic counselling*, (2nd ed.). Wright, Bristol.
Royal College of Physicians (1989). *Report on prenatal diagnosis and genetic*

screening; community and service implications. The Royal College of Physicians of London.

WHO (1985*b*). *Community approaches to the control of hereditary diseases.* WHO Advisory Group on Hereditary Diseases. Unpublished WHO document HMG/WG/85.4. May be obtained free of charge from: The Hereditary Diseases Programme, WHO, Geneva, Switzerland.

10. Genetic diagnosis: the place of DNA

In this chapter we briefly summarize the laboratory methods used to diagnose genetic disease, with particular emphasis on the potential and limitations of DNA technology.

Karyotyping

Karyotyping is the diagnostic study of human chromosomes. It can be done using any cells providing that they can be induced to divide in tissue culture. Lymphocytes from peripheral blood or fibroblasts from a skin biopsy are often used for diagnosis in a child or adult. Fetal cells for prenatal diagnosis are usually obtained from amniotic fluid, fetal blood, or chorionic villus material.

It is usually necessary to grow the cells (under meticulously sterile conditions) to generate enough cell divisions for analysis. Lymphocytes from adult or fetal blood require only 3 days in culture since they can be stimulated to divide rapidly. Amniotic fluid cells are grown for 2 or 3 weeks in special flasks, where they slowly spread as a single-layered sheet. Chorionic villus material is usually cultured for 2 days or 2 weeks. When enough cell divisions are taking place, the culture is prepared for study.

Chromosomes can be examined most easily at the metaphase stage of mitosis after the nuclear membrane has disappeared and they are arranged in a ring round the edge of the metaphase plate (Fig. 3.3). To maximize the number of mitotic figures available for study, the culture is exposed to colchicine for a few hours. This prevents dividing cells from completing mitosis by interfering with the mitotic spindle, so cells that start to divide after the colchicine is applied fail to complete the cycle, and accumulate in metaphase. The culture is then fixed and stained using methods that give a consistent pattern of bands, specific for each chromosome (Fig. 2.13). The chromosomes of a given number of metaphase plates are counted and examined in detail for abnormalities. One way to ensure accuracy is to take a photograph, cut out the chromosomes and arrange them in pairs; but many experienced cytogeneticists can make an analysis by careful visual examination of the preparation.

Chorionic villus material is growing very fast at the time of sampling, and often produces enough mitoses to permit a diagnosis directly or after only

1–2 days incubation. Rapid methods may be suitable for counting chromosome numbers, but are usually inadequate for diagnosis of chromosomal rearrangements, because the banding pattern is inferior to that of cells grown in longer term cultures.

Physiological and biochemical tests

Though many inherited syndromes are clear-cut clinically, such as thalassaemia, sickle cell disease, cystic fibrosis, a clinical diagnosis should always be backed up by diagnostic testing when possible. When the underlying genetic abnormality is unknown, diagnostic tests depend on evidence of typically disturbed physiology, e.g. the sweat test in cystic fibrosis or an electromyogram in muscular disorders. However, such secondary tests often have limitations. Seven (4 per cent) out of 147 children with a diagnosis of cystic fibrosis referred to one reference centre in the UK had been misdiagnosed because the sweat test had not been performed properly (Shaw and Littlewood 1987). Tests should be done by adequately trained people, and more precise diagnostic methods based on protein or DNA analysis are desirable.

Biochemical diagnosis usually depends on assaying a protein gene product such as haemoglobin, or measuring the activity of an enzyme in tissues from the individual under study. Table 10.1 lists diagnostic tests used for some common genetic diseases described in this book.

Diagnosis at the protein level has the advantage that a single measure of abnormality — deficiency of the end-product — is sufficient for a diagnosis, even though a wide range of mutations often underlie a particular disease. It is unlikely that biochemical diagnosis will be completely supplanted by DNA diagnosis, but protein methods have considerable disadvantages.

1. To study the protein product of a gene directly, one must examine a tissue in which that product is found, and this may be difficult to obtain. For example, phenylalanine hydroxylase, the enzyme that is deficient in phenylketonuria, is found only in liver tissue. Many protein abnormalities are therefore studied through secondary effects, such as the plasma phenylalanine level in phenylketonuria. However, this is not possible in the fetus, because metabolites cross the placenta and are cleared by the mother, so fetal tissue biopsy and enzyme assay may be necessary for prenatal diagnosis.
2. Biochemical diagnosis of heterozygotes is only rarely possible. Even when there is no overlap between the ranges in normal and affected individuals, there is usually a large overlap between normals and heterozygotes. Tay–Sachs disease, α-1 antitrypsin deficiency and the haemoglobin disorders are exceptions.

Biochemical carrier diagnosis is particularly complicated in X-linked

Table 10.1 Conventional diagnostic tests for some common inherited diseases

Condition	Test	Distinguishes	
		Homozygote	Carriers
Cystic fibrosis	Sweat test	+	−
Haemoglobin disorders	Electrophoresis, red cell indices etc.	+	+
PKU	Phenylalanine assay	+	−
Tay–Sachs disease	Hexosaminidase A assay	+	+
Haemochromatosis	Serum Fe or ferritin level	+ (only when iron loaded)	−
α-1 antitrypsin deficiency	Electrophoresis	+	+
Huntington's disease	−	−	−
APDK (adult polycystic kidney disease)	−	−	+
Familial hypercholesterolaemia	Serum cholesterol	+	+
Haemophilia	Factor VIII etc	+	Most
Duchenne muscular dystrophy	Creatine phosphokinase	+	Most
G6PD deficiency	Enzyme assay	+	Most
Fragile X mental retardation	DNA analysis	+	+

disorders such as haemophilia. Classical methods can only give a risk that a woman is a carrier.

Diagnosis using DNA

The advantages of DNA for genetic diagnosis are:

1. It is easy to obtain, since the whole DNA programme is present in every cell of an individual or fetus.
2. Genes can be studied whether they are producing their gene product or not. For example, the phenylalanine hydroxylase gene can be studied using chorionic villi, even though the gene is only expressed in liver tissue.
3. Usually a yes/no diagnosis can be made, permitting definitive carrier diagnosis.
4. Essentially the same type of laboratory approaches can be used for carrier diagnosis and for prenatal diagnosis at any stage of pregnancy, for all conditions diagnosable by DNA studies.

The main limitation of DNA diagnosis is that several different mutations may underlie the same clinical disorder, and it can be difficult to identify them all, particularly the rarer ones.

For DNA diagnosis one needs to detect changes in one or a handful of base pairs in genes that are a few thousand base pairs long. But DNA extracted from human cells comes in sets of 6000 million base pairs, and less than one millionth is relevant for any diagnosis. Since the sequence to be studied is both extremely small and surrounded by massive amounts of 'contaminating' DNA, extraordinarily specific and sensitive methods are required. Fortunately, DNA can be used to study DNA: single-stranded DNA 'probes' with a specific sequence that bind to complementary sequences in the DNA being studied make genetic diagnosis possible.

DNA methods have become radically simplified over the past ten years, and are still evolving rapidly. At present DNA diagnosis of genetic disease depends on different combinations of the following steps, which are described in detail below.

1. Extracting and purifying DNA from a tissue sample – a time-consuming procedure needed to make DNA single-stranded and reactive.
2. Amplifying the relevant segment of DNA by the 'polymerase chain reaction' (PCR).
3. Digesting the DNA with restriction enzymes.
4. Separating the resulting DNA fragments by electrophoresis.
5. Localizing the fragments carrying the gene in question, often using labelled DNA probes.

Making cDNA probes

Specific 'complementary DNA' (cDNA) probes are essential for analysis or manipulation of DNA. A probe is a length of DNA with a specific sequence that is complementary to the DNA sequence being sought (the 'target sequence'). When the target sequence is single-stranded, its cDNA probe will pair with it, given the right conditions. The target sequence may be a gene or a neighbouring non-coding DNA sequence. So that they can be easily detected, probes are 'labelled' either with radioactivity or by binding a fluorescent or coloured molecule.

Once a suitable probe is available, the inheritance of a gene within families may be tracked, the gene itself may be isolated and sequenced, and mutations responsible for disordered functioning may be defined. The first (and often major) step in DNA analysis is to obtain an appropriate probe, which can then be multiplied either in bacterial culture or synthetically, and widely used.

The first human cDNA probes were made for haemoglobin genes in the following way. Haemoglobin is essentially the only protein synthesized in reticulocytes, so the only messenger RNA present in adult red blood cells is α- and β-globin mRNA. Although mRNA is only a tiny fraction of the total RNA in the cell, it can be isolated relatively easily because it binds to a number of ribosomes (Fig. 2.6) forming large units which can be separated from other cellular material by ultra-centrifugation. The mRNA and the ribosomes can then be separated quite easily. Messenger RNA

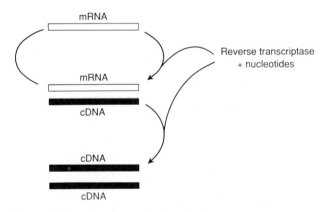

Fig. 10.1 Diagram illustrating the synthesis of complementary DNA (cDNA) probes from mRNA. The enzyme reverse transcriptase, made by some viruses, can synthesize a strand of DNA complementary to either mRNA or DNA. When the RNA–cDNA duplex is separated, the enzyme uses both the old mRNA and the new cDNA as templates for synthesizing more cDNA. DNA made using mRNA as a template lacks the introns that are present in chromosomal DNA.

obtained in this way can be used as a template for synthesizing cDNA (Fig. 10.1). It is more complicated to obtain cDNA for other genes or for non-coding sections of DNA, but the basic principles are the same.

When a specific cDNA has been made, it can be multiplied by inserting it into a plasmid (a self-replicating circle of DNA) which can be multiplied in bacterial cultures. cDNA probes are usually made radioactive after being harvested from the cultures. Large cDNA probes that represent all or part of a gene can, when sequenced, give the precise DNA sequence of the corresponding gene. This in turn allows smaller probes to be produced and simpler methods to be used for diagnosis.

Oligonucleotide probes

Once the precise DNA sequence of a gene and the surrounding DNA, and the mutations causing disease, have been worked out, short oligonucleotide probes complementary to specific sequences (usually about 19 base pairs long) can be synthesized easily. Automated oligonucleotide synthesizers are now available. Oligonucleotide or 'oligo' probes can be made radioactive when necessary, and have many uses in genetic diagnosis.

'Dot blotting' is a simple application of oligo probes to detect known point mutations. Two oligo probes are synthesized, one exactly matching the normal sequence and the other matching the mutated one – i.e. with only one base pair difference (Fig. 10.2). Spots of test and normal DNA solution are dried on to filter paper and exposed to radioactive normal and radioactive

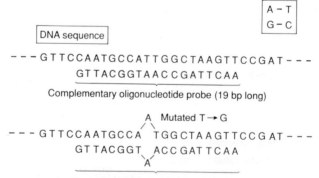

Fig. 10.2 Oligonucleotide probes will bind specifically to an exactly complementary DNA sequence. However, given specific laboratory conditions, if even a single base does not match, such a short probe cannot bind to otherwise complementary DNA. A single base mismatch does not interfere with binding of longer probes – genomic probes can be thousands of base-pairs long.

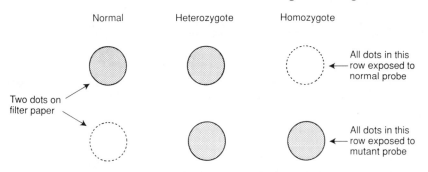

Fig. 10.3   Using radioactive normal and mutant oligonucleotide probes, normal, carrier, and homozygote individual may be easily distinguished by simple methods such as 'dot blotting'. For details see the text.

mutant probes. Normal probe binds only to a normal DNA dot, mutant probe binds only to a mutant DNA dot, and both probes bind to a heterozygote DNA dot (Fig. 10.3).

The polymerase chain reaction (PCR)

This is a simple method, summarized in Fig. 10.4, for producing a large amount of target DNA (Saiki *et al*. 1985). It solves the problem of the small quantity of target DNA versus the massive amount of unwanted DNA in DNA samples. Oligo probes are essential, so it is necessary to know the precise base sequence of the section of DNA to be amplified.

The length of DNA selected for amplification can be hundreds of base pairs long. Oligo probes complementary to sequences at either end are used to 'bracket' it, and also as 'primers' for DNA synthesis. PCR requires only a small amount of the target DNA, the appropriate oligonucleotide primers, the enzyme DNA polymerase and the four nucleotides adenosine, thymine, cytosine, guanine (A, T, C, and G). It does not matter how much contaminating DNA is present, and DNA need not be extracted first. The PCR cycle consists of three steps:

Step 1. The mixture is first heated to 92°C to separate the double-stranded DNA, then cooled to 56°C so that DNA returns to a double-stranded conformation. This allows the oligonucleotide primers to bind to complementary sequences, because they are present in excess. The original homologous DNA strand cannot compete with thousands of oligonucleotide molecules in binding to its complementary strand.

Step 2. The temperature is then raised to 72°C, the optimal operating temperature for the DNA polymerase (*Taq* polymerase) to synthesize DNA. To do so, it needs to start from small sections of double-stranded DNA.

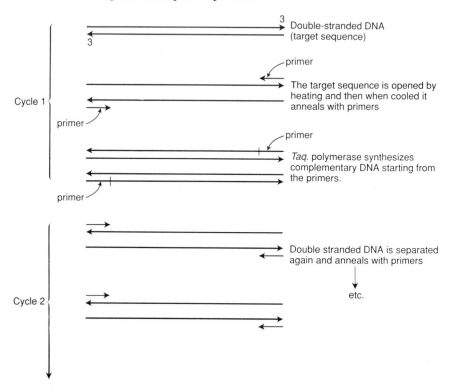

Fig. 10.4   The sequence of steps in the polymerase chain reaction (PCR) (from above downwards). The reaction mixture contains (1) the DNA sample including the 'target sequence', (2) a heat-stable DNA polymerase (*Taq* polymerase), (3) a supply of nucleotides for DNA synthesis, and (4) two kinds of oligonucleotide 'primers' which 'bracket' the section of DNA to be amplified. One primer is complementary to the nucleotide sequence at the 3′ end of the target *coding* strand, the other is complementary to the nucleotide sequence at the 3′ end of the target *complementary* strand. The 3′ end of each strand is indicated by the arrowhead. Cycle 1 is as follows: (a) The double-stranded DNA is separated by heating to 92°C to give two single strands. (b) The temperature is reduced to 56°C to allow the primers to anneal to complementary DNA: these short sections are now double-stranded. (c) The temperature is increased to 72°C, and *Taq* DNA polymerase uses the short double-stranded regions to start synthesizing the two complementary DNA strands in the 3′ to 5′ direction of the template. In Cycle 2, steps (a) to (c) are repeated. The number of DNA strands increases exponentially as the cycle is repeated: after 20 cycles the target DNA has been multiplied one million times. An amplified sequence of 300 base pairs will now constitute about 10% of the total DNA test sample (total length of the genome = 3000 million base pairs).

These are provided by the oligonucleotide primers, which are bound to complementary sequences on opposite strands, at either end of the target DNA sequence. The DNA polymerase binds at these short double-stranded regions and automatically replicates both DNA strands, using A, T, C and G as the raw material, moving in the 3′ to 5′ direction. It takes only minutes to duplicate the target sequence.

Step 3. Synthesis is then interrupted by heating to 92 °C again. This separates the DNA polymerase from the DNA, and the newly synthesized DNA from the template DNA strands.

On subsequent cooling to 56 °C (Step 1 of the cycle), the primers bind to both the original template DNA and to the newly synthesized DNA strands. Synthesis (Step 2) is then restarted, and the target DNA is doubled in each cycle. The mathematics are the same as in the Chinese story of the modest man rewarded by the Emperor. He asked only for one grain of rice to be placed on the first square of the chess board, two on the next, four on the next and so on. Since there are 64 squares, he needed a great deal of help to carry away his reward.

PCR machines now exist for repeating the cycle as often as necessary. Twenty to 30 cycles are easily executed in 3–4 hours and can amplify the target sequence to 10–30 per cent of the entire DNA sample, sufficient to be studied by basic methods such as electrophoresis.

PCR has made it possible to develop simple methods for identifying carriers of the common mutation for cystic fibrosis, in which 3 base pairs are

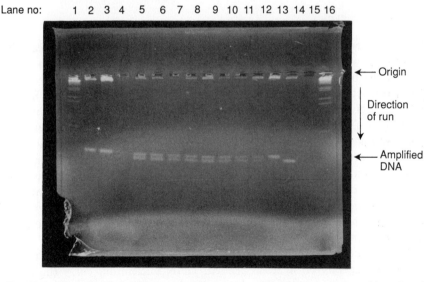

Fig. 10.5 Use of PCR for diagnosis of the common CF mutation. 16 'lanes' (numbered at the top of the figure) are shown. The samples are applied at the 'origin', indicated, and run towards the bottom of the figure by electrophoresis. Shorter, (mutated) CF DNA runs faster. Lanes 1 and 16, DNA standards. Lanes 2 and 3, normal DNA. Lane 4, blank control. Lanes 5–12, heterozygotes for the common CF mutation show 2 bands. Lane 13, normal DNA. Lane 14, homozygous CF. Lane 15, blank control. (Photograph kindly provided by Professor M. Bobrow.)

deleted from the cystic fibrosis (CFTR) gene (p. 213). A short length of DNA that includes the site of the mutation is amplified. In heterozygotes, 50 per cent of the amplified DNA is 3 base pairs shorter than the rest. When enough amplified DNA is present, the two types of fragment can be separated by electrophoresis, stained, and examined under ultraviolet light (Fig. 10.5). Alternatively, they may be distinguished by dot blotting using non-radioactive probes that develop a colour when suitably treated (Saiki *et al.* 1986).

When a point mutation is being sought, a modified form of PCR – the amplification refractory mutation system (ARMS) technique – can signal its presence or absence very quickly. If an oligo probe specific for the normal sequence is used as one of the primers, amplification will occur only if the normal sequence is present. If only the mutant sequence is present, the normal primer cannot bind and amplification cannot occur. Other simple approaches described below are feasible when PCR and restriction enzymes are used together.

PCR can even be used to amplify sequences from a single cell, starting from only two copies of the gene in question. This may make pre-implantation prenatal diagnosis a realistic possibility (Monk and Holding 1990) (p. 161).

Restriction enzymes

Restriction enzymes are bacterial enzymes that can cut DNA only at certain sequences, which differ with each enzyme. Their natural function is to protect the bacterium from virus invasion by cutting up foreign DNA. A given restriction enzyme cuts the 6000 million base pair sequences in the DNA of a given person consistently into a very large number of fragments, each of a different characteristic length (Fig. 10.6). Different enzymes produce completely different sets of restriction fragments. Most, but not all, fragments produced by a given enzyme are the same size in everyone.

Many restriction fragments do not include genes, but all genes are included in some of the restriction fragments (Fig. 10.6b). A whole gene and some flanking DNA may be included entire in a single restriction fragment. However, when there is more than one restriction site in coding sections or introns the gene will be divided up among two or more fragments.

Once produced, restriction fragments of different sizes can be separated by electrophoresis on agarose gel. All DNA nucleotides carry the same negative charge, so fragments separate strictly according to size, the smallest moving fastest. The distance a given fragment travels from the point where it was applied is inversely proportional to its length, which is expressed in thousands of base pairs (= kilobases = kb). When whole DNA is digested, innumerable fragments of many different lengths are produced and DNA is smeared all along the gel. The smear can be seen by staining the DNA with

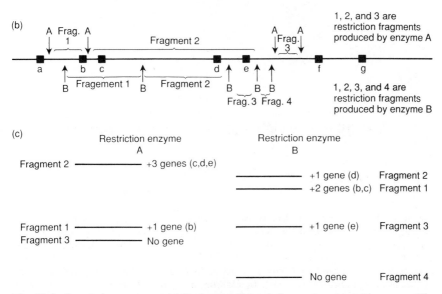

Fig. 10.6 Restriction enzymes. (a) Each type of restriction enzyme recognizes a specific base sequence in the DNA. These recognition sites are usually palindromic i.e. the sequence is the same in both directions. Most restriction enzymes cut the strands in a staggered way as shown, leaving two 'sticky ends'. (b) A section of human DNA carrying seven genes (a–g), with four restriction sites for enzyme A and five for enzyme B (indicated by arrows) in non-coding inter-gene segments. (c) Location of DNA fragments after digestion of the DNA with enzyme A (left) or enzyme B (right), followed by electrophoresis. If the length of DNA shown above is cut by enzyme A or by enzyme B the DNA fragments bearing the genes will be different sizes, and so reach different positions on electrophoresis. If a labelled cDNA probe specific for one of the genes is applied, a band will show the position of that gene. The position of the bands marking a gene are characteristic for each restriction enzyme. Usually only one gene is being sought, and only one probe is used at a time. DNA sections for which no probe is available or used, cannot be localized seen on the gel.

ethidium bromide and looking at the gel with ultraviolet light. However, all DNA fragments of a given length are situated in a band at a specific location in the smear and can be identified either by locating them with a labelled cDNA probe (Fig. 10.6c), or because the relevant fragment has been amplified by PCR or the ARMS technique.

To use a cDNA probe to locate the band that contains the target DNA sequence, the preparation is exposed to a solution containing the appropriate radioactive probe, which binds specifically to the sequence. Unbound probe

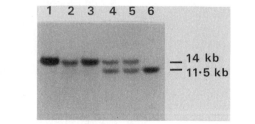

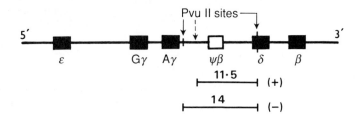

Fig. 10.7   Typical Southern blot, as used for prenatal diagnosis for thalassaemia. The diagram in the lower half of the figure shows the β-globin gene complex. A radioactive probe for the pseudo-β gene (indicated as a hollow box) has been used, following digestion of DNA with the restriction enzyme *Pvu* II. Everyone has a site in the δ globin gene that is cut by this enzyme (indicated by a bar in the δ gene), and another just to the right of the Aγ gene (indicated by a bar). In addition, many people carry a neutral mutation in the non-coding DNA between the Aγ and δ genes, which creates a new site for this restriction enzyme (indicated by an arrow). The pseudo-β gene is therefore found in a fragment of DNA that is either 14 or 11.5 kb long, which can be separated by electrophoresis. Individuals can be homozygous for one or the other form, or heterozygous. Lanes 1 and 2 in the above photograph show results in control individuals homozygous for absence of this restriction enzyme site. Lanes 3 and 4 show results in the parents of a child with thalassaemia major: one is homozygous for absence of the site so it is not possible to distinguish the thalassaemic and normal chromosomes: however, the other is heterozyogous for the mutation; in this parent the thalassaemia gene must be associated with presence or absence of the mutation. Lane 5 shows the result in the affected child; clearly in the heterozygous parent the thalassaemia gene is associated with the presence of the restriction enzyme site. Lane 6, the result in a chorionic villus sample from a fetus, shows that this chromosome is not present, so the fetus must be unaffected. (Photograph kindly supplied by Dr John Old.)

is washed off again. Classically, the bound probe is visualized by placing a photographic plate over the filter and leaving it in the dark for few days, until the probe's radioactivity causes a band to appear on the plate (Fig. 10.7). In its original form, this method was known as 'Southern blotting'.

If PCR is first used to amplfiy the target DNA sequence, electrophoresis concentrates the target DNA band and spreads out the background DNA, and the restriction fragments can be observed directly without radioactive probes (Fig. 10.8).

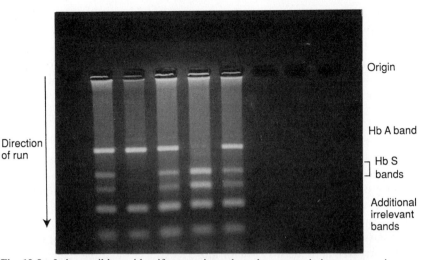

Origin

Hb A band

Hb S bands

Additional irrelevant bands

Direction of run

Fig. 10.8 It is possible to identify mutations that alter a restriction enzyme site very simply, using PCR, without using radioactivity all. The example shows detection of the haemoglobin sickle mutation, which creates a new restriction site for the enzyme *Mst* II within the β-globin gene. The section of DNA to be examined, in this case the β-globin gene, is amplified. (This takes about 3 hours). The relevant restriction enzyme, in this case *Mst* II, is then added to the reaction mixture, and cleaves the amplified DNA into two shorter bands if a β–S mutation is present. The DNA is run on an electrophoretic gel, stained with ethidium bromide, and examined in ultra violet light. A slow migrating (i.e. larger) band indicates the β–A gene, while two faster migrating bands show the presence of the β–S gene. The presence of both types of bands together indicates a heterozygote. The Hb S bands are twice as thick in Hb SS individuals compared to people with Hb AS. When this method is used for prenatal diagnosis after chorionic villus sampling, a result can be obtained on the day after sampling.

Restriction enzymes can be used to detect mutations (e.g. deletions) that change the length of a gene, and so alter the position on the gel of the restriction fragment(s) that contain the gene (e.g. α-thalassaemia, many cases of Duchenne muscular dystrophy, and the common CF mutation) (Fig. 10.5). They can also detect mutations that introduce a new or abolish an existing restriction site, such as Hb S, Hb E, and Hb D. However, only a limited number of the mutations that cause disease fall into these groups. Variations in the sites cut by restriction enzymes in non-coding sections of DNA are therefore more often used for carrier and prenatal diagnosis.

Restriction fragment length polymorphisms (RFLPs)

Coding sequences of DNA are usually highly 'conserved', because most mutations are removed by naturally selection, so restriction enzymes usually cut coding sequences at the same sites in everyone. However, many harmless mutations in the non-coding introns or flanking DNA are not eliminated, and these regions can differ between individuals. Some of these mutations can create new restriction enzyme sites, or eliminate usual ones so a particular enzyme may produce fragments of different length in different individuals. These can be separated by electrophoresis as described above (Fig. 10.8, Fig. 10.9). The band carrying a particular gene (or part of a gene) will appear in a different position, or may be split into two or more bands, according to whether a specific restriction enzyme site is present or absent. Restriction sites that can differ between individuals are called 'restriction fragment length polymorphisms' (RFLPs). A better name might be 'polymorphic restriction enzyme sites'.

There are many RFLPs scattered along all the chromosomes, and many genes have an RFLP in an intron or nearby in a non-coding region of DNA. Because of their proximity, a gene and any adjacent RFLPs are inherited together, i.e. they are linked (p. 39). If the inheritance of the RFLP can be tracked in a family, the inheritance of the linked gene is also traced. Genetic diagnosis using linked RFLPs depends on classical genetic family studies.

The first step is to identify one or more RFLPs close to the relevant gene, in the population as a whole. When the gene is known (e.g. haemoglobin

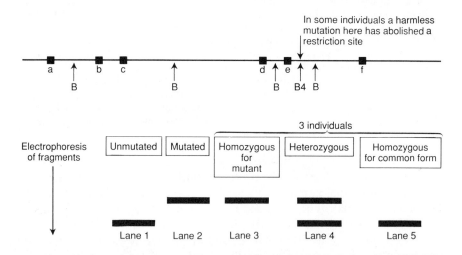

Fig. 10.9 Use of restriction fragment length polymorphisms (RFLPs) for genetic diagnosis. The figure shows the same length of DNA as in Fig. 10.6, with the same genes, but shows restriction sites for enzyme B only, and only a probe for gene e is used. Lane 1 shows the usual position of gene e in most of the population. However, in this case a proportion of the population carries a mutation which abolishes a restriction enzyme site (B4) adjacent to the e gene. In these people, the e gene is carried on an unusually long DNA segment, which moves more slowly on electrophoresis, as shown in lane 2. The population includes three types of individuals (lanes 3, 4, and 5): those who are homozygous for the mutant form show only one, slow band with the e probe (lane 3); heterozygotes have different restriction enzyme sites on each of a pair of homologous chromosomes and so show two bands with the e probe (lane 4); homozygotes for the common form show a single fast band (lane 5).

disorders, factor VIII), this is usually done by studying the length of fragments bearing the gene, after digestion with several different restriction enzymes, in a randomly selected sample of the population. When the site and nature of the gene are still unknown (e.g. in Huntington's disease), it can be done by finding a RFLP that is consistently inherited with the disease in several large families transmitting the disorder. Only RFLPs very close to a gene will remain associated with it through several generations. Numerous RFLPs spaced at intervals on every chromosome have now been defined, and this makes the search much easier. When a disorder is found to be linked to one of these RFLPs, the chromosome, and even the region of the chromosome, carrying the gene responsible is also known.

Some RFLPs are uncommon. The fact that the Hb S mutation first occurred in a person with an uncommon RFLP next to the β globin gene allowed the first use of RFLPs for genetic diagnosis (Kan and Dozy 1978). This RFLP eliminated a restriction site usually cut by the restriction enzyme *HpaI*. On electrophoresis the DNA fragment bearing the Hb S gene was therefore visibly longer than that carrying the Hb A gene, and this allowed prenatal diagnosis of sickle cell disease using DNA. However, the example also illustrates one of the principle limitations in using linked RFLPs for genetic diagnosis. With the passage of time, the Hb S gene and the RFLP have become dissociated due to crossing over in a significant proportion of cases. Therefore DNA studies could not be used on their own. Back-up conventional family studies were needed to discover whether the Hb S gene and the RFLP were linked in that particular family. However close an RFLP is to a gene, there is a small chance that crossing over will occur between the two sites in each generation, and this can occasionally lead to diagnostic errors.

Common RFLPs that occur in up to 50 per cent of the population are particularly useful in genetic diagnosis, because there is a good chance that family members will carry a different RFLP linked to the relevant gene on each of the two chromosomes, which will therefore be distinguishable from one another. RFLPs closely linked to the gene in question often allow a diagnosis to be made, even when the gene which has mutated to causes disease has not yet been identified, or when the gene is known but the specific mutation responsible for the disease in the family is not. Such linkage studies require two steps.

1. DNA samples from family members are examined for common RFLPs known to be close to the gene concerned (both normal and mutated). When a linked RFLP is found for which the parents are heterozygous, it can then be used to trace the inheritance of their chromosomes in their children. If differences are found the family is described as 'informative'. If not, they are described as 'uninformative'.
2. Assuming a difference is found, e.g. in each of a couple at risk for having

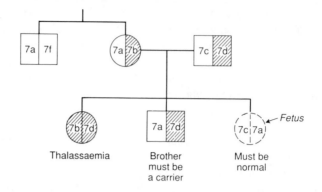

Fig. 10.10 A common way of showing RFLPs in a family. Each 'haplotype' (characteristic pattern of RFLPs) identifies the region of one chromosome carrying the gene that is being tracked in the family. Each haplotype is given a number in the family tree. It is easy to see that in this case, the gene in question (the thalassaemia gene) is associated with chromosome sections 7b and 7d, that the brother of the affected child is a carrier, and the fetus diagnosed prenatally is normal.

a child with cystic fibrosis, the second step is to find out which band (i.e. which RFLP) is linked with the normal gene and which with the abnormal one. This can easily be done by family studies if some DNA is available from a previous affected child of the same couple (Fig. 10.10). RFLPs found in the child are associated with the disease gene, and RFLPs not found in the child are associated with the normal gene. (However, it is important to remember that the linkage of RFLPs with a particular gene falls short of 100 per cent.)

Once the family pattern has been worked out and 'informative' RFLPs have been identified, carrier and prenatal diagnosis become relatively simple, using the methods described above.

Diagnosis using RFLPs is greatly simplified when PCR is used to amplify the DNA around the restriction site being sought. The amplified DNA is cut by the relevant restriction enzyme and examined by ultraviolet light after electrophoresis and staining. Enough DNA is present for the positions of different bands to be directly visible, without needing to use radioactive probes as markers.

Usually more than one restriction enzyme site is examined during family studies and a pattern of RFLP sites around the gene of interest becomes known for both chromosomes, for each family member. This set of known characteristics close together on a single chromosome and usually inherited together is known as a 'haplotype'. It can take some weeks to obtain enough information on a family to permit prenatal diagnosis. This is one strong reason why such studies should be carried out prior to pregnancy wherever possible.

Not surprisingly, it can be difficult to explain diagnosis based on linked RFLPs to families. A simple explanation for the at-risk couple shown in Fig. 10.10 might go roughly as follows: 'Everything that we inherit from our parents is determined by thousands of factors called genes. Genes are carried in 23 pairs of little rods called chromosomes, which are present in all the cells of our body. Each inherited characteristic or disease has to have two genes, one on each of a pair of chromosomes. You both have one thalassaemia gene, and one normal gene that prevents you from getting the disease. We are not able to detect these genes directly, but we can detect some factors associated with them. These are indicated in the diagram (Fig. 10.10) using the letters a, b, etc. Using these factors we can find out whether your child has inherited disease genes or normal genes'. With such an explanation, and the appropriate diagram, most people can understand how the inheritance of the disorder can be tracked in their own family.

The RFLP method is very widely used for DNA-based genetic diagnosis, because it does not require detailed knowledge of the mutation or even the gene responsible for the disease in every family under study. However, it is cumbersome and also involves a small but unavoidable risk of misdiagnosis due to crossing over. It can also produce unwanted information, for example on paternity. As technology progresses it is tending to be replaced by simpler and more reliable methods based on definitive detection of mutations.

'Reverse genetics': defining genetic disease by working backwards from DNA to protein

RFLPs make it possible to try to locate genes for inherited diseases, even when their protein product and biochemical effects are still unknown. The approximate location of the gene mutated in Huntington's disease was discovered by this method in 1983 (Gusella *et al.* 1983), and those for Duchenne muscular dystrophy (DMD) and cystic fibrosis (CFTR) in 1985 (Bakker *et al.* 1986, Knowlton *et al.* 1985).

Once closely linked RFLPs are identified, the gene itself may be sought by 'walking' along the chromosome, i.e. by identifying and studying adjacent sections of DNA sequentially. This is more difficult than it sounds: the CF gene turned out to be half a million base pairs away from the two RFLPs where the search was started. The dystrophin gene (DMD) was identified in 1987 (Koenig *et al.* 1987) and the CFTR gene in 1989 (Riorden *et al.* 1989). The gene for Huntington's disease had still not been identified at the time of writing.

Once a gene has been identified and sequenced, (1) the amino acid structure of its normal protein product can be worked out using the genetic code, and (2) the protein itself can be made *in vitro* by inserting the gene (or mRNA made from it) into a protein synthesis system. Its physiological function may then be deduced, and this may open the way to better treatment. Ultimately, it may even be possible to treat some diseases by correcting the mutation responsible.

Role of the primary care worker

The success of DNA diagnosis often depends on having some DNA from an affected individual in the immediate family available for testing. It is a part of the primary health care worker's responsibility to ensure that DNA from people with inherited diseases is preserved. This can be arranged through the local clinical genetics service.

Key references

Holtzman, N. A. (1989). *Proceed with caution. Predicting genetic risks in the recombinant DNA era*. The Johns Hopkins University Press, Baltimore.

Galjaard, H. (1980). *Genetic metabolic diseases, early diagnosis and prenatal analysis*. Elsevier/North Holland, Amsterdam.

Weatherall, D. J. (1991). *The new genetics and clinical practice*. (3nd ed). Oxford University Press (for the Nuffield Provincial Hospitals Trust).

11. Obstetric aspects of prenatal diagnosis

Prenatal diagnosis aims to answer specific questions about the genetic make-up, form, function, and development of the fetus, ideally in time for the option of termination of pregnancy to be considered when there is severe abnormality.

Though the dominating ethical issue in this area is the acceptability or not of therapeutic termination of pregnancy, we assume that most professionals involved in genetic counselling will agree that termination of pregnancy is ethically justifiable under certain circumstances. The UK 1967 Abortion Act contains a clause that allows a doctor to decline to participate if he or she has a 'conscientious objection'. However, the doctor then has a duty to refer the patient for help elsewhere. In the UK, the upper gestation limit for a termination was recently reduced from 28 to 24 weeks, to allow for the fact that some very premature babies are viable. However, the 24 week limit does not apply if the fetus has a lethal condition such as anencephaly or agenesis of the kidneys, or there is a serious risk of major handicap.

Ultrasound examination

Ultrasound is a method for producing an image by sound reflection, like deep sea sonar. It is increasingly used in many fields of medicine, and being non-invasive, is a popular investigation with family doctors.

Ultrasound waves are produced by passing an alternating electric current through a crystal of a material that has the natural tendency to distort with electric impulses. The resultant rapid vibrations produce high frequency sound waves that pass through tissues, but are partially reflected by surfaces. Reflected waves can be detected by the same crystal, since it also generates an electric pulse when distorted. A number of crystals arranged in a linear or revolving scan-head produce rapid sequential images, giving a moving real-time picture of deep structures, including the contents of the pregnant uterus. The picture can be 'frozen' on the screen, to allow detailed examination.

Ultrasound scanning is painless, and is usually a pleasant experience for the mother, unless a serious problem is suspected. The image can be quite detailed, and moves as the fetus moves. Couples often like to share the experience of seeing their child for the first time. Fig. 11.1 shows a first

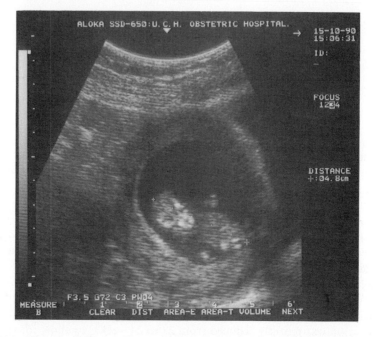

Fig. 11.1   Ultrasound picture of a normal pregnancy in the first trimester (10 weeks). The whole gestation sac and the fetus can be seen. (Kindly provided by Sue Price, Obstetric Ultrasound Department; University College Hospital.)

trimester fetus, and Fig. 11.2 a second trimester one. The gestational sac can be visualized at about 5–8 weeks' gestation, depending on the equipment and the position of the uterus, and the fetal heart-beat can be visualized from about 7 weeks onwards. From about 7–12 weeks of pregnancy, gestational age is assessed by measuring the length (crown–rump) of the embryo, and thereafter by measuring the maximum diameter of the skull (the bi-parietal diameter) of the fetus.

It is unsatisfactory that there were no controlled studies of the safety and efficacy of obstetric ultrasound before it became established in practice. There is no evidence of any harmful physical effect (Royal College of Obstetricians and Gynaecologists 1984), but it will probably be a long time before we can be sure of this. Reliable ultrasound examination depends on the skill and experience of the operator and the quality of the equipment. Its main known risk is misinterpretation of the image, which can lead to failure to detect abnormalities, misdiagnosis of abnormalities, or abortion of a healthy fetus.

At present, ultrasound is used for detecting fetal anomalies at three stages of pregnancy, at various levels of expertise.

Basic ('level 1') scanning is available for pregnant women at most district hospitals. In the UK it is usually carried out by radiographers, and takes only

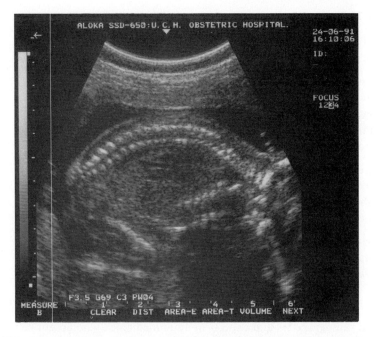

Fig. 11.2 Ultrasound picture of a normal pregnancy in the second trimester (16 weeks). The spine, base of the skull and the right femur can be seen. (Kindly provided by Sue Price.)

a few minutes. It is useful for diagnosing pregnancy (from 5 weeks of gestation onwards), and to confirm gestational age, fetal viability and number. It can detect 80 per cent of pregnancies that will go on to abort spontaneously, so a dating scan at booking (in the first trimester) is particularly valuable for older women, who have a high risk of spontaneous abortion (Fig. 21.3). A basic scan is also usually done in connection with maternal serum AFP screening (p. 147) because interpretation of the results is very dependent on gestational age. Basic scanning is not a primary screen for congenital anomalies; only the grossest malformations can be reliably detected at this level.

'Fetal anomaly scanning' (level 2) is a primary screen for congenital malformations. It is done at a limited number of centres, supervised by obstetricians, and takes about 20 minutes. As yet it is generally available only for women in recognized risk groups, for example, with diabetes or a raised serum AFP, or a history of a fetal abnormality, or of exposure to teratogenic agents such as anti-epileptic drugs. Fetal anomaly scanning is particularly important in twin pregnancies, as twins are more likely than singletons to have a congenital malformation.

'Level 3' (tertiary level) scanning is performed at expert centres by highly

Table 11.1   High risk groups of pregnant women referred for fetal anomaly scan

Those with raised maternal serum AFP.

History of:
 previous malformed child
 previous chromosomal anomaly
 familial genetic disease
 exposure to mutagens — radiation, chemotherapy, drugs

Maternal disease:
 infection (TORCH, p. 304)
 diabetes (insulin dependent)
 epilepsy (on drug treatment)

Other possibilities
 consanguinity
 advanced paternal age

trained radiographers and experts in 'fetal medicine', who accept referral of women in high risk groups (Table 11.1) or with findings suggestive of an abnormality on basic or fetal anomaly scanning. Fetal medicine experts also perform fetal sampling procedures and operations.

Ultrasound examination in early pregnancy is indicated in primary care under the following circumstances:

1. For mothers over 35, at about 8 weeks' gestation, to confirm that the pregnancy is viable, and to open the discussion about possible karyotyping for chromosomal anomalies.
2. To diagnose pregnancy at any time from 5 weeks of gestation onwards, if there is uncertainty from biochemical tests.
3. If there is bleeding in early pregnancy. If scanning shows an empty sac or a dead fetus, the mother can have an elective dilatation and curettage (D and C) if she wishes, rather than waiting for a spontaneous miscarriage and possible emergency admission to hospital. A pregnancy test on an early morning specimen of urine is a relatively unsatisfactory way to determine whether a pregnancy has become non-viable, as a positive test depends on the production of human chorionic gonadotrophin (hCG) from the placenta, but placental tissue may outlive the fetus by some weeks.
4. For women who have recently miscarried, and are worried about the viability of the current pregnancy.
5. When there is uncertainty about the dates.
6. When there is any possibility of an ectopic pregnancy, e.g. a missed

period with lower abdominal pain, more often one-sided. However, under these circumstances it is imperative to refer for an urgent gynaecological assessment as a simple ultrasound examination is not always helpful.

7. For detection of multiple pregnancies, which is possible in the first trimester and has replaced radiology. Early diagnosis of multiple pregnancy helps the parents to prepare psychologically, and by helping to ensure that the mother delivers with the support of neonatologists, is likely to reduce the incidence of perinatal problems.

Fetal anomaly scanning

This is best carried out at around 19 weeks' gestation when the fetal organ systems are sufficiently differentiated for major malformations to be apparent. At this stage an expert can detect over 90 per cent of major malformations in referred high risk pregnancies, with very few false positives (Campbell and Smith 1984). However, it is also possible to detect a surprisingly large range of severe malformations by ultrasound scanning in the first trimester of pregnancy.

Most congenital malformations occur in pregnancies in women who are not in known risk groups, and so may not be offered a fetal anomaly scan at present. At some centres more than 50 per cent of major malformations can be detected by offering routine level 2 anomaly scanning to all pregnant women. Table 11.2 shows the type of check-list used by ultrasonographers (Campbell and Smith 1984). Routine scanning is steadily becoming more widespread and efficient, though there is still some uncertainty about the proportion of severe abnormalities that can be detected in time for the option of termination of pregnancy. If ultrasound reveals that fetal growth is retarded (i.e. behind gestational age) in early pregnancy, there is an increased risk of chromosomal abnormality or congenital malformation, or that the fetus has been damaged by an environmental assault (infection, alcohol, smoking, or drugs).

As ultrasound equipment improves, it is becoming possible to use indicators such as the shape of the head, the shape of the cerebellum, the length of the femur in relation to gestational age, etc, during routine scanning, to pick out fetuses that may have chromosomal abnormalities and/or congenital malformations.

When of the fetus has a major abnormality, information on the karyotype is important in helping parents to decide whether or not to continue the pregnancy, as many malformed fetuses have an underlying chromosome anomaly. Nicolaides and co-workers (1986) studied 118 such fetuses diagnosed by ultrasound, using fetal blood sampling and rapid karyotyping. Thirty-eight (32 per cent) had a major chromosome anomaly and seven had Down syndrome. Many would have died during later pregnancy or at

Table 11.2   Fetal anomaly scan: example of check-list

1. Measurement of biparietal diameter, head and abdominal circumference, femur length
2. Head: anterior and posterior ventricular hemisphere ratios, cerebellum, palate, face and neck, the exclusion of intra-cerebral cysts and defects in the cranial vault
3. Spine: longitudinal and transverse sections
4. Heart and chest: four chamber view and confirmation of normal structure
5. Diaphragm: exclusion of herniae
6. Abdominal wall and cord insertion
7. Exclusion of abdominal masses and cysts
8. Kidneys and bladder
9. Genitalia
10. Limbs, fingers, and toes
11. Assessment of amniotic fluid volume
12. Placental site and appearance
13. Number of vessels in the cord

(After Campbell and Smith 1989)
Such an examination takes an experienced operator about 15 minutes.

birth. The mother may therefore be referred to a tertiary centre for rapid karyotyping by fetal blood or chorionic villus sampling.

Ultrasound is often used routinely to check the progress of the pregnancy and fetal growth at around 32 weeks' gestation, and congenital malformations are quite often detected at this stage, when selective abortion is no longer an option. This can be very worrying for parents when the abnormality is severe and they might have liked the option of abortion. However, diagnosis at this stage is useful for infants with less severe malformations, who will benefit from neonatal surgery or from surveillance, e.g. congenital heart disease, renal anomalies, or ureteric reflux (Allen *et al.* 1986; Thomas *et al.* 1985). The fetus is immersed in a water bath which makes scanning easy, so it can be examined for internal malformations more easily than the newborn can. It is possible that late intrauterine scanning could become complementary to the neonatal examination, as a form of newborn screening.

Fetal sampling methods

All fetal sampling or treatment procedures, including amniocentesis, should be done under ultrasound guidance.

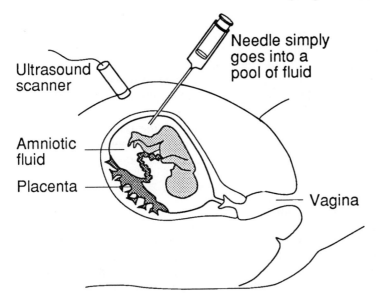

Needle simply goes into a pool of fluid

Ultrasound scanner

Amniotic fluid

Placenta

Vagina

Fig. 11.3 Amniocentesis.

Amniocentesis

Amniotic fluid is usually obtained by trans-abdominal puncture at 15–16 weeks' gestation (Fig. 11.3). The procedure involves a little discomfort when the needle penetrates the uterine wall. 15–20 ml of amniotic fluid contain enough viable cells to grow in tissue culture. It takes about two weeks of culture to get enough mitoses for reliable karyotyping and up to three weeks to be able to diagnose a metabolic disease. A DNA diagnosis can often be done directly on amniotic fluid cells using PCR, so some results may now be available much sooner.

Amniocentesis is now being done on an experimental basis as early as 9 weeks' gestation at some centres with improved ultrasound. The fluid contains enough cells for PCR, but not enough for karyotyping until 12 weeks' gestation (Rooney *et al*. 1990), so a chromosomal diagnosis cannot be obtained until about the 14th week.

There has long been uncertainty about the obstetric risk of amniocentesis. A recent, well-designed Danish study in which risk was compared between two randomly selected groups of women under 35 years old showed a 1 per cent increased risk of miscarriage in the amniocentesis group (Tabor *et al*. 1986). There is possibly a slightly increased risk of respiratory distress of the newborn and of common mild, correctable orthopaedic deformities such as club foot, after amniocentesis.

Fetal blood sampling

Fetal blood sampling at around 18–22 weeks of pregnancy was first used for prenatal diagnosis of blood disorders, but its commonest current application is for rapid karyotyping using fetal lymphocytes when a major malformation is detected by ultrasound. Fetal blood is obtained by ultrasound-guided trans-abdominal needle puncture of the fetal cord insertion (Daffos *et al.* 1985), which can be done safely only after the 17th week of pregnancy. It involves only a little more discomfort than amniocentesis, and the associated risk of spontaneous abortion is not much higher than that of amniocentesis, at about 1–2 per cent.

Chorionic villus sampling (CVS)

This can be done in the first trimester of pregnancy, using a catheter passed through the cervix (Fig. 11.4a). This approach is most successful and has fewest complications between 9 and 11 weeks' gestation. It may also be done through a needle inserted through the abdominal wall (Fig. 11.4b). This can be done at any stage of pregnancy, providing the placenta is accessible (Holzgreve *et al.* 1990). Trans-cervical CVS is often less uncomfortable, but may be more embarrassing, than trans-abdominal puncture. Many more fetal cells are obtained than with amniocentesis, so quicker and simpler diagnostic methods can be used.

When it was introduced, it was difficult to measure the risk of miscarriage associated with CVS, because of the high spontaneous abortion rate in early pregnancy. Randomized controlled comparisons of CVS with amniocentesis (in lower risk pregnancies) were initiated early (Modell 1985). The first results suggest that CVS carries an only slightly higher risk of fetal loss than amniocentesis (Canadian collaborative group 1989), but a Medical Research Council European trial of chorion villus sampling (1991) suggests a risk of 3–4 per cent. The risk seems to be highest among women over 37 years old, and when CVS is done before the 12th week of pregnancy (Cohen-Overbeck *et al.* 1990). Therefore in some centres, CVS is now offered to older women only at around 12 weeks of gestation.

The commonest current application of CVS is for prenatal diagnosis of chromosomal abnormalities. It was initially introduced for couples at high risk of inherited diseases, to use DNA methods for early prenatal diagnosis, so that termination of pregnancy could be done before 12 weeks' gestation simply, painlessly, and in privacy (Old *et al.* 1982). Prenatal diagnosis for such disorders is now regularly done in the first trimester. In practice, this does greatly reduce the emotional and ethical conflicts for couples at high genetic risk.

Table 11.3 summarizes the main clinical features of the current sampling methods for prenatal diagnosis.

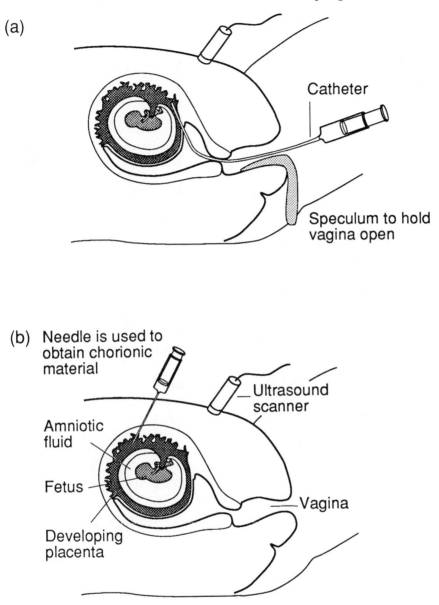

Fig. 11.4 Chorionic villus sampling. Two methods are widely used: (a) Transcervical. A soft plastic catheter is inserted gently through the cervix into the developing placenta under ultrasound guidance. Chorionic villi are aspirated using a syringe. (b) Transabdominal. A needle is inserted under ultrasound guidance directly into the placenta, and villi are aspirated using a syringe.

Table 11.3 Comparison of the clinical aspects of prenatal diagnosis methods

Sampling procedure	Obstetric aspects Gestation (weeks)	Risk to pregnancy (%)	Karyotyping Culture	Karyotyping Rapid	Biochemistry Culture	Biochemistry Direct	DNA Culture	DNA Direct
Amniocentesis	14–17	0.5–1	2–4 weeks	–	2–4 weeks	–	2–5 weeks	1–10 days
Fetal blood sampling	>18	1–7	–	3 days	–	2–7 days	–	1–10 days
Chorionic villus sampling	>9	1–4	2 weeks	2 days	2 weeks	1 days	–	1–10 days
Ultrasound	>19	+	Time to a definitive ultrasound diagnosis depends on many variables. Rapid karyotyping may be required					

(Based on Royal College of Physicians 1989)
- The risk to the pregnancy of specialized procedures such as fetal blood sampling depends on the characteristics of the patient and the experience of the operator.
- The risk of miscarriage following chorionic villus sampling rises with maternal age, and with gestation less than 12 weeks.
- The known risk to the pregnancy of ultrasound scanning is the possibility of a false positive diagnosis, leading to abortion of a healthy fetus.
- Until recently about 10 days have been required for a DNA diagnosis. The polymerase chain reaction often permits diagnosis within 24 hours

Twin pregnancies

Managing genetic risk in multiple pregnancies can be quite complex. All twin pregnancies in which prenatal diagnosis is contemplated should be referred early to a tertiary fetal medicine centre. With modern ultrasound guidance it is usually possible to obtain material from both fetuses. Nevertheless, counselling prior to prenatal diagnosis must cover the possibility of discordant results in the two fetuses, and of failure to obtain a diagnostic sample from one fetus, and the choices that will then have to be made.

In the past, when results were discordant, couples had to choose between aborting both fetuses even though one was healthy, and continuing a pregnancy that was certain to produce one seriously affected infant. Most couples at risk for severe abnormalities chose to terminate the pregnancy, a really heart-rending decision. Selective fetocide of an affected fetus is now possible in both the second and in the first trimester of pregnancy (Rodeck *et al.* 1982; Evans *et al.* 1988), but the pregnancy loss rate can be high unless the procedure is carried out by an experienced specialist. The dead affected fetus becomes a fetus papyraceous, and the healthy fetus often comes to term.

Termination of pregnancy

Despite recent advances, most prenatal diagnosis is still done at 16–20 weeks' gestation, and the decision whether or not to terminate a pregnancy on account of fetal abnormality is often made at about 20 weeks' gestation. Mid-trimester abortion of a wanted pregnancy is painful and distressing both for the mother and for the attendant medical and nursing staff. It is usually done either by inserting vaginal pessaries containing prostaglandin, or by injecting prostaglandin into the uterine cavity. This induces a labour which takes, on average, about 14 hours and ends in the delivery of a (usually) dead fetus.

Methods now exist for safe removal of fetal and placental material via the cervix in the mid-trimester of pregnancy, under general anaesthesia. This approach may spare the mother much physical suffering, but the emotional distress caused by termination of a wanted pregnancy remains. In addition, the method has the disadvantage that post-mortem examination to confirm fetal abnormality is impossible. Mid-trimester abortion by any method is often associated with distressing and painful filling of the breasts; most women need to be given bromocriptine immediately afterwards to suppress lactation.

Termination of pregnancy after CVS, providing the diagnosis is reached before 13 weeks' gestation, is done quickly and painlessly under general anaesthesia by the standard suction method. As 'privacy' can mean that friends and relatives did not even know the woman was pregnant, the couple may suffer from having no-one to talk to after the termination. Abortion can now also be induced medically in the first trimester (Rodger and Baird 1987). Some placenta from the abortus should be saved, since, as it was affected, it often provides the key to the essential DNA family studies.

Women undergoing this distressing procedure at any stage of pregnancy require constant, expert, and sensitive physical and psychological support. Mid-trimester genetic abortion needs to be done in the care of trained mid-wives. The parents often benefit from seeing and handling the fetus, which to them is already their child; and many are glad to have a photograph. They also need to be seen some time later for a follow-up discussion, and to be told when they can start another pregnancy, while information on the recurrence risk needs to be reinforced. They may also benefit from being put in touch with a support association. SATFA (Support After Termination for Fetal Abnormality) provides literature and the opportunity for contact with other women who have had the same experience — see p. 368 for contact address.

Treating the fetus

The commonest form of fetal treatment is intra-uterine transfusion for severe rhesus haemolytic disease, which is fortunately now becoming rarer, as prevention improves. There has been moderate success in treating female fetuses with congenital adrenal hyperplasia by treating the mother with dexamethasone, to suppress testosterone formation in the fetus (p. 228).

Possibilities for fetal surgery are limited to a few conditions with a single treatable malformation, that would progress to cause irreversible consequences prior to birth (Harrison *et al.* 1984). For instance, urethral valves are a simple obstructive cause of severe, often fatal, urinary tract malformation. Four out of eight fetuses treated recently by opening the bladder on to the abdominal wall at about 16–20 weeks of pregnancy, survived and developed into normal babies. The 'marsupialized' bladder is closed when the infant is a few months old (M. Golbus, personal communication). Aortic stenosis, an isolated malformation which would certainly lead to neonatal death in most cases, can now be diagnosed by ultrasound in pregnancy. Recent attempts to catheterize and dilate the aorta of the fetus at around 20–30 weeks have been encouraging, though as yet there have been no survivors. Though fetal surgery is very expensive, if it could be made safe and reliable it may produce a healthy baby, and is less costly than the intensive care needed by an affected one.

Role of the primary care worker

If couples are to benefit from genetic diagnosis early in pregnancy, pregnant women in risk groups must be referred to an appropriate centre as early as possible. This can be achieved only if primary care workers are alert to genetic risks, aware of the importance of early diagnosis of pregnancy, and refer women at risk directly to the appropriate diagnostic service, rather than waiting for a routine appointment at the local antenatal clinic.

When first trimester prenatal diagnosis is possible but a couple at risk have been identified too late to benefit from it in the presenting pregnancy, they need to be informed that CVS is available for future pregnancies. Family practitioners should encourage couples to contact the diagnostic centre as well as the practice immediately a pregnancy is suspected, in order to ensure early prenatal diagnosis and avoid mid-trimester abortion. It is equally important for the diagnostic centre to keep the family practitioner informed of the results of investigations.

Close liaison between the obstetric unit and the practice ensures that women who have had an abortion for a genetic reason are visited early by a member of the primary care team, for example the health visitor, for emotional support and to discuss practical questions, e.g. about future pregnancies.

Key references

Ferguson-Smith, M. A. (ed.) (1983). *Early prenatal diagnosis*. British Medical Bulletin; Vol. 39.

Fraccaro, M., Simoni, G., and Brambati, B. (eds) (1985). *First trimester fetal diagnosis*. Springer-Verlag, Berlin.

Rodeck, C. H. and Nicolaides, K. H. (eds) (1983). *Prenatal diagnosis*. Proceedings of the Eleventh Study Group of the Royal College of Obstetricians and Gynaecologists. John Wiley and Sons, Chichester.

Royal College of Physicians (1989). *Report on prenatal diagnosis and genetic screening; community and service implications*. The Royal College of Physicians of London.

12. The future: new genetic methods

Here we discuss some current research that may find future application for treatment or diagnosis of genetic disease. However, there is a long way to go before most of the methods being explored will match established conventional approaches in terms of reliability and safety.

Tissue and organ transplantation

Tissue transplantation is a first step towards gene therapy, because it provides experience of the possibilities and limitations of replacing genetically defective tissues and proteins. In the future it is likely that tissue transplantation and gene therapy will be combined.

Tissue transplantation for treating genetic disease is limited to disorders that affect transplantable tissues such as bone marrow, heart, lung, liver, or kidney. Transplantation of brain, which is inaccessible and unique, or muscle, in which a large mass of tissue would have to be replaced, seem more unlikely.

In cystic fibrosis, death is usually due to pulmonary fibrosis leading to heart failure. Heart/lung transplantation is being attempted increasingly for older CF patients. There were some early failures, possibly because the first patients treated were already very sick, but results seem to be improving. The operation offers the possibility of 'piggy-backing', i.e. a transplant of the heart removed from the CF patient, which is not diseased, to another recipient who needs only a heart.

Bone marrow transplantation is already used to correct genetic disorders of blood-forming tissues such as thalassaemia, sickle cell disease, severe combined immune deficiency (SCID), and disorders of reticulo-endothelial cells such as Gaucher disease. Only about 30 per cent of patients can have a marrow transplant for one of these conditions because a completely HLA compatible donor is required (except for immune deficiency syndromes, where affected children cannot mount an immune reaction), and because the procedure involves risk of infection, graft rejection, or graft-versus-host disease. These risks increase with age, but are diminishing as doctors obtain experience and new techniques are introduced.

Technically, bone marrow transplantation is relatively simple. First, the recipient's bone marrow is destroyed by cytotoxic drugs in order to remove diseased cells and make room for new ones. These drugs operate by killing rapidly dividing cells. They also destroy white cells (reducing defence against infection), make the patient feel ill, and usually cause the hair to fall out. Multiple samples of normal marrow are then removed by needle, usually from the sacrum of the anaesthetized donor, who may even be a small infant. These donor cells are then transfused into the recipient as in a normal blood transfusion, and a proportion settle in the now empty marrow. There follows a period of about six weeks during which the recipient must be kept isolated in sterile conditions because of the risk of infection. Immune competence returns as the graft takes, but is not fully re-established for about two years. Since there may be long-term complications (for example, an increased incidence of malignancies of the lymphoid system), it is very important that children who have had a transplant are followed up regularly.

Genetic engineering

In 'genetic engineering' the DNA code of living organisms is changed. This is commonplace in experiments with viruses and bacteria. DNA encoding for the relevant protein is inserted into a plasmid or viral vector, and cloned in bacteria or mammalian cells. Engineered bacteria are already used commercially to manufacture some proteins used in patient treatment like growth hormone, insulin, factor VIII, erythropoietin and interferon, and vaccines such as hepatitis B vaccine that are difficult to manufacture by traditional methods. Engineered bacteria have already been tested to protect crops against frost damage, and may be used to digest oil spills. Genetic engineering will probably contribute to improved treatment for genetic disease in several ways in the near future.

Transgenic animals

Animal models, i.e. animals suffering from, or with an inherited tendency to, a particular disease (e.g. cancer or diabetes) are invaluable in research on treatment, but animal models do not exist for most human inherited disorders. It is now possible to create 'transgenic animals' with one or more additional copies of a gene from either the same species, or a different species such as man. Thus abnormal human genes may be inserted into research animals such as mice to study their effects.

Transgenic animals are made by washing fertilized ova out of the genital tract of females, observing them until they reach the blastocyst stage, injecting the desired DNA into the blastocyst cavity, and returning the treated blastocysts to the animal's uterus. This is a rather hit-and-miss affair. The DNA may integrate into the nucleus of some cells of some blastocysts ('transfect' them). If some transfected cells happen to be in the inner cell mass that will form the embryo (rather than the trophoblast), the effects of the gene in the animal that develops from that embryo can be studied. When the experimenter is very lucky, the transfected portion of an embryo includes the germ cells. When such transgenic animals mate, the additional gene is passed on to some offspring, giving rise to a line of transgenic animals all of whose tissues carry the new gene.

In one of the first successful experiments, rat genes for growth hormone were introduced into mice, and, as expected, the result was giant mice (Palmiter *et al.* 1982). But results can also be unexpected. When the human gene for Lesch–Nyhan disease (a severe disorder of uric acid metabolism) was introduced into mice, the mice had the abnormal human gene without the pathology. To create a successful disease model, it may therefore be necessary to remove corresponding normal genes before inserting the disease gene.

Gene therapy

The objective of gene therapy would be to cure genetic disease either by introducing a functioning gene when one is absent, or by replacing a defective gene with a normal one. The first attempt to cure an enzyme defect causing immune deficiency, has already been started in the USA, but gene therapy is unlikely to be a simple solution for the problems of most genetic diseases. For success the following will be necessary:

1. Full understanding of the DNA sequence involved, the exact genetic defect in the patient, and the way that it causes the disorder.
2. Regulatory regions of DNA affecting the functioning of the gene must be known, and may need to be included in the 'transplant'.
3. There must be a safe and efficient way of inserting the gene into the host cell DNA. Chemical methods are very inefficient, so it will probably be necessary to insert the DNA into the genome of a modified retrovirus 'vector', capable of integrating itself into cellular DNA.
4. There must be a defined, accessible, and appropriate target cell that the gene can be inserted into; stem cells rather than terminally differentiated ones are required.
5. Transformed cells must be able to survive and function in the body in competition with unmodified ones. It may be necessary to introduce genes that confer a proliferative advantage together with the transplanted DNA.

Where a specific protein is missing, or is present in inadequate amounts, the aim will be to introduce a normal gene that will boost production of the protein by the appropriate tissues. If altered genes are introduced indiscriminately into the organism, it is uncertain whether they will function in all the cells, or only in those where they are normally expressed. There is evidence from transgenic mouse experiments that, providing their controlling regions are included, globin genes will function only in bone marrow.

However, dominant disease genes that produce an abnormal protein that interferes with normal processes (e.g. osteogenesis imperfecta, p. 57) offer a more complicated challenge. In addition to inserting a normal gene, the abnormal one may have to be destroyed. Even this may not be quite out of the question. It is possible to 'target' transplanted DNA to specific host DNA sequences by (1) combining it with a retrovirus, that normally integrates randomly into host DNA, and (2) adding sequences at either end of the 'virus' that are complementary to the two halves of the abnormal gene. In a small proportion of cells this causes the synthetic virus to pair with the abnormal gene as it integrates into the chromosome and split it in half. However, this approach is (as yet) extremely inefficient in terms of the number of cells that can be modified.

Thalassaemia and sickle cell disease may be early candidates for gene therapy. The globin genes and their regulation are particularly well understood. The target cell into which a normal gene must be inserted is the haemopoietic stem cell, the original source of the red, white, and platelet cell lines. In adult bone marrow these stem cells are relatively uncommon, cannot be recognized by their appearance, and are usually dormant, but when marrow is grown in culture, some stem cells are likely to be included.

Two approaches might be visualized. In the first and most probable scenario, a bone marrow sample would be taken from an affected person and grown in culture. It would be 'transfected' with the normal gene incorporated into a retrovirus that can enter the cultured cells and insert itself into their DNA, but is incapable of reproducing further. As only a small proportion of target cells become transfected, the transfecting virus is usually also tailored to contain a gene that confers resistance e.g. to a cytotoxic drug. After transfection the culture is exposed to this drug, which kills untransfected cells, leaving only the transfected ones. Mouse experiments have shown that DNA inserted into stem cells in this way functions, and can produce large quantities of globin.

It is necessary to demonstrate that the normal gene has been taken up and is functioning by studying the proteins synthesized in the culture. Once the presence of the normal gene has been confirmed, the marrow would have to be grown for some time to obtain enough cells to transplant back into the donor. For the transplant to take, probably at least part of the donor's marrow would have to be eliminated, so the recipient would run some of the risks of conventional marrow transplantation. From the patient's point of view, the procedure might be more like a bone marrow transplant than anything else — but it would be available to anyone, not just those with an HLA compatible donor.

A second possibility would be to introduce the gene directly into the stem cells of a person's bone marrow without tissue culture. This will require a far more sophisticated level of control over biological systems than we have today. It may be necessary to create an entire artificial virus-like entity containing the normal gene, and covered with surface proteins that could recognize and selectively attach to marrow stem cells. The particle would then inject its DNA and integrate it into stem cell DNA. Synthetic targeted viruses could become a reality. Vectors with a functioning CFTR gene have been successfully introduced by insufflation into the lungs of rats. Cystic fibrosis might be treated in some such way.

Other types of stem cells can be grown in culture: liver cells (hepatocytes) for instance, grow well. Since they make most of the clotting factors, corrected hepatocytes could correct bleeding diseases such as haemophilia, especially as massive quantities of clotting factors are not required: promising experiments have been carried out in mice.

Might it be possible to introduce genes, for example for circulating factors such as Factor VIII (missing in haemophilia), into cells that do not normally express these genes? Large sheets of epidermal skin cells (keratinocytes) are regularly grown in culture for use in covering burns. Is it possible to introduce genes into them and persuade them to secrete the wanted factors into the circulation? Animal experiments with factor IX (Christmas disease) and insulin (diabetes) show that though these genes can be inserted into keratinocytes, they do not function well, and in particular it is hard for the protein product to enter the circulation. This kind of work is just beginning.

Treating the fetus

In utero stem cell transplantation promises to be an interesting possibility for treating some genetic disorders in the fetus. Haemopoietic stem cells injected intra-peritoneally into monkey embryos at a stage before they are capable of mounting an immune response are accepted as 'self'. After birth, treated baby monkeys were found to be chimaeras, 5–12 per cent of their red blood cells originating from the graft and persisting (Harrison *et al.* 1989). This approach might be successful for human inherited diseases that could be corrected by the presence of a small proportion of normal cells: in conditions like thalassaemia and haemophilia, even as little as 10 per cent of normal cells might convert a severe syndrome into a mild one. Attempts to treat inherited diseases in the fetus using stem cells obtained from aborted fetuses, are now starting. If they are successful, attempted treatment could become an alternative when a fetus is affected and the parents cannot accept abortion. Treatment would have to be done before 20 weeks of pregnancy, when the fetus starts to become immunologically competent.

Pre-implantation genetic diagnosis

Prenatal diagnosis is most distressing for couples at high genetic risk, as they suffer great anxiety in every pregnancy, and some have two or three abortions before obtaining a healthy child. People often ask for diagnosis before the embryo implants, so that they could know their child was unaffected from the start of the pregnancy, and they could avoid the issue of abortion altogether. With PCR (p. 131) it is now possible to achieve a DNA diagnosis on a single cell (Monk and Holding 1990). If these methods are combined with the *in vitro* fertilization (IVF) techniques, that have been developed for helping infertile couples to have children, the idea of pre-implantation diagnosis seems realistic. Several approaches are being considered for couples at risk, for disorders with Mendelian inheritance.

In vitro fertilization (IVF)

The usual procedure in IVF is for the woman to be given injections of gonadotrophins to stimulate her ovary to produce a number of follicles simultaneously (super-ovulation). Her response is monitored by ultrasound scanning. When the follicles are ripe, a further dose of gonadotrophin is injected to release the ova, which are then collected with a needle under ultrasound guidance. In the laboratory the ova are mixed with sperm from the husband and observed for a few hours. Most will be fertilized, but by no means all divide normally. At about the 4 cell stage, a maximum of two of the healthiest looking eggs are returned to the mother by means of a fine tube passed through the cervix into the uterus. A major problem with IVF at present is its low success rate in older mothers, the ones who are most likely to seek it (Fig. 12.1). Attempting to achieve a higher pregnancy rate by returning more fertilized ova to the uterus can lead to multiple pregnancies, with serious medical and practical difficulties (Wagner and St Clair 1989).

A higher success rate may be achieved by the gamete intra-fallopian transfer (GIFT) technique, unless the mother's fallopian tubes are blocked (a common cause of infertility). Here the eggs are collected and placed with the sperm directly into the inner end of the Fallopian tubes, so fertilization takes place in a natural environment, as egg and sperm pass down the fallopian tube together. GIFT is done under direct observation with a laparoscope (like tubal sterilization) and requires a general anaesthetic. There is no evidence that embryos or eggs treated by regular IVF techniques have an increased chance of producing an abnormal baby.

Some spare fertilized eggs are often left. In the UK it has recently been agreed that with the parents' permission 'spare embryos' can be used for research, as long as they are not grown for more than 14 days in culture, and several possibilities for pre-implantation genetic diagnosis using a

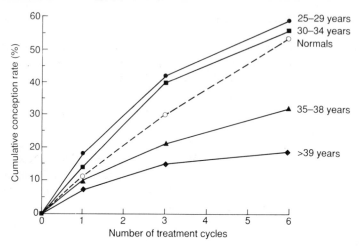

Fig. 12.1   Pregnancy rate in IVF related to the number of treatment cycles. The success rate varies with maternal age. In younger women it compares favourably with natural conception shown by the dolted line, but success falls off markedly after the age of 35 years. (Based on Tan *et al.* 1990; with thanks to Dr Stephanie Taylor.)

combination of DNA technology and innovative approaches to IVF are being investigated.

Using one cell removed from a fertilized egg

At about the 2–8 cell stage, a nick is made in the zona pellucida of each viable fertilized egg, and one cell is withdrawn using a fine needle. There is no evidence that removing a cell at this stage interferes with the normal development of the embryo. A genetic diagnosis is obtained for each ovum. Two of those diagnosed as unaffected are returned to the uterus in the usual way; any diagnosed as affected are discarded or used for research. Unused healthy ones can be frozen for possible use in the future, though the pregnancy rate with frozen ova is considerably lower than with unfrozen ones. At the time of writing, the first successful procedures had been carried out to select female ova for couples at risk for having sons with severe sex-linked disorders. Male ova were identified with Y-chromosome specific DNA probes and excluded (Handyside *et al.* 1990). However, the laboratory methods are far from simple. Contamination with even a single cell from a laboratory worker can lead to a mis-diagnosis.

Using the polar bodies

The first polar body may be removed and tested. When the aim is to diagnose a single gene disorder for which the mother is heterozygous, in theory if the polar body were found to contain the abnormal gene, the ovum must contain the normal one, and vice versa (Strom *et al.* 1990). This approach is attractive because it seems to offer the possibility of diagnosis before fertilization, and the GIFT technique could be used to return normal ova and husband's sperm to the mother, so increasing the chance of a successful pregnancy. However, it is not quite as simple as it may sound.

The description of meiosis in the female (p. 38) emphasizes that each ovum contains four sets of genes (in two chromosomes, each consisting of two chromatids) and crossing-over has already taken place. At ovulation, one complete set of chromosomes (each consisting of two chromatids) is extruded in the first polar body. When the woman is heterozygous there is a 50 per cent chance that the two chromatids of the first polar body will contain the two different versions of the gene (in which case the ovum would also be heterozygous) and a 25 per cent chance that it will contain two normal or two abnormal genes. Thus there is only a 25 per cent chance for each ovum of being diagnosed as normal at this stage, and returned to the mother using GIFT.

If only heterozygous polar bodies are found, it will be necessary to fertilize the eggs and examine the second polar body, which contains only one set of chromatids and is extruded immediately after fertilization (p. 39). In 50 per cent of cases it would contain the abnormal gene, in which case the fertilized ovum containing the normal gene could be returned to the mother. However, a stepwise procedure of this kind, involving such sensitive laboratory methods, multiplies the possibilities for mis-diagnosis.

Flushing normally fertilized eggs from the uterus

A third possibility is either to let the woman ovulate normally or to induce superovulation, leave fertilization to occur in the normal way, and after 4–5 days, wash out the ova from the uterus. They will now have reached the blastocyst stage. Then a slit would be made in the zona pellucida opposite the inner cell mass (Figure 4.2, p. 150), allowing some trophoblast cells to herniate out which would then be used for diagnosis. Blastocysts diagnosed as healthy would be returned to the uterus. This approach would be less invasive than those described above, and might lead to a higher pregnancy rate. However, it may be very difficult to retrieve all the fertilized ova produced, and any left behind could produce an affected child. There is also a risk of causing an ectopic pregnancy by accidentally washing a fertilized egg back up the fallopian tube.

Assuming that a pregnancy is achieved by one of these methods, it will be necessary to check the progress of the pregnancy by regular ultrasound scans, and the diagnosis will need to be confirmed by chorionic villus sampling. This should, however, be much less worrying than a regular prenatal diagnosis, because the fetus is already expected to be healthy.

Despite the complicated technology involved, this is a promising area: a great deal of research will be required to define the best methods, and their reliability and risks.

Safer fetal sampling

Another possibility is to obtain fetal cells circulating in the maternal blood, in order to achieve safer and simpler fetal diagnosis.

It has been known for a long time that some trophoblast cells enter the maternal circulation, and several efforts have been made to obtain them for diagnosis, until recently without conclusive results. However, it now seems that small numbers of nucleated fetal red blood cells, and of trophoblast cells, may be obtained from maternal blood. At present this is a very high technology approach (Bianchi *et al.* 1990; Meuller *et al.* 1990).

For such an approach to be useful for genetic diagnosis, cheap and simple ways of obtaining the cells would have to be found. Adequate cells would have to be obtained reliably at a stage that would allow a diagnosis before 12 weeks of pregnancy (to preserve the advantage of first trimester diagnosis). It seems rather unlikely that fetal red cells will be obtained reliably, since they are cleared from the circulation in the 15 per cent or so of pregnancies in which mother and fetus are ABO incompatible (p. 337). In addition, up to 5 per cent of pregnancies start as twins (Robinson and Caines 1977), so it would have to be shown that the trophoblast cells retrieved truly represent the viable fetus: an error rate of this order in prenatal diagnosis would be unacceptable.

Another possibility is to by-pass the need to separate out the fetal cells, simply by using the polymerase chain reaction (p. 131) on whole maternal blood, to amplify sequences characteristic for paternal chromosomes, and so make a genetic diagnosis in the fetus (Lo *et al.* 1989). This may be feasible, but would permit less than 50 per cent of genetic diagnoses. It may be possible to tell whether the fetus has inherited the father's normal or abnormal chromosome, e.g. by using RFLPs (p. 137): this would give a clear answer only for male sex, and for dominant conditions carried by the father. With recessively inherited disorders, 50 per cent of fetuses at risk would be found to have inherited the father's normal gene and so must be unaffected. However, if the father's abnormal gene is found, it would be necessary to proceed to a conventional diagnosis, now with a 50 per cent risk of fetal abnormality (since the fetus will have inherited either the mother's normal or mutant gene) and later than would otherwise have been the case. Overwhelming maternal contamination makes it impossible to define maternal genes in the fetus, excluding diagnosis of X-linked disorders (carried by females).

With such a proliferation of technology it is certain that new combinations of methods will become available for early prenatal diagnosis, and for treatment. However, it is not yet clear which approaches will be really useful.

Ethical problems associated with the new technology

The new developments in reproductive and genetic technology are not greeted with universal enthusiasm. Some people feel they involve a threat to human identity, and others have religious objections to the manipulation of potential human beings. All technological advance is associated with a fear that it could be used coercively to the disadvantage of individuals. Could genetic technology be exploited to satisfy the curiosity or ambition of individual doctors, or could politicians or big business use it to manipulate populations for political or economic purposes? Might people find they were no longer free to choose their own partner, have the number of children they want, or continue a wanted pregnancy if the fetus has a disability?

We cannot deal with this subject extensively. Ethical issues are best discussed in the context of practical experience, and we have done our best to deal with them briefly as they arise, throughout the book. The ethical dilemmas geneticists can face in practice have been described in detail by Czeizel (1988*a*), and Fletcher (1982) has discussed some issues from a religious point of view. Holtzman (1989) has considered the social and economic implications of screening for genetic predisposition to common diseases (see Chapter 13). Here we raise only two general issues.

Can doctors be trusted? Doctors are like any other group of people. Though most are concerned with good clinical practice, some are relatively insensitive to human feelings and needs, and some are egocentric and ambitious. So even if doctors are trustworthy as a group, not every doctor can be trusted. It is clearly necessary both to develop and publicize defined standards of practice, and to establish mechanisms to ensure that they are followed, and are monitored. For very sensitive areas like embryo research and termination of pregnancy, it is also desirable to establish independent bodies with powers of inspection and licensing, like the licensing authority recently established by the British Government, for IVF and embryo research.

Can politicians be trusted? For dictatorships, the answer is no. In a democratic society the government can define what is acceptable in that society at that particular time, and this can result in liberal and humane legislation.

In our view, genetic technology is acceptable as long as it is used to meet the expressed needs of informed people at risk (p. 181). Medical genetics services are not forced on the population: on the contrary, they are in very short supply. The central ethical issue in medical genetics is that these services, which can be critical for the happiness of entire families, are not yet either adequately or equitably delivered to the population (Royal College of Physicians 1989).

Key references

Fletcher, J.C. (1982). *Coping with genetic disease. A guide for clergy and parents.* Harper and Row, San Francisco.

Harrison, M.R., Slotnick, R.N., Crombleholme, T.M., Golbus, M., Tarantal, A.F., and Zanjani, E.D. (1989). *In utero* transplantation of fetal liver haemo-poietic stem cells in monkeys. *The Lancet*, **ii**; 1425–7.

Holtzman, N.A. (1989). *Proceed with caution. Predicting genetic risks in the recombinant DNA era.* The Johns Hopkins University Press, Baltimore.

13. Screening

First do no harm

Comprehensive delivery of the services described in the last two chapters depends on *screening* entire healthy populations for latent genetic problems. These services are not yet delivered as well as they should be. Many pregnant women in their late thirties are not offered CVS or amniocentesis, opportunities to identify people at genetic risk by taking a family history are missed, and members of ethnic minorities are not screened for haemoglobin disorders. If services are to be universal, an infrastructure is needed, much of it within the primary care system.

Screening differs from acute medicine in many ways. It is a large-scale exercise, aimed at detecting potential abnormalities in people who consider themselves healthy. It may lead to damage of people's healthy self-image, and unless it is appropriate could do more harm than good. To be delivered equitably, it should be provided through national health services. It requires an outreach approach to the population, including adequate information and counselling when necessary. It can generate false positive and false negative results, and so risks making things worse rather than better. Screening must therefore be justified both in terms of medical benefit and of cost. Methods evolve continuously, so long-term follow-up and continuing evaluation of technology, acceptability, and non-financial and financial costs and benefits is necessary.

General principles of screening

Established screening programmes for identifying genetic risks and congenital disorders are listed in Table 1.3. Most involve an initial 'primary screen' that selects the smallest possible group of people who *may* have the condition, but includes the maximum possible proportion of those who actually *do* have it. Definitive diagnosis then usually depends on a stepwise series of further tests, each involving a smaller group of people than the previous one (e.g. Table 13.1). The primary screen is usually the most problematic of these tests, because of the large scale, the need for cheapness and simplicity, and the fact that it can generate large numbers of false positive results.

Primary screening may be:

1. *Universal*: Screening is performed routinely at some point where everyone comes into contact with the health system. The only true example is

Table 13.1 Steps in the 'screening cascade' for haemoglobin disorders in the UK compared with cystic fibrosis

Step	Haemoglobin disorders	Cystic fibrosis
1. Identification of person in risk group	'Not North European' (10% of births)	Whole population
2. Carrier tests/year	60 000	600 000
3. Results positive	*ca.* 10% (6000)	*ca.* 4.5% (27 000)
4. Tests of partners (uptake >75%)	At least 4500	At least 20 250
5. Partner positive = 'at risk couple'	*ca.* 10% (At least 450)	*ca.* 4.5% (At least 910)
6. Prenatal diagnosis requested Most thalassaemias 30–50% of Sickle Cell Disease	*ca.* 150 } 300 *ca.* 150 }	? Uptake: ? (At least 455)

Estimates can be based on screening a single annual cohort of the population, such as all pregnant women, or all newborns. Estimates are given for the UK, with about 600 000 births per year. In reality, 2–3 times this number of tests are needed annually, to allow for pre-pregnancy screening, testing other family members when a carrier is found, etc. We have as yet no real idea of the uptake of prenatal diagnosis by couples at risk for cystic fibrosis.

neonatal screening. Routine antenatal clinic screening involves only the female half of the population.

2. *On request*: People are motivated, e.g. by direct recruitment through a patient register or through media publicity, to come voluntarily for a specific test. This is usually the least successful approach for genetic screening in terms of population coverage, though one can be fairly sure that the people who come really want the result.

3. *Incidental or opportunistic*: People are offered screening when they attend a health service for other purposes. This is one of the main approaches to screening in general practice, although a number of studies have shown systematic recruitment to be more effective than opportunistic.

These approaches are complementary rather than mutually exclusive. Screening on request tends to favour higher social classes, and opportunistic screening is limited to the sample of the population attending. At present, (nearly) the whole target population can be reached only when antenatal or neonatal screening is used either as the primary approach, or as a back-stop for people who have slipped through other nets.

The two main objectives of screening for genetic and congenital disorders are:

1. To identify a risk to personal health in time for treatment or prevention. This is the main objective of neonatal screening (Chapter 22).

2. To identify a reproductive risk in time to avoid it. This is the main objective of antenatal and pre-pregnancy screening (Chapters 20 and 21).

3. To reassure people who are not at risk. In general, people seek testing in the hope of finding they are not at risk, so a negative result is as important as a positive one.

Counselling is inseparable from genetic screening. It has been shown many times that it is useless to screen for carriers of genetic disorders without adequately informing people (Harper 1988; Rowley 1984). They cannot take the steps necessary to avoid their potential problem without full information.

Requirements for acceptable tests

It is important to be able to assess screening tests critically, and to understand the chance that a positive result indicates that the subject really has the condition being screened for, and that a negative one excludes it. This aspect of screening is described with clarity and in some detail by Cuckle and Wald (1984). Important considerations are as follows:

1. **The problem must be important**. Some conditions are important because they are common. Others (though rare) may be important because they

cause severe disease unless treated early or prevented.

2. **The test should be reasonably accurate**. A high standard of laboratory practice is essential, and a quality control system is necessary for reliable results to be obtained at different centres. The more screening methods can be developed from quantitative towards yes/no methods, the better (see below).

3. **There must be an effective solution**. Treatment should be effective, and early diagnosis must be necessary for its success (e.g. phenylketonuria and congenital hypothyroidism). Abortion may be the chosen solution for fetuses with severe disorders if detected early enough in pregnancy. However, this criterion is not as clear-cut as it sounds (see Chapter 22, on neonatal screening).

4. **Information and counselling are inseparable from screening**. Primary care workers can contribute enormously by informing and motivating the target population, collecting and sending samples, providing initial counselling, and helping people who have had false positive results.

5. **Facilities for information, testing, and counselling should be generally available**. Screening should ideally be provided equitably throughout the country and without relation to social class. In reality the standard will vary from one place to another, but such differences should be minimized as far as possible.

6. **The tests must be acceptable to the target population**: i.e. they should be convenient, as non-invasive as possible, and culturally acceptable.

7. **The cost per test should be affordable**. The cost of screening needs to be related to both non-financial and financial benefits. It can be difficult to equate the costs for a family of looking after a severely handicapped child with the expense of screening large numbers of mainly healthy people (p. 176).

Qualitative yes/no screening tests (most DNA methods, electrophoresis for abnormal haemoglobins, or assay for presence of IgM) produce clear-cut results. However, many tests (such as assay of maternal serum AFP level) are quantitative, and there is a substantial grey area between 'normal' and 'abnormal' results (Fig. 13.1). Such tests miss some affected individuals (false negatives) and include (often a large number of) others, who later turn out to be unaffected (false positives). But the concept of a 'false positive' is not clear-cut. People may fall into this group because they have a significant medical condition other than the one explicitly screened for. For example, a pregnant woman with a raised serum AFP level may have a non-viable pregnancy, or twins, and people with microcytosis may have iron-deficiency anaemia, rather than a thalassaemia trait.

The proportions of false negatives and false positives in a screening test depend on the cut-off point selected to separate 'normal' from 'abnormal' results, and this judgement is difficult to make. Attempts to base it on finan-

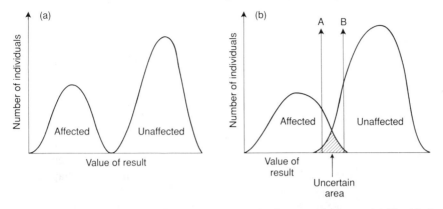

Fig. 13.1 Problems in the interpretation of quantitative screening tests. (a) The ideal quantitative screening test would discriminate clearly between affected and unaffected individuals. However, this is rarely possible. (b) In most tests there is considerable overlap between unaffected and affected ranges. A 'cut-off' point has therefore to be chosen to select individuals for further testing. A cut-off point indicated by line B, which includes almost all affected cases, also includes a significant proportion of unaffected people, who will be subjected to further unnecessary and probably costly testing. A cut-off point indicated by line A excludes almost all unaffected people but will miss a significant proportion of affected cases.

cial considerations are criticized on p. 176. It is also important to know, and to communicate, the extent to which a negative result *excludes* the condition being screened for. False negative results lead to false reassurance and may actually delay diagnosis. The terminology generally used in describing screening tests is summarized in Table 13.2.

Table 13.3 illustrates one application of these concepts in using maternal age as an initial screen for chromosome disorders in the fetus. As screening is extended to younger women, (1) the false positive rate rises, so the complication rate of the obstetric procedure may exceed the number of abnormal fetuses detected, and (2) the proportion of conditions of intermediate severity diagnosed rises — the test becomes less specific.

The fact that screening tests aimed at specific conditions may also detect a spectrum of abnormalities with a less predictable prognosis, or where no positive action can be taken, can create problems. Detection of sex chromosome aneuploidies can cause parents great anxiety and presents them with difficult decisions. Counselling should be available to cope with the psychological as well as the physical consequences of these possibilities.

Much clinical screening is also carried out during pregnancy and in the newborn period. Though ultrasound examination for fetal abnormality relies more and more on quantitative measurements, it ultimately depends on the clinical judgement of the ultrasonographer. The same is true of newborn screening for congenital malformations, such as congenital dislocation

Table 13.2 Key terms used in population screening

Detection rate (= 'sensitivity')	=	Proportion of those who have the condition being sought, who are positive on the test (e.g. detection rate for Down syndrome using maternal age >35 = *ca.* 30%)
False positive rate	=	Proportion of those who have do not have the condition being sought, who are positive on the test (e.g. false positive rate for maternal age >35 for Down Syndrome = 97%). These values can be expressed as (a) a proportion (of 1), e.g. detection rate = 0.3, or (b) as a percentage; e.g. detection rate = 30%
Odds ratio	=	$\dfrac{\text{Proportion affected among those with a positive result}}{\text{Proportion unaffected}}$
		This gives the probability that a person with a positive result is affected. e.g. 3/97 for maternal age >35 for Down Syndrome = 0.03 or 3%.

Other terms (unnecessary when above are used).

Specificity	=	Proportion of true negatives who are negative on test (= 1 – false positive rate)
False negative rate	=	Proportion of true positives who are negative on test (= 1 – detection rate)

(Bold type shows terms used in this book).
Note:
All of these terms are population-specific. They depend on the prevalence of the condition being sought in the particular population studied.

Table 13.3 Results of karyotyping for chromosomal disorders at different maternal ages

	Maternal age			
	35 years		40 years	
	% of fetuses	% of abnormalities	% of fetuses	% of abnormalities
Severe	0.64	52	1.91	72
Intermediate	0.27	22	0.44	17
Harmless	0.3	24	0.3	11
Total	1.24	100	2.65	100

'*Severe*' abnormalities include autosomal aneuploidies and mosaics, and unbalanced rearrangements. '*Intermediate*' abnormalities include sex chromosome aneuploidies and mosaics. '*Harmless*' abnormalities are balanced rearrangements.

of the hip. Training and quality control are just as important in clinical as in laboratory screening, so it is advisable for two skilled clinicians to agree on a positive diagnosis of an important abnormality in the fetus or in the newborn, before proceeding to intervene.

Timing the offer of screening

Reproductive risks fall into two groups (p. 6). *Sporadic* disorders, like chromosomal non-disjunctions or many malformations, arise during the process of conception or embryonic development. They occur with relatively low frequency and usually little or no warning. Everyone is at risk for them, though some groups which are at higher risk, such as older or diabetic mothers, can be identified. Women possibly carrying an affected fetus can be detected only during pregnancy. At present, the greater part of screening for fetal abnormality is aimed at sporadic conditions, but often without adequate information or counselling.

Several screening tests have been an obligatory part of pregnancy care for a long time. Blood is automatically taken for ABO and Rh grouping, haemoglobin level, and serology (for syphilis, rubella, and hepatitis B antibodies). Recently, in the UK anonymous testing for HIV has been added. In France, screening for *Toxoplasma* infection is also routine. Women are not usually informed about these tests, which became standard clinical practice when medical decision making was more authoritarian than is considered acceptable today.

By contrast, *genetic* risks can often be predicted before pregnancy, either from the family history, or if carriers can be identified, by screening. However, screening for inherited conditions is also usually left until women reach the antenatal clinic. This may seem efficient, but in fact antenatal screening for a genetic risk is a 'last ditch' approach, for several reasons:

1. It may be too late. Family studies are often needed and the tests required to establish risk take time and often cannot be completed in time for first trimester prenatal diagnosis. Some couples who would have been glad of prenatal diagnosis in the first trimester find second trimester testing unacceptably late. There may not be time to repeat a test which has produced an inconclusive result, and some women present too late for prenatal diagnosis to be possible at all.
2. Some tests such as carrier screening for Tay–Sachs disease and haemophilia give less clear results during pregnancy.
3. Usually women are inadequately informed about the implications of routine genetic screening tests in pregnancy, prior to the test being carried out, and so may be ill prepared to make the serious decisions needed if a test turns out positive.
4. Even when mid-trimester prenatal diagnosis is acceptable, identification of a major genetic risk for the first time at this stage of the pregnancy is extremely harrowing because of the extent of the parents' commitment to the child, and the unpleasantness of mid-trimester abortion.
5. Screening during pregnancy does not allow the couple an informed choice among the full range of possible options (see p. 187).
6. Screening only women reinforces the misleading notion that only women have genetic problems.

Genetic screening and counselling before pregnancy will have to become part of primary care if the large numbers of people at risk are to be identified and informed in time. However antenatal screening will also be needed as a 'backup' service.

The infrastructure for genetic screening

The requirements include:

1. An informed primary care system.
2. Information for the population at risk.
3. A system for collecting samples from the population at some point prior to reproduction, and delivering them to a laboratory.
4. A network of diagnostic laboratories, and a quality control system.
5. A system for reporting results and their meaning to doctors and the people concerned.
6. An information storage and retrieval system.
7. Information and counselling for people with positive results.

8. Adequate expert centres for counselling at-risk couples and providing prenatal diagnosis.
9. A system for monitoring the service.

Screening in primary care

In the UK, the organization of primary health care is ideally suited for primary screening and surveillance of vulnerable groups before pregnancy. Apart from the 'homeless and rootless' of the inner cities, nearly everyone is registered with a general practitioner, and 30 000 family practitioners cover the whole country. The structure of general practice is very variable, but in 1986, 50 per cent of family practitioners were working in groups of three or more, many as part of a primary care team including administrative staff, practice nurses, health visitors, district nurses, and sometimes midwives. In future, many practice teams will include specialist nurses and counsellors. At present the majority of community nurses are not attached to practices but are primarily responsible to the community unit of the District Health Authority (DHA). They look after patients in a specific geographical area, and may relate to several family practitioners.

An age/sex register is an essential tool for enabling the practice to provide such services. At present, many General Practices have either a manual or computerized age/sex register, and many others are working towards it. It contains the name, sex, and age of each patient, their address, and the date on which they registered with the practice. The register is used as a database for screening procedures, and also as an aide-memoire, e.g. for ensuring immunization of all the babies registered with the practice.

A rapidly increasing number of practices have microcomputers which will hold the age/sex register, store details of patients' medical histories, and issue recall reminders. They will eventually be linked with (1) the local hospital computer for results of investigations, (2) the screening and immunization records of Health Authorities, and (3) the Family Health Service Authorities (FHSAs), so that it can gather local health statistics. When fully established, this system will be ideal for storing results of screening tests. It will also be possible to generate lists of women of child-bearing age with specific 'risks, and to check that the relevant tests have been performed.

Most practices have a stand where leaflets about important general or local health problems are available free of charge, and health information posters are displayed on waiting-room walls. Health education units usually ensure these are available in a range of languages to suit the local population.

Some screening procedures such as cervical smears and mammography are organized through the community services on a district or regional basis often using computerized patient registers held at the FHSA as a database. In keeping with national policy, primary care teams, especially practice nurses, are becoming increasingly involved in screening.

The practice team can participate in the following ways in a screening programme to identify individuals with a reproductive risk, e.g. for a haemoglobin disorder or cystic fibrosis

(1) informing the practice population and motivating them for testing;
(2) sample collection and dispatch;
(3) interpretion of results for patients (see below) (ideally, the diagnostic laboratory should send out written information for the doctor to give to the patient together with the diagnosis);
(4) arranging testing of the spouse and other family member when necessary;
(5) ensuring referral when necessary;
(6) supporting the woman or couple through any procedures they may undergo;
(7) discussion of family planning and support in future pregnancies.

Confidentiality

While confidentiality of genetic, as of other information is essential, information on a genetic risk should usually be entered in a prominent place in the primary care record, as well as given to the patient. This will ensure that it accompanies patients who move away, and prevent its being forgotten with the passage of time. However, specially sensitive genetic information should not be easily accessible to other people through a patient's notes, and on rare occasions a patient may request that an item of genetic information, such as risk of a late-onset genetic disorder, be omitted from the records entirely. Naturally the doctor should be prepared to comply.

Though practice infrastructure is steadily becoming more appropriate for community genetic screening, many tools for carrying out the task are not yet available. Posters, information leaflets, training for health workers in genetics and in counselling techniques, training video tapes, courses, books, reference materials, and computer software are all still needed.

'Costs' and benefits of genetic screening

Most attempts at an economic appraisal of genetic screening services compare the cost of screening with the cost of lifelong treatment of an affected child. However, analyses confined to *financial* costs and savings are not acceptable for an economic appraisal (Drummond 1980) and can be very misleading. Costs and benefits to the family are often dismissed as 'imponderables' (i.e. impossible to put a true financial price on), even though many benefits of screening can be measured. For example, Fig. 13.2 summarizes the costs of Down syndrome to the family and society. It can cause great

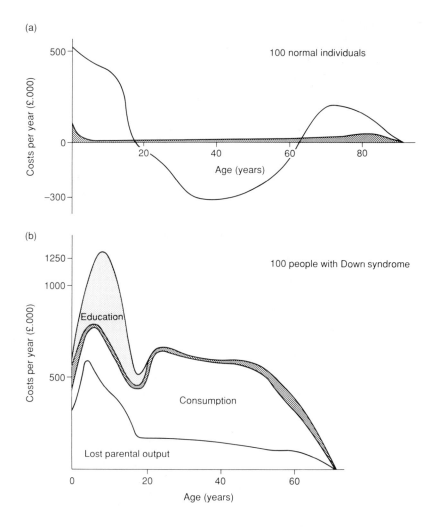

(a)

Costs per year (£.000)

500

0

−300

100 normal individuals

20   40   60   80

Age (years)

(b)

1250

1000

Costs per year (£.000)

Education

500

Consumption

Lost parental output

100 people with Down syndrome

0   20   40   60

Age (years)

Fig. 13.2 Analysis of financial costs related to people with Down Syndrome, by comparison with non-Down individuals (Gill *et al.* 1986). Costs are broken down so that those which fall on society and those which fall on the family can be seen. Financial costs to the Health Service are shown in black. (a) Summary of average life-time costs and contributions to society of 100 individuals without Down Syndrome. Costs, shown above the line, are greatest during childhood and schooling, and after retirement. However these costs are more than offset by net input during adult life, shown below the line as 'negative costs'. Costs to the Health Service are greatest in early infancy and old age. (b) Summary of average life-time costs of 100 individuals with Down Syndrome. Costs in childhood exceed those of non-Downs' children, and costs to the family, society, and the health service continue throughout the relatively shortened life. They are not offset by input. Note that costs to the Health Service, though considerably higher than for non-Down individuals, represent only a very small proportion of the total. It is unrealistic to evaluate costs and benefits of genetic services only in terms of financial costs to the health and social services. (After Gill *et al.* 1986. Reproduced from Royal College of Physicians 1989, with permission.)

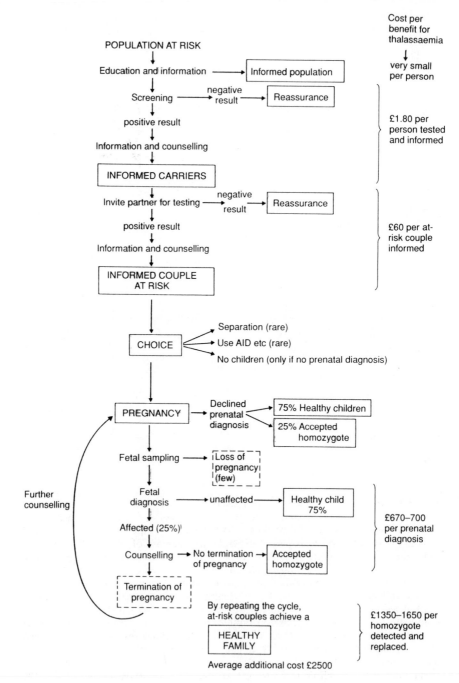

distress to patients and their families to see the price of their life counted out in public, and constantly hear themselves described as a burden rather than a benefit to society, especially when this is done by members of the medical profession.

It is not difficult to design a more realistic approach based on Fig. 13.3, which summarizes costs and benefits of screening for the haemoglobin disorders (Royal College of Physicians 1989). It identifies the main problem for couples at genetic risk as the unexpected or unwanted birth of a seriously affected child, and the main benefit of screening programmes as informed parental choice. Informed parental choice embraces both the decision to replace an affected fetus with a healthy one, and the decision to accept an affected child. In a society where people are free to do what they feel is right, the birth of a wanted affected child is a benefit, not a cost. In fact, parents rarely express regret after an affected child has been born as the result of an informed choice, made at any point in the flow chart.

Coronary heart disease: an example of screening for risk factors

Screening for coronary heart disease (CHD) is complicated; one does not screen for the disease itself but for factors that increase the risk of getting the disease (p. 112). Screening for risk factors is screening once removed (but similar to maternal serum AFP screening).

In England and Wales, people consult their family doctor on average three and a half times a year. This frequency is ideal for regular weighing, measurement of blood pressure, testing the urine for sugar and protein, and a discussion of smoking and drinking. A brief family history (which needs to be brought up to date every few years) is part of the introductory consultation following the registration of an individual with a family doctor. The above is good medical practice and will incidentally identify some people at increased risk of CHD.

However, screening the *practice population* for biochemical as distinct from clinical risk factors is controversial and not established practice at the moment. For example, measurement of serum cholesterol is not part of a routine medical check. We agree with the recommendation of 'cholesterol testing for those with a family history of premature coronary disease, clinical

---

Fig. 13.3 A flow chart summarizing some of the real costs and benefits of screening for prenatal diagnosis (From Royal College of Physicians 1989). Benefits are in a solid boxes. The main costs, termination of pregnancy, or miscarriage as a complication of prenatal diagnosis, are in dashed frames. The birth of an accepted affected child to informed parents is a benefit. For the sake of clarity, some costs such as the consequences of false positive and false negative results in screening tests have been omitted. The estimated average financial cost per person for each step in screening for the haemoglobin disorders (in the UK in 1987) is indicated in the right-hand column. (Reproduced from Royal College of Physicians 1989, with permission.)

features of hyper-lipidaemia, those with manifest coronary disease, those under treatment for diabetes and high blood pressure and those with a long history of heavy smoking' (Kings Fund Centre). The situation will inevitably evolve over the next few years.

Key references

Drummond, M.F. (1980). *Principles of economic appraisal in health care*. Oxford University Press.

Modell, B., Kuliev, A.K., and Wagner, M. (1991). *Community genetics services in Europe*. WHO Regional Office for Europe, Public Health In Europe Series. (In press.)

Wald, N.J. (ed.) (1984). *Antenatal and neonatal screening* Oxford University Press.

14. Counselling

General principles of genetic counselling

Counselling forms part of most consultations between patients and their medical advisers. Basically it consists of encouraging people to voice their concerns and feelings, and giving them enough information to help them decide on a course of action, or to understand the likely origins of their symptoms, or why they are behaving in a particular way. This chapter is concerned with *genetic* counselling in primary care before pregnancy, during pregnancy, and after the baby has been born.

Genetic counselling requires specific knowledge, training, adequate time, and the ability to communicate. Whilst all members of a primary health care team who come into contact with parents or potential parents must be able to give basic information, it may be advisable for larger practice teams to arrange for one member, perhaps a health visitor or practice nurse, to receive additional training. He or she will then receive intra-practice referrals.

Many of the relevant topics are still inadequately taught to medical students or nurses. In addition, the field is rapidly developing and knowledge needs to be continually updated, not least because the intense media interest in reproductive medicine leads people to ask their doctors for a fuller explanation of items that they have read about or seen on television, and think may be relevant to themselves.

Fletcher and co-workers (1985) carried out an international study of counselling by clinical geneticists, and concluded that they practise according to three key ethical principles:

(1) the autonomy of the individual or couple;
(2) their right to complete and accurate information;
(3) the highest standard of confidentiality.

Counsellors need to be aware of their own limitations and to be able to admit uncertainty instead of, for example, giving 'off the cuff' figures for percentage risks of producing an abnormal child, which may turn out to be quite wrong. They also need help with explaining the central concepts of inheritance to people of varied educational and cultural backgrounds.

It is widely accepted that genetic counselling should be 'non-directive', but this does not mean simply giving people the facts and leaving them to make up their own mind. It involves actively helping them to reach the decision that they feel is right for themselves, in the light of their unique social and moral situation. Decisions should be made by the parents according to their own, and not the doctor's or nurse's moral code. However, the counsellor also has responsibilities, particularly if a couple are uncertain or confused. Especially if he or she has known the family for several years and has previous experience of their ability to cope with stressful events, it can be appropriate to focus on the long-term implications of having a child with chronic disease, or the advisability of a diagnostic procedure if a pregnancy has been preceded by several years of infertility. It may be legitimate to put particular emphasis on certain items of information, or show somewhat less readiness to support a request for termination of pregnancy, if the mother shows a significant degree of ambivalence.

Table 14.1 shows the circumstances in which reproductive counselling may be needed in primary care. Each practice will find it useful to develop a *pre-pregnancy counselling policy* involving the general practitioner, health visitors, and practice nurses. Patients come into contact with each of these groups independently, so all three groups have an equally important role, though the health visitors usually concentrate on preventive care in young families. Many of the topics can usefully be discussed with women who visit for contraceptive advice, but wish to become pregnant at some time in the future. Many couples will find the booklet by Macgregor (1990) a useful source of information.

Counselling may include the following:

1. Discussion on methods of increasing the likelihood of conception. Ovulation usually occurs 2 weeks after the onset of a period.
2. The possible deleterious effects of smoking, drinking, and drugs on the fetus can be discussed opportunistically. Future parents may be concerned about environmental hazards such as radiation (p. 321).
3. Discussion of the change in risk factors with increasing age of the parents.
4. The summary of the persons' family history at the front of their notes (p. 93) should be reviewed, to see if they need a referral for genetic counselling.
5. Non-pregnant female teenagers who are unsure of their rubella immunization status can be offered immunization, as can women in the postnatal period whose antenatal test show they were not immune to rubella. A more recent policy of measles/mumps/rubella (MMR) immunization in the second year of life or before school entry will increase the population level of immunity.
6. Certain groups of women can be encouraged to report pregnancy as soon as a period is missed. Women over 35 years old and couples with specific risks need to be informed about the possibility of prenatal diagnosis. People at special risk, for example diabetic women and women on anti-

Table 14.1 Reproductive counselling in primary care

Pre-pregnancy

Previous obstetric history.

Discussion of life-style. Smoking, alcohol, and diet. Discussion of vitamin supplementation.

Need for screening test because of belonging to an 'at risk group' (haemoglobin disorders, Tay–Sachs disease, cystic fibrosis).
Explanation of the test result.
If positive, desirability of investigating other members of the family.

Implications of advancing maternal age.

Consequences of maternal disease e.g. diabetes, certain genetic disorders.

Effect of drugs at the time of conception.

Need for referral to a geneticist because of
  • family history;
  • infertility;
  • previous abnormal child or still-births;
  • recurrent abortions;
  • if screening shows couple to be at genetic risk.

During pregnancy

Above if not already done

Desirability of continuing the pregnancy if the woman wishes to raise this issue.

Explanation of the standard prenatal tests, including the possibility of 'false positives'.

When a fetal abnormality is found, collaborating with the geneticist and/or obstetrician in explaining the likely effects of the particular disorder in a child, and of a termination of pregnancy for the mother.

After pregnancy

Explanation of neonatal screening tests.

If a child has a congenital abnormality:
  • Can its cause be found?
  • What are its likely consequences?
  • What treatment is available?
  • Is referral for genetic counselling indicated?

Discussion of family planning.

If there has been a still-birth or neonatal death, or a pregnancy has been terminated for fetal abnormality, the support of the practice team is needed.

---

epileptic drugs, need to be told of the risk and of the need for special care in pregnancy.

7. A special policy may be developed for any ethnic minorities in the practice population. Most have specific genetic risks (e.g. of haemoglobin disorders or Tay–Sachs disease), and should be informed and offered testing before pregnancy if possible. Some ethnic groups have a convention of consanguineous marriage. It is desirable to refer such couples to a clinical geneticist if there is any suggestion of an inherited disease in the family.

8. In the very near future screening will be available for cystic fibrosis, and people of reproductive age will be the first target. They will need advice in primary care.

Some of the commonest genetic issues are counselling of single heterozygotes for common recessively inherited conditions, pre-test counselling, counselling people with false positive results on tests, and helping people who are worried about genetic risks. When a couple at risk of having a child with an inherited disease is identified, they should always be referred for specialist genetic counselling.

The following paragraphs were written with counselling for single carriers of common inherited conditions such as the haemoglobin disorders, Tay–Sachs disease, or cystic fibrosis in mind. However, much of what is said also applies to counselling for e.g. older mothers, or those with a raised maternal serum AFP.

Though genetic screening often relieves anxiety, it also often involves telling people about risks that they were previously unaware of. A positive result is disturbing, can lead to moral and psychological problems, and sometimes arouses anger and disbelief. Even if a person's initial reaction is hostility to the counsellor, it is important to maintain a calm and supportive attitude so that they will feel able to come back for further discussion in the future.

It is often said that genetic counselling takes too much time to be possible in primary care. Counselling single carriers often takes longer than the basic ten-minute consultation. Whichever practice member is involved needs to set aside enough time for people to absorb information and to make an informed decision when necessary, for instance on whether to have other family members tested. A great deal of time can be saved if appropriate leaflets are handed or sent to people prior to the consultation: the discussion will be more fruitful if they have had time to decide on the questions they would like to have answered. Many excellent leaflets can be obtained from support associations (see Appendix).

In many parts of Europe at present, ethnic minorities are the groups with the greatest need for genetic counselling before or during pregnancy. Some practices with many patients in these groups will have made special arrangements to ensure that they feel welcome when they consult, by, for example, employing receptionists and other members of staff from the relevant groups, making arrangements for interpreters to be available when necessary, referring patients to special ethnic counsellors when these exist, and making sure that the practice leaflets are available in the relevant languages.

Specific points for discussion with carriers of common recessively inherited disorders are as follows.

1. The high incidence of the trait, and the fact that carriers are healthy, and will not develop the major disease.
2. The pattern of inheritance. It is not necessary to mention genes and statistics. Most people are satisfied with a simple explanation such as the following. 'Cystic fibrosis is an inherited characteristic, like skin or eye colour, that is passed on from parents to their children. It is common. About 1 in 20 of the population is a healthy carrier of CF, like yourself. In fact, you are half CF and half normal, but the normal half protects you. When you have children, you can only hand on either CF or normal, but not both. If your partner is a non-carrier, some of your children will be carriers and some not, but no-one will be ill. However, if by chance your partner is another carrier, one of your children could inherit CF from both of you, and that child would be sick. But if you know about the risk beforehand, you can avoid it.' This should lead on to the offer of testing for the partner.
3. The main points about the major disease.
4. The fact that the carrier state may be advantageous (p. 363).
5. Many close relatives are likely to be carriers.
6. Knowledge of carrier status very rarely influences people's choice of partner, but may affect reproductive behaviour.
7. Further information can be obtained from Support Associations.

Before ending the consultation, it is important to check that the person really understands that they are a 'healthy carrier', and do not have, and never will have, the disease. They should be given time to reflect back what they have heard, and specifically asked if they have any further questions. Verbal counselling should be supplemented by written information booklets.

Couples at risk

If the partner is tested and found also to be a carrier, the couple are 'at risk', and should be referred for counselling and the offer of prenatal diagnosis.

Counselling for at risk couples should include:

1. Full information on the nature and prognosis of the disorder involved, and available treatment.

2. The chances of an affected infant in *each* pregnancy, and the reasons for the risk given.
3. Possible ways to avoid the birth of an affected child, including different types of assisted reproduction, and adoption.
4. Prenatal diagnosis. Obstetric and laboratory techniques, problems, risk of obstetric complications or laboratory error. Methods for termination of pregnancy.
5. No couple can be guaranteed a healthy child. Even if they can be assured that they are not overtly at risk of producing an infant with an inherited disease, 2 per cent of all babies are born with a significant congenital anomaly. This risk also applies to children born after prenatal diagnosis for specific conditions, or as a result of successful insemination by donor semen.

People who find they are at risk of bearing an abnormal child, and pregnant women who find they actually are carrying an abnormal fetus, cannot escape from choosing among the options listed in Table 14.2. These are important and difficult decisions. All possible choices affect other people, and none is simple (Modell 1988). If prenatal diagnosis is not available, many couples at high genetic risk choose not to have children, but this can be very painful and may involve aborting accidental pregnancies. Most parents who have requested prenatal diagnosis will also consider abortion if the test result is abnormal.

When a woman has undergone prenatal diagnosis and has been found to carry an affected fetus, primary care workers are likely to be involved in helping her make her decisions, and it is important to give up to date information about the relevant disease. In this book we have included clinical descriptions of the commoner congenital disorders to help primary care workers to do this. The mother should be encouraged to come to a joint decision with her partner, but if there is disagreement, usually it is the mother's decision that should be supported.

Once a couple have made up their minds whether to terminate a pregnancy or continue it, it is appropriate then to support them in their choice and assist them through the associated problems, e.g. getting up the courage to start a new pregnancy, or ensuring optimum care for an affected child. If a woman decides to continue a pregnancy with an affected fetus, it can be very helpful to arrange an early meeting between the parents and the paediatrician who will be looking after the child when it is born. Naturally, it is important to confirm the diagnosis made prenatally, at birth.

The way a couple will behave in the next pregnancy cannot be deduced from the way they behave in the present one. Once a couple have terminated a pregnancy for a particular disease, they are likely to do so again if the next pregnancy is also affected. However, those who have voluntarily given birth to a child with a known abnormality are often interested in prenatal diagnosis in the next pregnancy.

Table 14.2 Possibilities open to carriers of an inherited disease, to avoid having affected children

Time of discovering risk	Possible action
Before marriage (uncommon)	1. Remain single (uncommon)
	2. Avoid selecting another carrier as partner (very uncommon)
	3. Select as partner in the usual way (the commonest choice)
After marriage (more common)	4. Remain childless (common only for severe disease when PND impossible)
	5. Risk having affected child (common for less severe diseases)
	6. Use prenatal diagnosis (very common)
	7. Use AID or other form of 'assisted reproduction' (very uncommon)
	8. Adoption (very uncommon)
	9. Separate and find another partner (very uncommon indeed)
After birth of an affected child (commonest)	Options 4–8 above for further reproduction, plus:
	10. Accept infant and treatment (usual)
	11. Accept infant, but reject treatment (sometimes)
	12. Reject infant (can happen)

From Royal College of Physicians 1989
PND = Prenatal diagnosis
AID = Artificial insemination by donor

Ethical issues

Occasionally there is a conflict between the ethical views of the professionals providing the service and those of the person coming for advice. For example, in the UK it is unacceptable within the NHS to provide fetal sexing with a view to terminating a pregnancy purely on grounds of sex (Royal College of Physicians 1989). Prenatal diagnosis laboratory staff can be very upset if their work is used in this way. The rare couples requesting fetal sexing should be informed of the reasons for refusal.

It is important to stand back and try to view requests for fetal sexing objectively. It is often assumed that female fetuses will be preferentially aborted, but when the (uncommon) request originates from European couples there is no sex bias evident from statistics (A. Czeizel, personal communication). The request is most often made by couples with two to four children, all boys or all girls, the pregnancy under discussion is accidental, and the couple have decided to have an abortion unless the sex of the fetus is different from that of their existing children. In face-to-face discussion the counsellor is likely to feel considerable sympathy with the request. Occasionally a couple of Asian origin ask for fetal sexing with the intention of aborting a female fetus. This may generate more hostility, until it is recalled that the request reflects the weak position of women in some social groups, including the woman involved in the request. Fetal sexing occurs in India and occured in China. It is now forbidden by law in both countries.

There can be a conflict between a patient's right for genetic information to remain confidential, and the need of relatives to know that they may be at risk of having a child with an inherited disease. A patient carrying a mutant gene may not wish the information to be passed on to a spouse or close relative. In these cases, on balance, the responsibility to a future child and the welfare of a whole family usually outweighs the principle that information given to a patient should not be given to a third party. However, every effort should be made to persuade the carrier to allow the doctor or counsellor to divulge the information when necessary. A similar problem arises when a man (or woman) who is discovered to be sterile refuses to tell their partner, who may then decide to be investigated for infertility, especially as the investigations can be unpleasant and invasive.

Cziezel (1988*a*) has written a thoughtful and objective book, about ethical aspects of genetics practised in Hungary and elsewhere.

How genetic screening is used in practice

It is often asked how severe a condition must be to justify abortion, and whether, assuming it becomes technically possible, prenatal diagnosis should be permitted if it is known that the fetus is at risk of mental disease or diabetes? According to the ethical principles of medical genetics, such decisions should be made by the couples directly concerned, the health worker's main responsibility being to provide accurate, clear and comprehensible information. In practice therefore, as new tests becomes feasible, the best way to assess their value to the community is to offer them to families with affected members, and to couples at risk. Further decisions may be based on experience with these pilot groups. Recording the choices that people make in practice could be a key determinant in the future development of community genetics services (Modell *et al.* 1991). Though this general principle has yet to be widely accepted, the information given below about the choices people actually make supports its validity.

It is often also assumed that the ability to predict a wide range of genetic characteristics will lead to terminations of pregnancy for minor or even frivolous reasons. However, this assumption gravely underestimates the seriousness with which people view abortion of a wanted pregnancy. There is no evidence of abuse or over-use of prenatal diagnosis and selective abortion by the population. In the case of thalassaemia, the effect of informed choice has been a major reduction in the number of affected births, but the same is not true for all inherited diseases. In the USA and the UK, less than 50 per cent of couples at risk of having children with sickle cell disease choose prenatal diagnosis at present (Anionwu *et al.* 1987), and there has so far been relatively little demand in Western Europe for selective abortion for phenylketonuria, where dietary treatment, though burdensome, is relatively simple and is thought to be effective (H. Goldstein, personal communication). When congenital malformations are detected by ultrasound scanning in early pregnancy, parents usually request abortion only if the future for the fetus appears extremely grave.

There is little evidence that people take genetic risk in to account in choosing a partner (Angastiniotis *et al.* 1986), and there is no evidence of an increased divorce rate among couples who discover a genetic risk after marriage.

All available studies show that most couples wish to know of, and if possible avoid, any risk they may run of having children with a serious congenital anomaly, and that less than 10 per cent have absolute objections to termination of pregnancy. Unless prenatal diagnosis is available, couples at risk of having children with severe genetic diseases often decide to have no (further) children and abort accidental pregnancies. When the fetus has a severe disorder, couples usually request abortion, and most ask for prenatal diagnosis again in subsequent pregnancies. Thus prenatal diagnosis may actually reduce the number of abortions done for fear of genetic disease (Modell *et al.* 1980). Other solutions such as artificial insemination by donor or adoption have not proved popular. In the UK most pregnant women accept maternal serum AFP screening when it is offered (Ferguson-Smith 1983), and between 60 and 80 per cent of counselled women in risk groups for chromosomal disorders of the fetus request fetal karyotyping (Knott 1988).

When a family decides to sue the health service, usually it is not because of a complication of a diagnostic procedure, but because a woman with a detectable risk was not informed and offered testing, and subsequently had an affected child.

Key references

Czeizel, A. (1988). *The right to be born healthy. The ethical problems of human genetics in Hungary.* Akademiai Kiado, Budapest.
Fletcher, J.C., Berg, K., and Tranoy, K.E. (1985). Ethical aspects of medical

genetics. A proposal for guidelines in genetic counselling, prenatal diagnosis and screening. *Clinical Genetics*, **27**, 199–205.

Harper, P.S. (1988). *Practical genetic counselling*, (2nd edn). Wright, Bristol.

Holtzman, N.A. (1989). *Proceed with caution. Predicting genetic risks in the recombinant DNA era*. The Johns Hopkins University Press, Baltimore.

Modell, B. (1988). *Ethical and social aspects of fetal diagnosis for the haemoglobinopathies: a practical view*. In '*Prenatal diagnosis thalassaemia and the haemoglobinopathies*' (ed. D. Loukopolous.) CRC Press Inc. Boca Raton, FL.

Rothman, B.K. (1988). *The tentative pregnancy*. Pandora, London.

Part 3

Specific congenital and genetic disorders

15. The haemoglobin disorders: an example of the community approach

The major haemoglobin disorders are listed in Table 15.1. About 260 000 affected children are born annually worldwide (Fig. 15.1), a global birth incidence of about 2 in a 1000 (Modell and Bulyzhenkov 1988). It has been possible to detect carriers simply, cheaply, and reliably for many years, so when prenatal diagnosis became feasible, it was possible to set up programmes for *community control* of these disorders. A 'control' programme for an inherited disease is defined by WHO (1987) as 'an integrated strategy combining the best possible treatment for affected individuals, with community education, population screening, genetic counselling, and the availability of prenatal diagnosis'. Control programmes for haemoglobin disorders provide the most fully developed example so far (WHO 1983, 1985a, 1988).

Table 15.1 The major haemoglobin disorders

Thalassaemias	Sickle cell disease
β-thalassaemia major	HbSS = sickle cell anaemia
β-thalassaemia intermedia	HbS/C disease
Hb E/β-thalassaemia	HbS/D disease
α^0-thalassaemia hydrops fetalis	HbS/β-thalassaemia
Haemoglobin H disease	

Haemoglobin disorders are indigenous in southern Europe, and are the main genetic problem of the ethnic minorities that now constitute about 5 per cent of the population and contribute up to 9 per cent of births in much of North-West Europe. Table 15.2 gives figures for the populations at risk in the UK, and Table 15.3 gives minimum estimates of the number of affected infants born annually. In many industrial 'inner city' areas, haemoglobin disorders are now the commonest major genetic disease; for example, their birth incidence is more than 0.85 per 1000 in Greater London.

Most populations at risk include carriers both of thalassaemia trait and of sickle cell trait. In the UK there are more thalassaemia carriers, but sickle cell

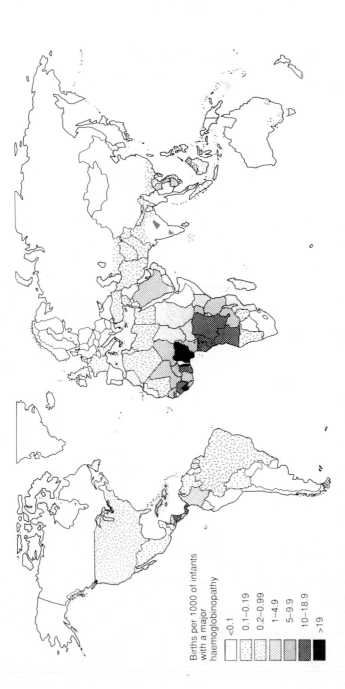

Fig. 15.1 World distribution of the haemoglobin disorders, expressed in terms of births of affected infants/1000, by country. (Reproduced from WHO 1985*a*, with permission).

Table 15.2 Ethnic groups at risk of haemoglobin disorders in the UK: number, carrier (=heterozygote) frequency, birth rate of affected infants (=homozygotes)

Ethnic group	Numbers in UK (thousands)	Common haemoglobinopathy genes	% heterozygotes	Estimated number of UK carriers	% Males G6PD deficient	Estimated no. of UK male hemizygotes
Afro-Caribbean	750	S > C > βth > D	11-14	94 000	12	45 000
West African	98.5	S > C > βth	20-25	23 000	20	10 000
Indian	873	βth > D, E, S	3-10	44 000	7	30 000
Pakistani	370	βth > D, E, S	6	22 000	5	9000
Bangladeshi	87.2	HbE > βth > S	5	2500	3	1250
East African Asian	221	βth > D, E, S	6-10	13 000	??	7750
Cypriots	180	βth > αth > S	17	31 000	7	6300
Middle Eastern (Iranian, Lebanese etc.)	?200	βth > S	?	?	5-10	5000
Chinese (Hong Kong Singapore)	>250	αth > βth	6	15 500	3	3750
Italian	>250	βth > αth > S	1-6	7500	0.6	750
TOTAL	>3280			452 500		118 800

(From Modell and Modell 1990)
βth = β-thalassaemia trait; αth = α-thalassaemia trait. S, C, D, E, = Hb S trait etc.
Precise information on the frequency of these traits in several ethnic groups in the UK is still unavailable.

Table 15.3 Estimates of birth incidence of major haemoglobin disorders in the UK

Ethnic group	Estimated		Major haemoglobin disorders			
	Birth rate/1000	Annual births	Births/1000[1]	Annual births		
				Potential	Actual	
Sickling disorders						
Afro-Caribbeans	13.6	10 150	4.5	46	Neonatal screening in London suggests 50% more, i.e. about 150/year	
West African	27	2660	19.0	50		
Total				ca. 96	ca. 150?	
Thalassaemia disorders						
Indians	21	18 330	?0.3	6	Reduced by parental diagnosis by about 50%	
East African Asians	21	4640	0.9	>4		
Pakistanis	36	13 300	2.0	28		
Bangladeshis	46	4010	0.3	>1		
Cypriots	15	2700	7.0	19		
Chinese	20?	4000	0.6	2.5		
Total				60	<30?	

[1] Estimated from the % heterozygotes listed in Table 15.2.
In the absence of a patient register, there is considerable uncertainty about the true number of births

disease is commoner than thalassaemia major because sickle cell genes are so common among Afro-Caribbeans and Africans (Figure 5.5).

Molecular basis

Adult haemoglobin (Hb A) consists of two α and two β globin chains, controlled by corresponding genes in chromosomes 11 and 16 (p. 49). Haemoglobin disorders may therefore be caused by two groups of mutations, affecting either α or β genes (Table 15.4). The inheritance of two α or two β chain mutations can cause clinical disease, but the combination of one α with one β chain mutation is harmless. Both groups include 'severe' and 'mild' mutations, and much of the clinical heterogeneity of the resulting disorders is due to different combinations of mutations of different degrees of severity. The following summary shows how different mutations affecting the same protein can interact to give different clinical pictures ranging from still-birth to the essentially normal. Similar genetic and clinical correlations will probably be discovered for many of the other genetic disorders that are now becoming understood at the molecular level.

The commoner major haemoglobin disorders are caused by β chain abnormalities. The clinically important **abnormal haemoglobins** (Hb S, Hb C, Hb E and Hb D) arise from different point mutations in the coding sequences of the β-globin gene that cause structural and charge changes in the haemoglobin molecule. 'Sickle cell disease' includes, in decreasing order of severity, homozygous Hb SS (sickle cell anaemia), Hb S/D disease, sickle cell/β thalassaemia, and Hb S/C disease. However, even the apparently clear-cut SS disease includes mild and severe types, as the expression of the S gene can be influenced by adjacent DNA polymorphisms that have no observable effect in the absence of the disease (Kulozic *et al.* 1986).

Thalassaemias are disturbances of haemoglobin synthesis that may be caused by a wide variety of mutations, many in the introns or flanking sequences of the α- or β-globin genes. The inheritance of two severe β-thalassaemia mutations causes thalassaemia major. Haemoglobin E results from a point mutation in the coding sequence of the β-globin gene, which leads both to a single amino acid substitution, and also to moderately reduced β chain synthesis. Inheritance of one severe β-thalassaemia mutation and one Hb E gene causes a less severe anaemia. People who inherit two HbE genes are usually quite healthy.

A similar situation occurs in α-thalassaemia (reviewed by WHO 1990b). Normally there are two functioning α genes in tandem on each chromosome 16 and there are two main forms of α-thalassaemia. In α^+**-thalassaemia,** one of the two genes is deleted, but the condition is nearly always harmless, since even in homozygotes plenty of α-globin is still made and there is only a slight reduction in haemoglobin level. By contrast, in α^0**-thalassaemia,** both α genes on one chromosome are deleted; this mutation involves a

Table 15.4   Harmful and harmless combinations of haemoglobinopathy genes

Group	Effect	α-chain mutations	β-chain mutations
1 severe	Cause serious disease in homozygotes, or when combined with a Group 2 mutation *of the same globin chain*	α^0-thalassaemia	β-thalassaemia Hb S
2 mild	Cause serious disease only when combined with a Group 1 mutation *of the same globin chain*	α^+-thalassaemia Hb Constant Spring	Hb E Hb C
3	Nearly always harmless	Numerous rare forms	Hb D Numerous rare forms

α chain mutations and β chain mutations do not 'interact' to cause pathology.

serious genetic risk. Carriers are healthy, but homozygous α^0-thalassaemia causes still-birth or neonatal death, because fetal haemoglobin (Hb F) is $\alpha_2 \gamma_2$, so when no α-globin is made, no Hb F can be made. The combination of α^0- with α^+-thalassaemia leads to a chronic, usually fairly mild anaemia, haemoglobin H disease.

α^+-thalassaemia is extremely common, being carried by up to 30 per cent of people originating from Africa or the Indian sub-continent. It is essentially the only form of α-thalassaemia found in these populations, so α-thalassaemia carriers of such ancestry have practically no genetic risk.

Both α^0- and α^+-thalassaemia occur among East Asians and Mediterraneans, and can be distinguished definitively from each other only by DNA studies. α-thalassaemia in people of such ancestry may carry a significant genetic risk.

The major disorders

Diagnosis

β-thalassaemia major and sickle cell disease rarely present before 4–6 months of age, because the predominant haemoglobin at birth is Hb F ($\alpha_2 \gamma_2$). In normal infants this is replaced by Hb A ($\alpha_2 \beta_2$) by six months of age (p. 50). In β-thalassaemia major, the Hb F level remains high (more than 10 per cent, usually over 80 per cent) after 6 months of age, and β-thalassaemia trait is found in both parents. In sickle cell disease there is usually anaemia, blood tests usually show Hb S and Hb F, and both parents are found to carry a trait that can contribute to a sickling disorder (Table 15.4).

β-thalassaemia major

Children with β-thalassaemia major are healthy at birth, but develop a severe intractable anaemia between 6 months and 2 years of age (haemoglobin level = 4–7 g/dl). They present with failure to thrive, pallor, lethargy, and sometimes an enlarged spleen. Without diagnosis and treatment, most die from anaemia or infections before 2–5 years of age, though a few with a milder disease, thalassaemia intermedia, survive into adult life.

Thalassaemic patients are very hospital-dependent. Most require regular (approximately monthly) blood transfusions to maintain a mean Hb level of about 12 g dl^{-1} (WHO 1982). This preserves excellent health in the short term but leads to severe iron overload. After 11 years of age this can cause endocrine disturbances such as growth retardation, failure of puberty, and diabetes, and can cause death in early adult life from intractable heart failure. This outcome can be avoided, but only by regular subcutaneous infusion of the iron-chelating agent desferrioxamine (Desferal) from a small portable syringe-driver, over 8–12 hours on at least five nights a week. When treatment is started early and faithfully maintained, quality of life can be good and the prognosis is hopeful. Many patients now live well into their 30s, and it is reasonable to hope for a near normal life expectancy. However there may be complications of transfusion such as hepatitis B or C. Patients should be immunized against hepatitis B at an early stage. Thalassaemic patients are also at risk of overwhelming infection (1) because many are splenectomised and (2) because Desferal treatment increases the risk of infection (usually severe gastroenteritis) with *Yersinia enterocolitica* (Piga *et al.* 1991), which uses the iron chelate as a source of iron for growth (Chiu *et al.* 1986).

Desferal is too expensive for many countries to afford, and in developed countries adolescents and young adults often find it difficult to comply with the treatment. Failure to take 'the pump' is now the main cause of death in thalassaemia in most of Europe (Borgna-Pignatti *et al.* 1988). A cheap and effective oral iron-chelating agent (Herschko 1988) would probably solve the problem of compliance, and could make treatment possible in more countries.

Bone marrow transplantation from an HLA compatible sibling is increasingly popular and successful (Lucarelli *et al.* 1985), and is far less expensive than the cumulative cost of conventional lifelong treatment. An important problem is that less than 30 per cent of patients have a suitable donor in the family. The hope that a new child might be HLA compatible with the affected one helps some couples to work up the courage they need to start a new pregnancy, and undergo the rigours of prenatal diagnosis.

Sickle cell disease

This includes a wide spectrum of illness (Serjeant 1985). The following description covers problems that may arise in all sickling syndromes. People with SS disease have lifelong chronic anaemia with an Hb level around 8 $g\,dl^{-1}$, but the main problems are risk of sudden death in childhood, and painful crises and other complications lifelong. These problems arise from the tendency of the red blood cells to sickle and block capillaries at low oxygen tension. Patients with sickle cell disease should attend a sickle cell clinic if possible. If one is not easily accessible, they should be seen regularly by a specialist.

In children, sickled red cells often become trapped in the spleen, leading to a serious risk of death before the age of seven from an acute 'splenic sequestration crisis' (sudden profound anaemia associated with rapid splenic enlargement), or because functional asplenia permits an overwhelming infection. It is estimated that about 10 per cent of undiagnosed affected infants die in this way in developed countries (*Neonatal screening for sickle cell disease* 1989), but in rural Africa the figure is almost 100 per cent. In the absence of neonatal screening, affected children most often present between 6 and 18 months of age with painful swellings of the hands or feet (hand–foot syndrome). Once a diagnosis has been made, prophylactic penicillin and folic acid (5 mg) should be given daily. Since these precautions are most effective if susceptible infants are identified at birth, neonatal diagnosis is recommended for the sickling syndromes (p. 350).

The spleen and its associated risks usually disappear before adulthood due to auto-infarction, so later in life the main problems are due to trapping of sickled cells in other organs. At this stage sickle-cell disease is sometimes viewed as a relatively mild disorder, but one of its worst aspects is its highly unpredictable course, which varies with the gene combination, ethnic origin, environment, level of awareness of the disease and its hazards, and other unknown factors. This places a great strain on the patient and the family, who need a good understanding of the disease and easy and open lines of communication with primary care workers and a specialist centre.

Affected people may suffer recurrent and unpredictable severely painful crises, and other complications such as 'acute chest syndrome' (representing either pneumonia (more common in children) or pulmonary infarction (more common in adults)), bone or joint necrosis, priapism, or renal failure. The incidence of complications can be reduced by avoiding excessive heat or cold, dehydration and stress, and minor crises may be managed at home with rest, warmth, adequate analgesia, adequate fluid intake, and antibiotics if indicated, but the average 'sickler' requires one or two hospital admissions a year for more severe episodes, and occasional transfusions.

About 10 per cent of patients suffer from particularly severe disease and may have catastrophic complications such as stroke, blindness, or total marrow necrosis, leaving severe residual disability. They may need regular exchange transfusions, or maintenance transfusion and Desferal treatment as in thalassaemia, at least for several years. If there were some way to identify these patients at an early stage, they might be suitable candidates for bone marrow transplantation. Sicklers are often transfused throughout pregnancy to minimize risks to the mother.

Many families with haemoglobin disorders get considerable help from the Support Associations, and from the staff of special sickle-cell centres (see appendix to this chapter).

α^0-thalassaemia: Hb Bart's hydrops fetalis

When both partners carry α^0-thalassaemia, there is one in four chance in each pregnancy of still-birth or neonatal death due to Hb Bart's hydrops fetalis. The syndrome has this name because the fetus cannot make the α chains of Hb F but produces γ chains normally. These aggregate in fours to make the abnormal haemoglobin Bart's, easily identified by electrophoresis. The syndrome used to be very rare in Europe, but is now seen increasingly in families of Chinese, Vietnamese, or Eastern Mediterranean extraction. It is dangerous to the mother, as well as to the fetus, and couples at risk of homozygous α^0-thalassaemia are particularly interested in prenatal diagnosis and selective abortion.

α-thalassaemia hydrops fetalis can cause a wide range of fetal and maternal pathology. A study from Bangkok, where its birth incidence is 0.3 per 1000 (Fucharoen *et al.*, submitted) showed that 25 per cent of affected infants died *in utero*, 18 per cent during delivery, and 54 per cent soon after birth. Pathological findings in the fetuses include gross anaemia with enlargement of the heart, liver and spleen (where red cells are made in fetal life), and severe retardation in brain growth. Hypoplasia of the lungs accounts for death soon after birth.

Up to half the mothers needed 'assisted delivery', including Caesarian section, and sometimes embryotomy (cutting up a fetus that is causing an obstructed labour). Maternal complications include pre-eclampsia and eclampsia, antepartum haemorrhage and post-partum haemorrhage. Many mothers carrying a hydropic fetus may die when obstetric assistance is not available.

Diagnosis of carriers

Tests to identify carriers of haemoglobin disorders (Fig. 15.2) are, in general, simple and accurate (British Society for Haematology 1988). Carrier screening is one of the simplest forms of genetic screening, but can sometimes present quite difficult problems: similar problems will certainly be encountered as other forms of genetic carrier screening become established.

Since the abnormal haemoglobins have a charge change, carriers can be identified by electrophoresis. In the thalassaemias there is decreased production of normal haemoglobin, so in carriers the amount of haemoglobin per red cell is greatly reduced. However, the haemoglobin level is maintained in the normal range by the normal erythropoietin mechanism: the kidney secretes erythropoietin if it does not receive enough oxygen; this stimulates the bone marrow to produce more red cells until the Hb level is high enough to ensure adequate oxygenation of the kidney, and other body tissues. The primary screen for a thalassaemia trait is thus for microcytosis rather than anaemia.

Samples with a low mean cell haemoglobin level (MCH) should be investigated further. Diagnoses will include iron deficiency, α-thalassaemia trait and β-thalassaemia trait. Both the incidence of β-thalassaemia trait and the proportion of people with microcytosis differ greatly between populations. In Mediterranean populations, β-thalassaemia trait is the commonest cause of marked microcytosis, but in Indians and Pakistanis, iron deficiency and α-thalassaemia predominate and often occur together, particularly in

For thalassaemias

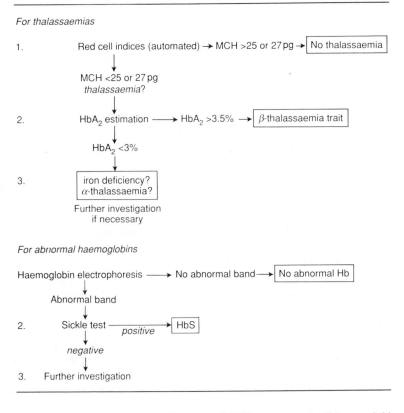

Fig. 15.2 The haemoglobinopathy screen. MCH = mean red cell haemoglobin.

pregnant women, over 25 per cent of whom may be iron deficient (L. Tillyer, personal communication). Iron deficiency does not interfere with the diagnosis of β-thalassaemia trait: but measurement of serum iron or ferritin is necessary to differentiate α thalassaemia from iron deficiency in 'at risk' groups.

For genetic purposes, the prime targets of screening are β- and α^0-thalassaemia traits. The definitive diagnosis of β-thalassaemia trait is made by measuring the haemoglobin Hb A_2 level; a value over 3.5 per cent indicates β-thalassaemia trait. But Hb A_2 estimation is expensive, and the cut-off value of the MCH below which Hb A_2 estimation is recommended can vary with the population being screened, the need to diagnose *all* or just obvious carriers, and local facilities. The vast majority of carriers of β- or α^0-thalassaemia trait have an MCH of less than 25 pg, so for opportunistic screening, measurement of Hb A_2 on samples with an MCH of less than 25 pg may be sufficient. However, a few β-thalassaemia carriers have a MCH between 25 and 27 pg, so for genetic screening, e.g. in the antenatal clinic or when patients ask for screening, Hb A_2 measurement on samples with an MCH of less than 27 pg is recommended.

Once β-thalassaemia has been excluded, a number of people with micro-cytosis and a normal Hb A_2 level remain, and the differential diagnosis can be difficult. The next step is to diagnose and treat iron deficiency. Once this is excluded, remaining possibilities are (1) one of the two forms of α-thalassaemia trait, (2) a combination of α- and β-thalassaemia traits, or (3) (in some populations) 'normal Hb A_2 β-thalassaemia' trait. As it can be very difficult to establish a definitive diagnosis, the next step in counsell-ing depends on the ethnic origin of the subjects. If they originate from the Indian subcontinent or Africa, they almost certainly carry α^+-thalassaemia and can be told that they have a harmless characteristic of the blood with no known clinical importance. But if they originate from the Mediterranean or South China, Hong Kong, Singapore, Malaysia, etc., they should be advised to bring their partner for testing before marriage, or before starting a family. If the partner is haematologically normal, the couple can be reassured, but if there is *any haematological abnormality at all*, they must be referred for further investigations and expert genetic counselling.

Thalassaemia trait occurs very infrequently (about 1–2 per 1000) in Northern Europeans, and is often mistaken for iron deficiency (Knox-Macaulay *et al*. 1973). Carriers may be incorrectly treated long-term with iron preparations, and especially in men this can lead to iatrogenic iron overload, clinically very similar to idiopathic haemochromatosis (p. 224). Thalassaemia trait is not an iron-loading condition: pregnant thalassaemia carriers should be given extra iron when indicated.

Prenatal diagnosis

Prenatal diagnosis for haemoglobin disorders can be done from 9 weeks of pregnancy onwards, by chorionic villus sampling (CVS) and DNA analysis (p. 150). In the UK a centre capable of CVS is within reach of most at risk couples. Because of the wide range of possible mutations DNA diagnosis for thalassaemia usually requires prior family studies, which may take up to two weeks, but diagnosis of sickle cell disease is relatively straightforward. In order to benefit from first trimester prenatal diagnosis, couples at risk need to be identified and studied either before pregnancy or immediately a pregnancy has started.

Prenatal diagnosis can also be done in the second trimester of pregnancy (at around 18 weeks) by fetal blood sampling, but this is less acceptable (p. 153). Though most couples at risk for thalassaemia request prenatal diagnosis and choose to terminate affected pregnancies, uptake depends on stage of pregnancy as well as on culture and religion (Modell *et al*. 1980). A majority of Muslim British Pakistanis find prenatal diagnosis in the second trimester unacceptable, but many request it in the first trimester (Darr 1990). When first trimester diagnosis was introduced in Sardinia, uptake by counselled couples rose from 94 per cent to 100 per cent (A. Cao, personal communication). About 50 per cent of couples at risk for having children with sickle cell disease in the second trimester of pregnancy request prenatal diagnosis, and about two-thirds of those found to have an affected fetus decide to terminate the pregnancy (Anionwu *et al.* 1987; Rowley *et al.* 1988). There are as yet no published studies on the relative acceptability of first, as opposed to second trimester prenatal diagnosis for sickle cell disease.

Whenever possible an experienced haemoglobinopathy counsellor of the appropriate ethnic group should counsel couples at risk.

Results of thalassaemia control programmes

In general, screening programmes can be monitored by comparing the number of affected infants born with the number expected, and following up new cases to find out if the parents were detected and counselled prior to the child's birth. When 193 parents of thalassaemic children born in Italy and Greece since the start of the control programme were interviewed, it was found that only a minority had been born as a result of their parents' informed choice. Most births were due to ignorance of risk and availability of prenatal diagnosis the part of the parents or their obstetrician (WHO 1985*a*). Since most couples at risk for thalassaemia request prenatal diagnosis, the effectiveness of whole programmes including public education and screening, can be monitored relatively simply by recording the changing birth rate of affected children. The results of several programmes, summarized in Fig. 15.3, show that it is possible to deliver these services to whole populations, and that thalassaemia may almost disappear as a result (WHO 1985*a*; Kuliev 1986).

The figure also shows that once prevention becomes possible, the problems of delivering the service are only just beginning. Thalassaemic births have fallen most rapidly where a programme for a small community at high risk is organized by motivated staff working from a single centre, with the help of an active Thalassaemia Association, as in Cyprus, Sardinia, and the Ferrara district in North-East Italy. The fall has been slower in large countries where the disease is endemic, such as Greece and Italy, because a great deal of time and effort is needed to integrate genetic screening into routine health services. Despite advanced medical services, the fall in thalassaemic births has been slower in the UK than in other countries. This is only partly because prenatal diagnosis is less acceptable to British Pakistanis than to other ethnic groups. It is mainly due to the additional difficulty of ensuring that the service is delivered to a range of ethnic minorities scattered in a larger population that is not at risk (Modell *et al.* 1985).

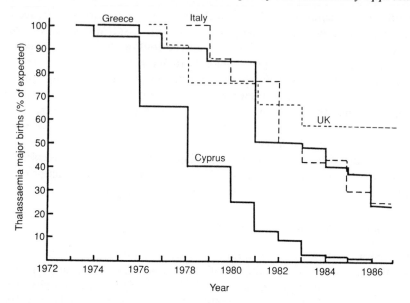

Fig. 15.3 Fall in the birth rate of infants with thalassaemia major associated with four national thalassaemia control programmes (Reproduced from Modell *et al.* 1991, with permission).

Important problems of communication associated with diagnosis of carriers of haemoglobin disorders in the UK (as in most of North-West Europe) include:

(1) the women are rarely informed that they are being screened, or why, or when their result is negative;

(2) carriers are asked to bring their partner for testing, but often without a comprehensible explanation or written information;

(3) results of screening are often filed in the notes without their genetic significance being explained to the patient;

(4) when carriers are informed, they are often told, for example, that they are 'a sickler', and can receive the impression that they have a serious disease when, in fact, they are perfectly healthy (Black and Laws 1986);

(5) people who know that they carry a haemoglobinopathy trait often get tested repeatedly, because of the lack of documentary evidence.

To try to deal with these problems, the UK Department of Health has recently provided cards to haematologists, giving the diagnosis, to send out to the family doctor to be handed to the person, when a carrier is found. But this is only a beginning. The accompanying short leaflet is inadequate, and handing a card to someone without an adequate explanation is of little use. Thalassaemia carrier couples who had both previously been diagnosed and issued with a card, but had not been informed of any genetic risk, are still being referred for counselling in the mid-trimester of pregnancy.

These problems are beginning to be resolved in the UK, because it has been recognized that trained counsellors and a network of haemoglobinopathy counselling centres are needed. Sickle cell and thalassaemia centres are gradually being established on a regional or district basis to provide counselling for patients and their families, and for carriers, and to support other primary care workers in screening and counselling (WHO 1988). Testing and counselling for haemoglobinopathy carriers can realistically become part of primary health care once such centres are available to provide the necessary back-up.

Role of the primary care worker

Treatment

Patients with thalassaemia major are usually managed at a hospital, but may call on their family doctor if they are unwell. If there is any suspicion of a serious infection, the patient should be treated promptly with a broad spectrum antibiotic and referred immediately to their treatment centre. If the patient develops gastroenteritis, Desferal should be discontinued. Even patients with symptoms of a relatively mild infection should be closely monitored, as deterioration can be sudden and severe.

Sickle cell disease may cause few problems over long periods and many patients are reluctant to go to hospital if it can be avoided. When problems arise they usually occur as emergencies. The family doctor is particularly likely to be involved in the treatment of painful crises and should not hesitate to give adequate analgesia (Pain in sickle cell disease 1983).

Carrier screening

People in the relevant ethnic groups should be informed of the existence of risk and offered carrier screening (Modell and Modell 1990). The information can be given verbally; a poster and simple leaflets are also available (See appendix to this chapter).

Haemoglobinopathy screening can be carried out either as an investigation in itself, or when blood is taken for some other reason from a person in a group at risk. In pregnancy it is obligatory as early as possible. Usually 5 ml of 'sequestrene' (EDTA) blood sent to the laboratory with a request for a 'haemoglobinopathy screen' (Fig. 15.2) will be adequate. Before starting screening for haemoglobin disorders in general practice, it is necessary to check the following with the local haematology laboratory:

1. Are they prepared to do haemoglobinopathy screening?

2. How should a screen be requested, in order to be sure that both haematological indices and electrophoresis will be performed?
3. Will they report a diagnosis (e.g. 'result compatible with β thalassaemia trait') rather than simply giving the test results, which are often hard for primary care workers to interpret?

Opportunistic screening can be done for thalassaemias, by scanning the haematology reports of people in risk groups, including looking through their records for previous blood test results. If a blood test shows microcytosis (MCH < 25 pg), an Hb A_2 measurement should be requested to confirm or exclude β-thalassaemia trait.

When a result is positive, the following action should be taken:

1. It should be recorded prominently in the notes, and noted in the records of other family members.
2. The person should be informed. Booklets explaining the implications of being a carrier are available (see appendix to this chapter).
3. Testing should be offered for other family members (there is a 50 per cent risk for each first degree relative of being a carrier) and for the partner, if there is one yet.

Counselling

Counselling for single carriers is most appropriately done in primary care (p. 184). Table 15.5 gives a check-list of topics that should be covered. Helpful leaflets and posters exist (see appendix to this chapter).

Table 15.5   Information check-list for people tested for haemoglobin disorders

---

- All people tested should be told (or, in the case of a child, the parents should be told) whether they do, *or do not* carry the trait tested for.
- Carriers should be given a booklet describing the implications of being a carrier (see Appendix to this chapter).
- Carriers must understand the difference between being a healthy carrier and having a disease.
- Carriers need to be reassured that there is no risk to their own health, though there could be a potential risk for their children.
- They must be aware that problems can be avoided, if they wish, but the first step is to arrange similar testing for their partner.
- They should be encouraged to tell other family members, and encourage them to get tested.
- They must know that the results are strictly confidential.
- A system must be developed for keeping the results in the primary care record.

One of the objectives of counselling is to encourage known carriers to invite their partner to have a blood test at some time prior to reproduction. If the partner has a completely normal blood test, the couple can be reassured: but if the partner has any red cell abnormality *at all*, the couple should be referred for definitive diagnosis. If the couple is found to be at risk, they should be offered genetic counselling at a specialist prenatal diagnosis centre or a sickle cell/thalassaemia centre, or by a clinical geneticist, as appropriate.

Key references

Loukopolous, D. (ed). (1988). *Prenatal diagnosis of thalassaemia and the haemoglobinopathies*. CRC Press Inc., Boca Raton, FL.

Modell, B. and Berdoukas, V. (1984). *The clinical approach to thalassaemia*. Grune and Stratton, New York.

Modell, M. and Modell, B. (1990). Genetic screening for ethnic minorities. *British Medical Journal*, **300**, 1702–4.

Serjeant, G. R. (1985). *Sickle cell disease*. Oxford University Press.

Vullo, C. and Modell, B. (1990). *What is thalassaemia*? The Cooley's Anemia Foundation, New York. (For the Thalassaemia International Federation). (A book for patients)

Weatherall, D. J. and Clegg, J. B. (1981). *The thalassaemia syndromes*, (3rd edn). Blackwell Scientific Publications, Oxford.

WHO (1983). Community control of hereditary anaemias: memorandum from a WHO meeting. *Bulletin of the World Health Organization*, **61**, 63–80.

WHO (1988). *The haemoglobinopathies in Europe*. WHO Regional Office for Europe, unpublished document IPC/MCH 110. (May be obtained free of charge from: Maternal and Child Health Division, WHO Regional Office for Europe, 8 Scherfigsvej, DK-2100, Copenhagen, Denmark.)

WHO (1990). *Hereditary anaemias: alpha thalassaemia*. Unpublished WHO document HDP/WG/HA/87.5. May be obtained free of charge from: The Hereditary Diseases Programme, WHO, Geneva, Switzerland.

Appendix

Aids for screening and counselling for haemoglobin disorders

Posters and simple leaflets to motivate people to be tested for the haemoglobin disorders can be obtained (in large numbers) from the Press and Public Relations Dept, North East Thames Regional Health Authority, 40 Eastbourne Terrace, London W2 3QR or (small numbers) from the George Marsh Sickle Cell and Thalassaemia Centre, St Anne's Hospital, Tottenham, London N-15. (Tel. 081-809-1797).

Leaflets for *carriers of haemoglobin disorders* can be obtained from the Sickle Cell Society, Green Lodge, Barretts Green Rd, London NW10 7AP, and (in English, Greek, Turkish, Urdu, Hindi, Gujarati, and Bengali) from the UK Thalassaemia Society, 107 Nightingale Lane, London N8 7QY.

A list of *sickle cell/thalassaemia centres* is given in Appendix 4 of the Royal College of Physicians Report on Prenatal Diagnosis and Genetic Screening (1989). Alternatively, addresses may be obtained by telephoning the George Marsh Sickle Cell and Thalassaemia Centre at St Anne's Hospital, Tottenham; London on 081-809-1797.

Specialist counselling and prenatal diagnosis for couples at risk for haemoglobin disorders is available at University College and Middlesex School of Medicine Perinatal Centre, 86-96 Chenies Mews, London WCIE 6HX, or the Department of Haematology, King's College Hospital Medical School, Denmark Hill, London SE5 9RS.

16. Common recessively inherited disorders

An average group practice in the UK serves a population of 8000–10 000 people, and is likely to include at least one family at risk for most inherited disorders that occur with an incidence of 1/10 000 or more. Our aim in this chapter and the next is to describe some inherited conditions which many primary care workers may meet in practice, and use them to illustrate the wide range of clinical pictures and problems presented by genetic disorders. These chapters may be viewed as reference material. To provide a background for advising families, we summarize the incidence, diagnosis, clinical picture, natural history, and possibilities for prevention for each condition, finishing with comments on the role of the primary health care team. Many of the points raised in Chapters 14 and 15 apply, and many of the problems discussed in Chapter 20 will occur in families with a child with one of the severe diseases discussed here.

This chapter concentrates on the common recessively inherited conditions shown in Table 16.1. Now that the gene for cystic fibrosis has been identified, primary care workers are likely to be involved in screening for carriers

Table 16.1 Carrier frequency and homozygote birth rate in some common recessively inherited disorders in Europe

Condition	% of the population heterozygous	Homozyogote births/1000
Cystic fibrosis	4–5	0.5
Haemoglobin disorders	0–25	0–20
Phenylketonuria	1–2	0.05–0.13
Lactase non-persistence	Most of some populations	Most of some populations
Tay-Sachs disease (in Eastern European Jews)	3–4	0.28
Idiopathic haemochromatosis	6–10	1.6–2.0
Congenital adrenal hyperplasia	c. 2	0.1–0.2
α-1 antitrypsin deficiency	1–4	0.025–2.4
Total in the population served by a general practice	14–>23	

of CF as well as of the haemoglobin disorders, which together are likely to affect more than 5 per cent of practice populations.

Cystic fibrosis (CF)

A disease causing recurrent respiratory infections and malabsorption: the commonest inherited disease in Northern Europe (Goodchild and Dodge 1985; Goodfellow 1989).

Incidence

In Europe, one in 2000–3000 children suffer from cystic fibrosis, correspon-ding to a carrier frequency of 3.5–4.5 per cent in different populations (Table 16.2). For this lethal gene to have become so common, carriers must have a considerable selective advantage. It has been suggested that they are protected against some forms of gastroenteritis, one of the major killers of young infants (Quinton 1982).

Table 16.2   Birth incidence of cystic fibrosis in different populations

Population	Homozygote births/1000	Calculated carrier frequency (%)
North Europe	$1/2200 = 0.45/1000$	4.3
Greece	$1/2600 = 0.38/1000$	3.9
Southern Europe	$1/3500 = 0.29/1000$	3.4
Middle East, Vietnam	$1/12\ 000 = 0.08/1000$	1.8
American Blacks	$1/17\ 000 = 0.06/1000$	1.5
Finland	$1/40\ 000 = 0.02/1000$	1.0
Japanese ⎫ Chinese ⎭	Very rare indeed	<1

Molecular basis

The molecular basis of CF is now close to being understood. The disorder decreases the normal transport of chloride (Cl^-) ions, and secondarily of sodium (Na^+) ions and water, through the epithelial cells lining sweat glands, and glands in the lungs and the intestinal tract (Cuthbert 1989). In lungs and gut, the reduced ability to secrete salt and water causes thickened mucous secretions, leading to most of the pathology of the disease. In sweat glands, reduced ability to reabsorb chloride ions leads to excessive loss of salt and water, which is not usually harmful (there is a small risk of heat stroke in exceptionally hot weather) but is useful for diagnosing the disease.

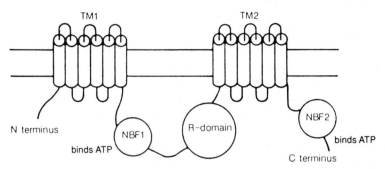

Fig. 16.1 Diagram of the probable tertiary structure of the cystic fibrosis trans-membrane conductance regulator (CFTR) protein. Sequences at either end of the protein are thought to span the cell membrane repeatedly, possibly creating a pore-like structure. Within the cell, two sections of the protein bind ATP, suggesting that the protein uses energy to transport ions across the membrane. (Reproduced from Ringe and Petsko 1990, with permission.)

The CF gene was traced to chromosome 7 by linkage studies in 1986 and was sequenced in 1989 (Riordan *et al.* 1989: see also p. 141). The precise molecular basis for the disease is being defined at the time of writing. The gene's DNA sequence suggests a protein resembling a membrane transport protein found in some bacteria (Fig. 16.1). As with most inherited diseases, a number of different mutations lead to the same end result: more than 90 mutations have been discovered since the gene was identified. The first to be defined is of a very unusual type — three base pairs (coding for one amino acid) have been deleted, leaving a structurally altered protein. This is the commonest CF mutation in Europeans, accounting for 50–80 per cent of CF genes in different populations.

Clinical features

Ultrasound examination shows that the majority of affected fetuses have signs of intestinal obstruction due to thickened intestinal secretions as early as 20 weeks' gestation: this has been confirmed by pathological examination of aborted affected fetuses (Boue *et al.* 1986). Obstruction usually resolves later in pregnancy, but about 15 per cent of affected infants are born with meconium ileus, i.e. intestinal obstruction with abdominal distension, bile-stained vomit, and constipation due to viscous meconium. It is usually necessary to remove the affected portion of the gut.

If cystic fibrosis is not diagnosed at birth, affected children usually present within the first five years of life, though about 5 per cent of patients with a milder form of the disease are only diagnosed in later childhood or adolescence. In general the course of the disease is fairly uniform. The main organs involved are the lungs, pancreas and liver, sweat glands, and, in males, the seminal vesicles. Coughing and wheezing usually begins within the first year and repeated infections cause progressive damage to the lungs. This ultimately leads to death after a long period of chronic illness. Many affected children have bronchial hyper-reactivity with recurrent episodes of wheezing. Air gets trapped in the bronchi producing over-inflated lungs with a resultant increase in the antero–posterior diameter of the chest. Bullae may rupture in about 20 per cent of older patients, causing a pneumothorax. Increasing episodes of wheezing may indicate an infection with the fungus *Aspergillus fumigatus*. As the child gets older, respiratory symptoms increase as more lung tissue is damaged, and bronchiectasis supervenes. Infections become less responsive to antibiotics and cor pulmonale develops. Finger clubbing may occur. Cyanosis is a late sign. Sinusitis and nasal polyps are common. Nasal obstruction may be relieved by corticosteroid nasal sprays.

Nearly all CF patients suffer from pancreatic insufficiency because thickened secretions reduce the outflow of pancreatic enzymes. In untreated children, this leads to diarrhoea with bulky foul-smelling stools, distention of the abdomen and failure to thrive due to malabsorption of food and fat-soluble vitamins. Rectal prolapse occurs in about one-third of untreated children in the first few years. Increased viscosity of the bile leads to biliary obstruction in some children, and ultimately to cirrhosis in those who live long enough. There is an increased incidence of diabetes in older patients. Other complications include gall stones and portal hypertension with a risk of haematemesis due to oesophageal varices. Arthritis may develop in adolescents. Most patients are underweight and growth tends to fall off before puberty, which is usually delayed. In the absence of treatment, death occurs early, usually as a result of a respiratory infection.

Management depends on preventing and counteracting the above effects as far as possible (David 1990). The parents are trained to give regular physiotherapy at home and to encourage vigorous exercise such as swimming, running, and cycling to help clear tenacious secretions from the lungs. Because of the sticky nature of the bronchial mucous, it is difficult to eradicate bacterial pathogens. The commonest are *Staphylococcus aureus*, *Haemophilus influenzae* and *Pseudomonas aeroginosa*. The latter, which colours the sputum green, is a particularly chronic infection. The bacterial population is monitored by regular sputum culture. Antibiotics are often needed in high doses and should be given early in infections. Severe infections require hospital admission for intensive treatment with intravenous antibiotics, though these can sometimes be given at home with the support of a specially trained domiciliary nurse. Supplementary pancreatic enzymes are scattered on meals to improve digestion, and a high calorie diet without fat restriction and with vitamin supplements and additional salt helps to improve growth.

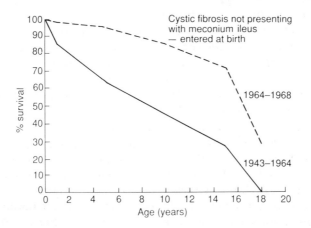

Fig. 16.2 Survival in cystic fibrosis (From Office of Health Economics 1986). Data are given only for infants not presenting with meconium ileus (this serious complication often leads to early death). Before 1964, 50% of patients were dead before the age of 10, and none survived past 18 years of age. Survival has improved greatly since then, and is now better than suggested by the figure. Such data are inevitably considerably out of data, as survival into the late teens and beyond can be assessed only for patients born more than 20 years ago.

The assiduous application of this approach from the time of diagnosis, aiming to maintain health rather than to treat disease, has led to an increase in mean survival after diagnosis from less than one year in the 1930s to 20–30 years in most developed countries today (Fig. 16.2). The number of adult patients with cystic fibrosis is increasing and many lead a good quality life and earn their living (Office of Health Economics 1986). Improved survival, however, introduces new challenges as existing physical problems get worse and new ones arise, including diabetes and sexual and psychosocial problems. Reproductive ability is severely limited. Ninety per cent of adult males are sterile because the vas deferens is absent, but the testes function normally and most are potent. Most females have a delayed menarche, and are sub-fertile due to the high viscosity of their cervical mucus. Maternal mortality is also increased, as in most other conditions where survival to adult life involves considerable organ damage. Heart–lung transplantation in CF is discussed on p. 156.

Diagnosis

CF may be diagnosed by neonatal screening in some places, or when a newborn has meconium ileus. More generally, initial suspicion must be based on the clinical picture. Most of the associated symptoms are non-specific, and CF is an infrequent cause of any of them. It is therefore not surprising that it usually takes time to reach a diagnosis, even when the parents have been convinced that something is wrong for some time. Most infants with CF have diarrhoea, may thrive poorly, and have recurrent chest infections. Prolapse of the rectum can occur. The mother may comment that the baby tastes salty when kissed. Cystic fibrosis may be suspected in the older child who fails to grow properly in spite of an excessive appetite, has fatty, foul-smelling stools, a chronic productive cough with persistent crepitations perhaps associated with bronchospasm on auscultation of the lungs, and a predominantly normal chest X-ray.

Hitherto, diagnosis of homozygotes has been by the sweat test, in which a patch of skin is stimulated with pilocarpine and sweat collected on a filter-paper pad. In cystic fibrosis, there are more than 60 mmol of sodium and chloride ions per litre, compared with less than 50 mmol in normal children. The method is not simple and depends on careful technique. A diagnosis should never be based on a single test, and babies less than 6 weeks old may not secrete enough sweat to give a reliable result. That we are still uncertain about the incidence of cystic fibrosis in many parts of the world is due to the difficulty of detecting all affected infants using these methods. Traditional approaches can now be supplemented with DNA methods.

Neonatal diagnosis is possible because obstruction of the pancreatic ducts leads to a raised plasma trypsin level in the first weeks of life, and this can be detected by radioimmunoassay using Guthrie filter-paper blood spots. The method has been used for establishing the true incidence of the disease, and for research on the value of very early diagnosis. However, neonatal screening is not yet widely practised because it involves a significant number of false positives and negatives, and there is not yet convincing evidence that very early diagnosis is advantageous (Chapter 22).

At present, definitive *carrier diagnosis* is possible for most relatives of known patients. It is also possible to screen for the commonest CF mutations by simple DNA methods (p. 133), but for the foreseeable future only about 85 per cent of carriers are likely to be identified in any screening programme. Such programmes are bound to involve the primary care service. Trials to assess the practicability and acceptability of CF screening are just beginning. Fortunately, experience with prevention of the haemoglobin disorders (Chapter 15) provide some useful hints on prevention services for CF.

The role of the primary care worker

It is important for primary care workers to be highly alert for CF and to refer children who present with relevant symptoms early for investigation, in order to ensure the best prognosis. As with all chronic conditions that involve gruelling treatment regimes at home, good communication between the practice and hospital unit is essential. These children need all their vaccinations, including MMR and pertussis, which should not be delayed because of a cough if the child is well. There is a case for an annual injection of influenza vaccine. Adolescents may need a lot of support at times, when they become demoralized by the disease or embarrassed at school because of their continual cough or offensive flatus. Patients with rare diseases requiring complex management including psychological and social support often do better and survive longer under the care of a specialist unit: it is up to the Primary Care Team to make sure that patients are referred appropriately (see p. 291).

Prenatal diagnosis for cystic fibrosis is possible using DNA methods. It is important to ensure that DNA from any living CF patient has been stored, and to be sure that couples with a CF child who plan to have another pregnancy have DNA family studies carried out before the pregnancy is started.

Phenylketonuria (PKU)

Phenylketonuria is a disorder of amino acid metabolism causing severe mental handicap. There are many inherited disorders of metabolism. Pathological effects may be due to deficiency of a particular product or to accumulation of toxic by-products (Fig. 16.3). Few metabolic diseases can be treated as satisfactorily or inexpensively as PKU.

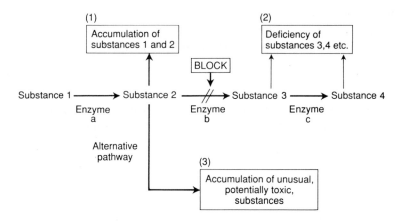

Fig. 16.3 Summary of some possible effects of an enzyme defect. Substance 1 is converted to substance 4 in three steps, mediated by enzymes a, b and c. When enzyme b is absent or deficient, substance 2 cannot be adequately converted to substance 3. This can cause (1) accumulation of substances 1 and 2; (2) deficiency of substances 3, 4 etc; (3) metabolism of substance 3 by alternative pathways, which may produce toxic or unusual substances.

Incidence

PKU occurs in from 1 in 8000 to 1 in 20 000 infants in different populations of European origin, corresponding to a carrier frequency of 1–2 per cent (Table 16.1). Heterozyogotes are thought to have some small unknown selective advantage.

Molecular basis

In most cases, the disease is due to a deficiency of the enzyme phenylalanine hydroxylase, which normally converts dietary phenylalanine to tyrosine (Fig. 16.4). In homozygotes the result is an abnormally high blood level of phenylalanine and its abnormal breakdown products, which can interfere with the development and maturation of the brain and with pigment metabolism. The fetus is not affected because the substances are cleared by the maternal circulation, but they begin to accumulate immediately after birth. Neonatal diagnosis is based on measurement of the plasma phenylalanine level in blood taken onto Guthrie cards 4–10 days after birth. Most infants with a persistent high plasma phenylalanine have classical PKU and require treatment. However, a minority have a milder disorder, with a plasma phenylalanine level up to $1.0 \, \text{mmol} \, \text{l}^{-1}$ (almost 10 times the normal level). These children develop completely normally without special treatment, and are said to have 'benign hyperphenylalanaemia', though some females have a reproductive risk later in life (see below).

DNA studies have revealed several different mutations that abolish the activity of the phenylalanine hydroxylase gene (see Table 2.3) or reduce it to different degrees, and this probably accounts for variations in clinical severity of disturbances of phenylalanine metabolism.

Clinical features

In untreated patients, mental retardation is usually first noted at around 4 months of age, and soon becomes profound: the IQ is usually around 20–40.

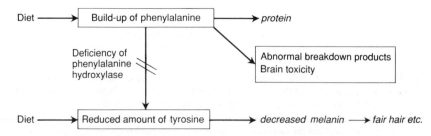

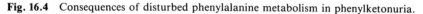

Fig. 16.4   Consequences of disturbed phenylalanine metabolism in phenylketonuria.

Patients have fair hair because they cannot make melanin pigment. In the past, most were looked after in institutions because their severe mental retardation was often combined with a hyperactive, irritable and apprehensive personality, temper tantrums, psychotic episodes, fits and other neurological abnormalities, and eczema. Half died before the age of 20 and three quarters before 30 years of age, mostly from infections (Knox 1966).

This unfortunate progression can be prevented when homozygotes are detected by neonatal screening, and kept from the first months of life on a low phenylalanine diet with just enough tyrosine for protein and hormone synthesis. Mental development is nearly normal and the IQ remains in the normal range as long as the diet is continued, despite some mild neurological impairment (Smith *et al*. 1990; Scriver *et al*. 1989). However, every silver lining has its cloud. In the past it was thought that the risk was over after 10 years of age, because brain myelination is completed by this time, and patients were allowed to discontinue the special diet. Longer-term follow-up now shows a gradual but steady deterioration of mental function with time (Fig. 16.5). In some patients, relaxation of diet in childhood or adolescence may be followed by upper motor neurone damage (Thompson *et al*. 1990). Efforts to get patients to resume the strict diet have not been very successful, and the current recommendation is to continue the diet lifelong.

Maintenance of a strict diet seems particularly appropriate for girls, because pregnant women with PKU who are not on the special diet have a raised serum phenylalanine level, which can damage the fetal brain and cause mental retardation and other congenital anomalies in their offspring (Lenke

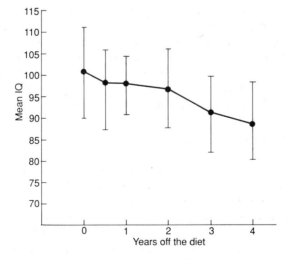

Fig. 16.5 If patients with PKU discontinue their diet, there is a slow and progressive decline in mean IQ. (After Scriver *et al*. 1989b).

and Levi 1980). When this happens, the net effect of treating the mother is only to postpone handicap by one generation. To avoid fetal damage, mothers who are not on a strict diet must resume the diet *prior* to conception. This is by no means easy: assessment of the effectiveness of strict dietary control in the mother is continuing.

In women who were diagnosed at birth as having benign hyper-phenylalaninaemia, the metabolic disturbance persists into adult life. As the fetal level of plasma phenylalanine tends to be higher than the maternal level, the amount of plasma phenylalanine in the fetus can sometimes reach a harmful level, and a clinically unaffected mother may give birth to a mentally retarded child. This outcome can be avoided by an appropriate diet started before conception and continued during pregnancy, providing the original diagnosis in these otherwise healthy individuals is not forgotten. It is now recommended to keep a special register of people with benign phenylalaninaemia in order to follow them up to provide genetic counselling in their teens.

Diagnosis

PKU is a target of neonatal screening in most developed countries, and most homozygotes are diagnosed in this way (p. 346). In families with an affected member, heterozygotes can be diagnosed by DNA methods.

Prenatal diagnosis

This is possible using DNA probes. There has so far been only a limited demand for it from parents with an affected child in Western Europe, in view of the ease of early diagnosis, the relative simplicity of treatment and the resultant good prognosis. There is a higher demand in Eastern Europe where it can be difficult to obtain the necessary special diet. Increased awareness of the reproductive problems of women with PKU is leading to increased interest in prenatal diagnosis with a view to selective abortion of affected female fetuses, because of their high reproductive risk.

The role of the primary care worker

Neonatal screening is the job of primary care workers, who must ensure that all newborns have a Guthrie blood sample taken during their first week of life. In the UK this is mostly done by community midwives, as most mothers are discharged early from hospital. Primary care workers need to be aware of the reproductive risks of women with PKU, and of those who were diagnosed as having benign hyperphenylalaninaemia at birth.

Lactase non-persistence (lactose intolerance)

This is a polymorphism of milk metabolism.

Molecular basis and frequency

The main carbohydrate in human and cow's milk is the disaccharide lactose, which is usually broken down to its constituents glucose and galactose by the enzyme lactase. This enzyme is regularly present in the intestines of infants and young children, but in many branches of the human race (Fig. 16.6) it is switched off after six years of age (Flatz 1989). Adults in these population groups (e.g. Africans, Chinese, and Afro-Caribbeans) experience indigestion and/or diarrhoea when they take more than a small amount of cow's milk, and often develop an aversion to it, though yoghurt or cheese, in which lactose is partially digested by bacteria or moulds, may be acceptable. Lactase persists in adults only in populations, like North Europeans, whose ancestors were pastoralists and lived off milk. Homozygotes for lactase non-persistence appear only rarely in these populations,

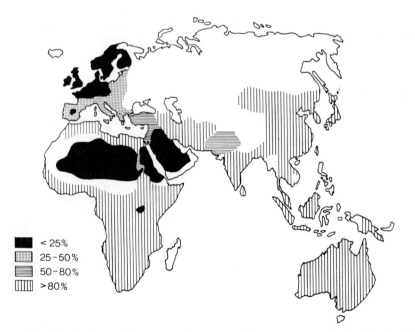

Fig. 16.6 Distribution of lactase persistence and lactase non-persistence in the old world. Shaded areas: per cent of the population with lactase non-persistence. Blank areas: no information. In Africa and the Middle East lactase non-persistance is associated with a milk-dependent pastoral way of life. (Reproduced from Flatz 1989, with permission.)

so the condition tends to be viewed as abnormal. However, lactase non-persistence is so common globally that it must be considered as a polymorphism. It is unlikely to have harmful effects except when milk is given as a food supplement to older children or adults from susceptible populations with diarrhoea and/or malnutrition.

In children of European origin, lactase deficiency causing diarrhoea is usually an acquired, transient phenomenon following an attack of acute gastroenteritis which damages the intestinal villi. Stopping milk for a short time may relieve the symptoms. It is often unnecessary to avoid all lactose-containing products.

Diagnosis

Diagnosis of homozygotes is by measuring a raised hydrogen excretion in the breath (collected by the subject blowing up a balloon) following a standard lactose load.

The role of the primary care worker

Aversion to milk can have a sound physiological basis. It is inadvisable to recommend milk to adults in relevant ethnic groups as a dietary supplement or as a source of additional calcium, e.g. during pregnancy or after the menopause.

Tay–Sachs disease

This is a lipid storage disease causing severe mental retardation and paralysis.

Incidence

In most ethnic groups, Tay–Sachs disease is very rare, but 3–4 per cent of Jews of Eastern European origin are carriers, and about 1 in 3600 of their infants suffer from the homozygous disease. The high frequency of the carrier state suggests a selective advantage in this population; increased resistance to tuberculosis has been proposed (Rotter and Diamond 1987).

Molecular basis

Lysosomes are cellular organelles that contain numerous enzymes and perform the vital task of degrading large molecules into smaller components that can be recycled or eliminated (Fig. 2–3). The lysosomal enzyme hexosaminidase A is essential to break down sphingolipids, which are an important component of most cell membranes and are especially abundant in the myelin sheaths of nerve cells. In Tay–Sachs disease, hexosaminidase A is absent or severely deficient. Undigested sphingolipids gradually accumulate in the lysosomes of neurones and interfere with their functioning. The retina becomes pale because it has neurones on its surface. The macula, which is not covered with neurones, stands out by contrast as a 'cherry red spot' easily visible with an ophthalmoscope. This specific diagnostic feature of Tay–Sachs disease is usually visible before the child is 1 year old.

Clinical features

Affected infants appear normal up to about 6 months of age, but then regress developmentally, becoming apathetic and hypotonic, with an exaggerated startle reaction and loss of head control. The infant becomes blind and deaf by 12–18 months of age, has frequent fits, and usually dies before 5 years of age from progressive paralysis leading to aspiration pneumonia. The disease is all the more heart-rending because of the early normal development of affected infants. In view of its remorseless nature, prenatal diagnosis and selective abortion are popular with couples at risk.

Diagnosis

Diagnosis of homozygotes is made clinically, and by finding a reduced white cell hexosaminidase A. Heterozygotes can be reliably and simply detected by measuring the level of hexosaminidase A in serum or white blood cells. Carrier diagnosis is rather more difficult in pregnancy.

Prenatal diagnosis is possible both in the first and second trimester of pregnancy by assaying the enzyme in chorionic villus or cultured amniotic fluid cells.

Effects of prevention programmes

Population screening and prenatal diagnosis for Tay–Sachs disease were introduced in the early 1970s in the United States (Kaback *et al.* 1974) and Canada (where the disease is common among French Canadians due to a founder effect (p. 121), and later in Israel and the UK. These programmes have provided much information on the social implications of genetic screening (Zeesman *et al.* 1984). Though precise statistics are not available, it seems that in the United States, the birth incidence of Tay–Sachs disease has fallen by over 90 per cent. There is no information on the relative roles of genetic screening and of intermarriage with the general population.

Role of primary care workers

Babies with Tay–Sachs disease can be distinguished from those with a non-progressive CNS disorder such as cerebral palsy because they are initially developmentally normal. Heterozygotes can be reliably detected even in the absence of a family history. Carrier testing should be offered to all people with a Jewish parent or grandparent, prior to reproduction. Further information can be obtained from the Tay–Sachs Centre, South East Thames Regional Genetics Centre, Guys Hospital, St Thomas Street, London SE1 9RT.

Idiopathic haemochromatosis

This is a disorder leading to damage from excessive gastro-intestinal iron absorption.

Incidence

1.6–2 per 1000 of the population are homozygous for the haemochromatosis gene, corresponding to a heterozygote frequency of 6–10 per cent in different populations. The condition is not lethal in early reproductive life, but may reduce reproductive efficiency among older men. Heterozygotes may have a selective advantage when iron deficiency is prevalent, because of increased ability to absorb dietary iron. The haemochromatosis gene is on chromosome 6, very closely linked to the HLA antigen A3 (and also to B7 and B14). Most people who carry HLA A3 also carry a haemochromatosis gene.

Though haemochromatosis is recessively inherited, there is a 6–10 per cent chance that a (pre-symptomatic) homozygote will select a heterozygous partner, in which case 50 per cent of the offspring will be homozygotes, and the inheritance pattern of the disease within the family may appear dominant.

Metabolic basis

Idiopathic haemochromatosis refers to iron overload without any external cause. It is a disorder of the intestinal mechanism that limits iron absorption, so that homozygotes absorb considerably more iron than usual. (Iron overload can also be caused by repeated blood transfusions or prolonged inappropriate administration of iron, or may occur in association with cirrhosis or alcoholism.)

Because of its ability to react with oxygen and oxidize other cell components, iron is usually tightly bound to proteins that control its activity, and hence cannot be excreted from the body when it is present in excess. To prevent excess accumulation, only about 10 per cent of dietary iron is absorbed. This is usually adequate for men, but iron balance is often precarious in growing children and women.

Clinical features

The increased iron absorption of haemochromatosis rarely harms homozygous children, or women before the menopause, because of losses through menstruation and pregnancy. However, homozygous men accumulate iron slowly and steadily throughout their adult life. Iron absorption is increased by alcohol taken with meals, so there is a large environmental factor in the development of the disease. A few grams of iron are normally stored in the liver, but when grossly excessive amounts are present, iron accumulates in and causes damage to several other vital organs. A store of more than 16 grams is associated with serious symptoms of iron overload.

Symptoms may appear in homozygous men in their 40s onwards. Patients usually present with weakness and lethargy and in about half the cases, loss of potency or libido. On examination there is abnormal skin pigmentation, which is particularly noticeable in exposed areas of the body. It is due to excess melanin rather than to increased deposition of iron. The liver is considerably enlarged and in severe cases there will be evidence of cirrhosis; this is especially likely with heavy drinkers. This is associated with a risk of primary liver cancer 200 times that of the rest of the population (Bothwell *et al.* 1989). There may be painful joints or symptoms of diabetes caused by a combination of iron overload and genetic predisposition. Untreated patients start developing cardiac arrhythmias in their mid-50s. They also have increased susceptibility to a variety of infections.

Treatment consists of removing one 500 ml unit of blood weekly until the iron load is normal, as indicated by the serum iron or ferritin level; blood is then removed regularly but less often. Since 5 units of blood contain 1 g of iron, this simple approach is highly successful when an early diagnosis is made and can also be socially useful as the blood is then available for someone needing a transfusion. In some instances the iron chelating agent, desferrioxamine, may be added to the treatment regime. However, even after iron overload has been corrected, irreversible organ damage (for example cirrhosis and diabetes) can significantly reduce a patient's survival in comparison with the normal population. In addition, an increased risk of various types of cancer, especially hepatoma, persists even after successful treatment. Hence the importance of early diagnosis and prevention.

Diagnosis

Homozygotes are often diagnosed only after the onset of symptoms, but the offer of serum iron or serum ferritin estimation and HLA typing to the patient's extended family usually reveals several pre-symptomatic cases.

They should be informed of their risk, offered dietary counselling, recommended to have their serum iron checked at 5-yearly intervals, and advised to become regular blood donors. Heterozygotes can be detected by HLA linkage studies. They often have a somewhat raised serum ferritin and an increased risk of developing iron overload, eg. in association with alcohol consumption. However, the cost of HLA typing precludes its use as a screening test at present.

Role of the primary care worker

Haemochromatosis may be suspected in a middle-aged man who presents with a combination of the symptoms described above. The disease may be confused with alcoholic cirrhosis, which can also result in excess deposition of iron in the liver. One way of screening for haemochromatosis is by incorporating iron evaluation into routine adult health check-ups. A serum iron level of $\geq 160\,\mu g$ dl^{-1} (the normal range in males is 50–160), or a serum ferritin of $> 550\,\mu g/l$ (normal range 16–540) should alert the doctor to the need for further investigation, which may include a liver biopsy. One of these blood tests may be more useful than many other tests done for a routine 'executive screen', which incidentally selects the risk group as it is primarily offered to well-off males over 35 years old, with a diet rich in red meat and alcohol.

Congenital adrenal hyperplasia (CAH)

The adrenogenital syndrome. This is a disorder of steroid hormone synthesis (New *et al.* 1989).

Incidence

1–2 per cent of European populations carry this disorder, leading to a birth rate of about 1 per 10 000. Deficiency of the enzyme 21 hydroxylase is closely linked to the HLA Bw27 gene.

Metabolic basis

More than 90 per cent of cases of CAH are due to a deficiency of the enzyme 21 hydroxylase, which is necessary for the adrenal gland to produce cortisol, aldosterone, and small amounts of other steroid hormones. Normally the hypothalamus stimulates the pituitary to secrete adrenocorticotrophic hormone (ACTH), which in turn stimulates the adrenal to synthesize steroid hormones. The cortisol in the circulation suppresses the hypothalamus by a feedback control mechanism, so that an optimum level of steroids is maintained. When cortisol production is decreased due to 21 hydroxylase deficiency, the ACTH level remains high and continually stimulates the adrenal to overproduce those steroids it can make. These include testosterone, which can virilize females. 21 hydroxylase deficiency also leads to the accumulation of 17 hydroxyprogesterone, which is easily measured in the blood.

Clinical features

In the fetus there is no effect on development of the internal sex organs. However, in females the external genitalia may be virilized, so that the baby is born with ambiguous external genitalia. In the absence of treatment there is progressive virilization, precocious puberty, and stunted growth. The severity of the disease is variable; some women who develop hirsutism and acne may have a mild form of CAH.

An affected female infant may have just an enlarged clitoris or partial fusion of the labia, but if virilization is very marked the baby's genitals can be mistaken for those of a male with undescended testes with or without hypospadias, leading to mis-assignment of sex. In boys the effects of androgen excess are likely to appear later with early growth of the penis, though the testes remain pre-pubertal. There will be premature development of pubic hair and rapid, early growth which can lead to early closure of the epiphyses. Some girls may present after infancy in the same way. About three-quarters of patients with this enzyme deficiency lose an excessive amount of salt due to deficient secretion of aldosterone. The baby usually presents in the second or third week of life with vomiting, dehydration, and lethargy and may die if untreated.

Treatment

Treatment with cortisol replaces the deficient hormone and controls the exaggerated ACTH stimulus to the adrenal. A sick infant with a salt-losing crisis needs intravenous saline and steroids including mineralocorticoids. Chromosomal studies are essential for correct gender assignment of females, who often need surgery to normalize their external genitalia. According to Hughes (1988), well-managed patients can be expected to live a normal life span and to reproduce.

Diagnosis

Confirmation of a suspected diagnosis is by chromosome studies (lymphocyte culture) and hormone assay. Neonatal screening is possible by assay of 17 hydroxyprogesterone on Guthrie blood spots. This has been advocated in order to forestall a salt-losing crisis, and to ensure that infants are assigned early to the correct sex. According to Pang *et al.* (1988) CAH is not diagnosed in up to a half of affected females, in spite of genital abnormalities, unless they are identified by neonatal screening. Heterozygote diagnosis will soon be possible.

Prenatal diagnosis

At present, linkage of CAH with the HLA system can be used for early prenatal diagnosis in families with an affected individual, or where a previous sibling has died unexpectedly in the neonatal period. Amniotic fluid can also be analysed for a raised level of 17 hydroxyprogesterone. The objective is not necessarily termination of pregnancy, but early detection of an affected female. Administration of dexamethasone to the mother during pregnancy appears to protect some female fetuses from virilization.

Role of the primary care worker

CAH should be considered in any infant with abnormal genitalia or who develops the non-specific symptoms of a salt-losing state. In general, the onset of puberty before the age of eight and a half in girls and slightly older in boys is considered precocious. Such children are usually referred to a paediatrician or endocrinologist and in about two-thirds of cases no obvious cause is found. There are many other reasons besides CAH for an abnormally early puberty. These include brain and ovarian tumours and exogenous oestrogens in girls, and cerebral tumours, some testicular growths and certain liver tumours in boys.

α-1 antitrypsin deficiency

This is a plasma protein disorder (Cox 1989).

Incidence

One to four per cent of different European populations are heterozygotes, the corresponding homozygote birth rate being 0.025–2.4 per 1000.

Molecular basis

The plasma proteins include α-1 antitrypsin, a small molecule which inhibits proteolytic enzymes, and helps to protect body tissues, especially the lung, when white cells release enzymes to destroy invading organisms. This is a fairly polymorphic protein with many different neutral alleles, some of which occur with particular frequency in particular populations. They are therefore often used as genetic markers in anthropological studies investigating the relationships between different ethnic groups. One variant, the pI Z allele, can lead to disease. Heterozygotes have about 60 per cent, and homozygotes about 15 per cent of normal antitrypsin activity. This variant is due to a single amino acid substitution and can be detected unambiguously by electrophoresis, or at the DNA level by oligonucleotide probes. The α-1 antitrypsin gene has been identified and sequenced.

Clinical features

The clinical picture in homozygotes is very variable; many may have no clinical problems, but 10–20 per cent of homozygous infants have juvenile hepatitis or cirrhosis. Affected babies present most commonly with a raised level of conjugated bilirubin and liver enzymes. Hepatosplenomegaly is often present. Many of the infants who develop hepatitis appear to recover or only have persistently raised liver enzymes. A small proportion of homozygotes who are apparently healthy in infancy later develop cirrhosis. Females seem to have less severe disease than males, and breast feeding may decrease the likelihood of severe liver disease.

In later life, α-1 antitrypsin deficiency is associated with emphysema, especially if the person smokes. Smokers tend to start getting breathless around 35 years of age, 10 years earlier than non-smokers (Fig. 16.7) (Wilson Cox, 1989). A small proportion of heterozygotes may be vulnerable to emphysema if they smoke.

Diagnosis

Homozygous infants can be identified accurately by measuring α-1 antitrypsin activity in the Guthrie blood spot. Both homozygotes and heterozygotes can be detected quite easily by plasma protein electrophoresis.

Prenatal diagnosis is possible both by measuring α-1 antitrypsin activity

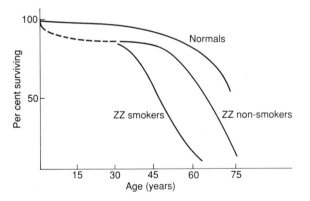

Fig. 16.7 Effect of smoking on survival in homozygous α-1 antitrypsin deficiency (after Wilson Cox 1989).

in fetal blood, and by DNA methods. There is a significant demand for prenatal diagnosis for couples who have had a previous child with juvenile cirrhosis.

Role of the primary care worker

In principle, it could be desirable to identify individuals with α-1 antitrypsin deficiency so as to advise them against smoking and exposure to polluted environments, and to offer monitoring of respiratory and possibly of hepatic function. It may be worth while to test people who develop irreversible obstructive airways disease at an early age for α-1 antitrypsin activity. If they are deficient, this can lead on to the discovery of other family members who are at increased risk of developing emphysema.

Neonatal screening has been attempted in Sweden where α-1 antitrypsin deficiency is particularly common, with these aims in mind. However, it was discontinued as a result of the problems encountered (see p. 351). It might be useful to associate screening in schools with an anti-smoking campaign.

Key references

For cystic fibrosis:
Goodchild, M. and Dodge, J. A. (1985). *Cystic fibrosis. Manual of diagnosis and management*, (2nd edn). Baillere Tindall, London.
David, T. J. (1990). Cystic fibrosis. *Archives of Disease in Childhood*, **65**, 152–157.
For others, the relevant articles in:
Scriver, C. R., Beaudet, A. L., Sly, W. S., and Valle, D. (eds) (1989). *The metabolic basis of inherited disease*, (6th edn). McGraw Hill Inc., New York.

17. Common conditions with dominant or X-linked inheritance

Dominantly inherited disorders

Huntington's disease

Incidence

This neurological disorder is a classical example of a severe dominantly inherited disease with onset of symptoms in mid-reproductive age. New mutations are very rare. Practically all cases are familial and have probably originated from a few foci, the gene having been disseminated by migration (Hayden 1981). The gene is highly penetrant, i.e. everyone who carries it will suffer from the disorder if they live long enough. About 1 in 3000 newborns in the UK and America carries the gene, and at any one time about 1 in 12 000 of the population suffer from the disease. This surprisingly high birth incidence suggests that the Huntington's disease gene has a selective advantage (p. 363), and in fact there is good evidence that carriers tend to have larger families than others (Walker *et al.* 1983). This tendency may now have gone into reverse: potential carriers have had significantly smaller families than the population norm since genetic counselling has been available (Harper *et al.* 1981).

Molecular basis

At the time of writing, though the gene was localized to the short arm of chromosome 4 in 1983, the nature of the problem in the brain is still not known, and there is no medical treatment.

Clinical features

Age at onset ranges from 20 to 60 years, with the median in the early 40s. The typical picture is of very gradual appearance of uncontrolled jerky movements which may affect speech. There may be subtle early changes in personality such as a decreased ability to function, irritability, impulsiveness, and blunting of emotions. Whilst the patient retains insight, depression is common and suicide a significant risk. Walking, talking, and feeding become increasingly impaired as they are interrupted by uncontrollable choreiform movements. Patients usually walk in an unsteady, swaying, jerky manner with sudden lurches, and danger of falling. Attempts to pick up objects may be hampered by unexpected lunges. Eventually speech becomes unintelligible and dementia becomes more marked, the proportion of chorea to dementia varying in different patients, and at different stages of the disease. Dementia, a decline in intellectual capacity, memory, and ability to develop new ideas, is very difficult to assess in a patient whose speech is grossly distorted.

The duration of clinical disease is usually about 17 years (Hayden 1981). The patient will have to be fed and taken to the toilet perhaps for some years before finally being admitted to hospital, where most patients spend the last two years or so of life. Death is usually due to pneumonia caused by aspiration of food or secretions because of poor neurological control. There are inevitably great stresses within the family, and social support and access to counselling are important. A Huntington's disease support association (see Appendix) may be very helpful to families.

Juvenile Huntington's disease can also occur. It usually begins in the second decade and differs from the adult form in that fits and muscular rigidity are commoner, and dementia develops earlier in the course of the disease and is more profound. Interestingly, the age at onset is related to the parental origin of the gene. Inheritance from the father may be associated with early onset, and from the mother with late onset. These were among the observations that have led to the concept of 'genomic imprinting' (p. 31) (Reik 1988).

There is a 50 per cent risk of the same disease developing in siblings and children of an affected person, and a lower risk to more distant relatives.

Diagnosis

Suspicion is aroused if a previously well co-ordinated person with a positive family history starts to become somewhat clumsy in movement and speech. An expert neurologist can make the diagnosis at this stage. However, since the gene for Huntington's disease has been located, it is now possible to a large extent to distinguish carriers and non-carriers by RFLP studies in families with the right pedigree structure. The first requirement is to have DNA from at least two family members who are definitely affected, or definitely unaffected (i.e. healthy and more than 60 years old). Even after testing, the residual probability that the diagnosis is incorrect ranges from 0.3 to almost 15 per cent (Brock *et al.* 1989), and some families are not informative at all.

The critical question is, how many family members will wish to know their status? Craufurd and co-workers (1989) actively offered testing to relatives

Table 17.1 Uptake of testing for the Huntington's disease gene in a group where testing was offered, compared with a group spontaneously seeking testing

	Total	Declined	Withdrew after counselling	Total not proceeding	Undecided	Tested
Test offered by centre	110	86	7	93 (85%)	9	8 (7%)
Spontaneously requested testing	78	–	12	18 (23%)	21	39 (50%)

(Based on Craufurd *et al.* 1989)

of patients on their register. Table 17.1 shows that only about 16 per cent of the whole group have taken up testing or are actively considering it, compared with over 60 per cent of those presenting with a request for testing. This contrasts with predictions from surveys of an uptake rate of 55–80 per cent, and is yet another illustration of the fact that what people think they will do is not a very good predictor of what they actually do, when faced with a difficult decision.

Carrier testing for Huntington's disease is done only with careful and repeated counselling and psychiatric support, and experience has revealed a number of unexpected problems (Morris *et al*. 1989). Those with a positive result usually seem to be able to return to their normal coping pattern after only a brief period of shock and depression, but those with a negative result are often not as relieved as they expected to be (Tibben *et al*. 1990). This may be partly because the residual uncertainty in diagnosis still leaves room for hope and fear, and partly because of complex reactions between family members who prove not to be carriers, and those who are.

Prenatal diagnosis

Prenatal diagnosis is possible for known carriers, and in some families can even be done by 'exclusion testing' (Quarrell *et al*. 1987) when the person at risk does not wish to know their own status (Fig. 17.1). The approach allows couples at risk a 50 per cent chance in each pregnancy of having a child that they know will be unaffected. Knowing that the fetus has a 50 per cent risk of disease does not alter the parent's own risk, which is already known from the family history.

Role of the primary care worker

It is particularly important to ensure that the family receive all the services they are entitled to. When the family doctor senses that the time has arrived when they can no longer look after the patient at home, it may be wise to suggest the possibility of hospital care rather than waiting for the request to come from the family, who may see it as a confession of failure.

A family doctor is more often involved with potential or known carriers. On occasions it may be the family doctor's responsibility to inform a relative of a Huntington's disease patient of their own risk and to ensure access to counselling with both genetic and psychotherapeutic input. Confidentiality about the family history and test results is essential, and the person's consent must be obtained before even another member of the practice team is told. People at risk will find it difficult to get insurance or to find a job if a prospective employer learns of the situation. The family doctor is the person most likely to be approached by a third party with a request for information, and this should naturally be refused, unless the person at risk consents. It

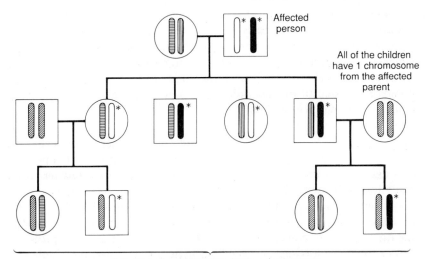

Only 50% of grandchildren have a chromosome
from the affected grandparent.

Fig. 17.1 'Exclusion testing' for risk of Huntington's disease. For prenatal exclusion testing DNA samples are needed from the affected grandparent, the parent who has inherited one chromosome 4 from the affected grandparent, and is therefore at 50% risk of developing Huntington's disease, and his or her partner. Offspring of this couple will inherit one chromosome 4 from each parent. The chromosome inherited from the parent at risk could have originated either from the affected grandparent or from the unaffected one. If the fetus has not inherited the chromosome from the affected grandparent, there will be no risk of Huntington's disease. However, a fetus which inherits a chromosome from the affected grandparent is at the same 50% risk as its parent, and in this case will usually be aborted. If the fetus is not aborted in this situation and the parent develops Huntington's disease, the offspring will be aware that he or she will definitely suffer from the disease.

is important to reply in a way that will not arouse suspicion, e.g. by giving a general statement that it is practice policy not to respond to requests for genetic information.

A positive family history may be relevant to many consultations, apart from those related to family planning and discussions of whether or not to get married. Any vague malaise may be perceived by a person at risk as the earliest stage of Huntington's disease, and they may often need reassurance. In this context it is helpful to remember that the risk falls with advancing age.

A very important practical step is to ensure that some DNA from the affected person has been stored.

Adult polycystic kidney disease

This affects about 1 in 1250 people born in Western Europe. It is a progressive disorder in which cysts form in both the kidneys and sometimes in other organs such as the liver, pancreas, seminal vesicles, and ovaries. Renal cysts are small and few in early life, but their size and number increases progressively, usually leading eventually to renal failure. The relatively high frequency of the disorder may be due to a high spontaneous mutation rate, as in one-third of cases there is no prior family history. Isolated renal cysts with no serious significance are quite often found incidentally on renal ultrasound or radiological examination.

Clinical picture

Mean age at presentation is around 40 years, usually with loin pain and urinary tract infections (Segal *et al.* 1977). There may be haematuria with or without pain. Renal colic may be due to stones. Patients may also present with hypertension secondary to renal complications. As 10–15 per cent of patients also have congenital cerebral aneurysms, they are prone to subarachnoid haemorrhage, especially if hypertension develops. Table 17.2 shows the main causes of death in adult patients (Dalgaard 1957). In the absence of treatment for end-stage renal failure, mean survival from the time of clinical presentation was about 4 years, shorter in earlier presenting (and therefore more severe) cases, and longer in patients presenting after 50 years of age. If they live long enough, practically all carriers have symptoms by their eighth decade, i.e. the gene has very high penetrance in developed societies.

The picture has changed considerably in recent years. Modern management of kidney disease, including dialysis and transplantation, can prolong and greatly improve the quality of life. Prompt treatment of urinary tract infections and raised blood pressure may slow the development of renal failure. Cerebral aneurysms can be sought by cerebral tomograms and radionuclide imaging, and prophylactic surgery may be carried out.

Table 17.2   Main causes of death in adult polycystic kidney disease (APKD)

Cause of death	Frequency (%)	Notes
Renal failure	60	5% of all patients with end-stage renal failure have APKD
CNS bleed	13	3% of deaths from ruptured berry aneurysms are due to APKD
Heart failure	6	Related to hypertension
Other	21	

(Based on Dalgaard 1957)

Diagnosis

The gene for adult polycystic kidney disease has been located on chromosome 16, close to the α globin genes. Linked RFLPs have been found (Reeders *et al.* 1985). Using a combination of clinical and DNA methods, 90–95 per cent of carriers related to an index case can be identified by the age of 20 years (Mulitinovic *et al.* 1980; Zerres and Stephan 1986).

Prenatal diagnosis

Prenatal diagnosis is also possible using RFLPs, but so far uptake has been relatively small, probably because of the late onset of the disease and hopes for treatment.

Role of the primary care worker

Most cases without a family history are likely to be diagnosed following investigations for the symptoms described above. In theory, the diagnosis in one family member should be followed by offering screening to other adult family members, as it may be possible to delay the onset of symptoms and moderate the course of the disease. However, as there is no cure and as there are already difficulties in obtaining life insurance for relatives of affected people (Dalgaard and Norby 1989), carrier testing is a delicate matter. The result of a study of the attitudes to presymptomatic diagnosis of 100 people with the condition and relatives at risk is summarized in Table 17.3 (Zerres and Stephan 1986). It confirms that young people fear that, as with other late-presenting, dominant conditions, early diagnosis may negatively affect employment or insurance prospects. For some young relatives, annual measurement of the blood pressure or renal function may be more acceptable than a definitive diagnosis. Surveillance can be offered by family doctors who share the care of patients with a specialist unit. The same comments as those made for Huntington's chorea apply with respect to confidentiality.

Achondroplasia

The commonest of many rare bone dysplasia syndromes.

Incidence

Achondroplasia is due to reduced growth of cartilage, particularly at the epiphyses of the long bones and leads to a highly distinctive form of dwarfism. The prevalence of achondroplasia is estimated to be around 1 in 3500–6600 and its birth incidence somewhat higher. Hitherto, over 80 per

Table 17.3 Pros and cons of early diagnosis of APKD, as seen by 21 patients and 79 relatives at risk

	Number of interviewees mentioning this point
For	
Family planning	86
Treatment possibilities	71
Reducing uncertainty	61
Planning work/finances	19
Against	
No cure	56
Problems with insurance	47
Fear of the future	45
Burden on the family	37
Difficulty in dealing with pain/premature death	30
Loss of self-esteem	25
Occupational difficulties	21
Desire to have children	15

91 of this group said their marriage plans would not be, or were not, influenced by early diagnosis.

46 said they would not have children, or would have no more children, because of the risk of APDK.

(From Zerres and Stephan 1986)

cent of cases have been due to new mutation. Improved management is increasing the proportion of familial cases (see below), but changes in paternal age distribution are thought to have lowered the mutation rate in this and other inherited diseases, possibly by as much as 50 per cent (p. 59).

Clinical features

The familiar clinical features are shown in Figure 17.2. Those achondroplasics who survive the first year of life have comparatively good health, normal intelligence, and a near-normal life expectancy. However, reproductive fitness is much reduced for a combination of physical and social reasons. Until the arrival of safe Caesarian sections, pregnancy was one of the main risks to life for achondroplasic females, because their narrow pelvis led to a high incidence of obstructed labour. Though male achondroplasics

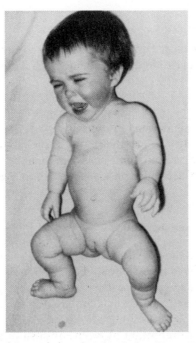

Fig. 17.2 Child with characteristic achondroplasia. (From Emery AEH and Mueller R.F. 1988, with permission)

married to normal females have been known to produce large families, social factors have been important in reducing fertility. A study of all known achondroplasics in Northern Ireland in 1955 located 37 individuals, 21 of whom were over the age of 25 (Stevenson 1957). Most lived an isolated and often withdrawn existence, and only one, a male, was married (he had eight children). By contrast, social factors which draw dwarfs together, like the court of Peter the Great (who actively sought out dwarfs for his court) or modern support associations, create an accepting social environment and the opportunity to find a partner of similar stature. In a study of achondroplasics carried out through the 'Little People of America' (Murdoch *et al.* 1970), 80 out of 92 individuals over 20 were married, 46 to another achondroplasic, 17 to another type of dwarf, and 17 to a person of normal stature. Sixty-five per cent of these marriages were fertile. Three-quarters of the surviving 50 children were dwarfs.

When two achondroplasics reproduce, in every pregnancy there is a 25 per cent chance of a child of normal stature, a 50 per cent chance of a classical achondroplasic, and a 25 per cent chance of an infant homozygous for the achondroplasia gene, who is almost bound to die before or soon after birth. The average number of living offspring in the American study was 0.82 per couple – considerably less than the population mean.

Diagnosis

Achondroplasia is the most typical of the 'chondrodystrophies'. As the gene involved has not yet been identified, diagnosis is clinical.

Prenatal diagnosis

Even though neither the affected protein nor even the chromosomal site of the mutation is known, it is possible to offer prenatal diagnosis by ultrasound scanning in the second trimester of pregnancy. This is particularly valuable to couples of achondroplasics who wish to avoid lethally affected homozygous infants.

Role of the primary care worker

When the child is born, or soon after, there will be obvious signs of a skeletal dysplasia. A head that is large compared with the rest of the body with a bulging forehead and depressed nasal bridge are common features. The baby will have shortening of the proximal part of the limbs, the fingers only reaching the iliac crest, and the limbs are covered with fatty folds of skin in infancy. However, it is important to refer any such infant to an expert, as a very similar clinical picture may be produced by other genes with other modes of inheritance.

These babies often have delayed motor development associated with hypotonia for the first year or two of life, and recurrent otitis media is a common problem. It can be helpful to put the families in touch with a Support Association (see Appendix).

Familial hypercholesterolaemia

This is a disorder of lipid metabolism.

Unlike the conditions discussed above, this is a treatable metabolic disorder. An elevated plasma cholesterol concentration with an increase in ratio of low density to high density lipoproteins is a major risk factor for coronary heart disease (Table 8.5). Many proteins, and therefore genes, are involved in the metabolism of the plasma lipoproteins, so many mutations can disturb their metabolism. One of the commonest and best defined of these disorders is familial hypercholesterolaemia (Goldstein and Brown 1989).

The disorder is due to a defect in a surface receptor for low density lipoprotein (LDL) on liver cells. Receptors are reduced to 50–60 per cent of normal in heterozygotes, resulting in a disturbance of lipoprotein metabolism and an increased level of LDL-bound cholesterol. One result is the deposition of plaques of cholesterol and other lipids in skin and other organs, including the lining of arteries.

Incidence

The gene is carried by 1–2 per 1000 people.

Clinical features

Patients usually present with a premature arcus senilis, present in half over the age of 30. Xanthomas (lipid nodules that contain deposits of LDL-derived cholesterol) are related to the severity of the disease. They are especially common on the extensor tendons around the knuckles, the Achilles tendon, the elbow, and the knee. Ninety per cent of patients have developed them by the age of 40. Raised yellow skin patches around the eyes (xanthelasmata) occur, but unlike xanthomas these are also often found in normal people. If untreated, the mean age of onset of coronary heart disease (CHD) is 43 years in men and 53 years in women (Slack 1969). Table 17.4 shows the percentage of over 1000 heterozygotes who have died from, or had symptoms of, ischaemic heart disease at various ages. However 6 of 214 treated heterozygotes aged 20–39 years, followed during 1980–9, died from CHD. The expected number was 0.06 (Scientific Steering Committee 1991).

If a couple are both heterozygotes, there is a one in four chance in each pregnancy of a homozygote. This is very rare, occurring in one in a million people. LDL lipoprotein receptors are reduced to 0–10 per cent, xanthomas develop by the age of 4, and death from myocardial infarction usually occurs before the age of 30.

Diagnosis

The LDL receptor gene has been located on the short arm of chromosome 19 (p. 00). Only about 1 in 20 people with type 2 hypercholesterolaemia has

Table 17.4 Familial hypercholesterolaemia: estimated percentage risk for heterozygotes of having symptoms of coronary heart disease and dying of myocardial infarction, at different ages

Age (years)	Male heterozygotes		Female heterozygotes	
	Coronary symptoms	Coronary death	Coronary symptoms	Coronary death
40	20	–	3	0
50	45	25	20	2
60	75	50	45	15
70	–	80	75	30

(From Goldstein and Brown 1989)

familial hypercholesterolaemia: in most others, it is multifactorial in origin. Unless patients with familial hypercholesterolaemia have a positive family history and typical signs it can be difficult to differentiate them from other patients with hypercholesterolaemia. In the familial form, the plasma cholesterol is particularly high. If there is a significantly raised cholesterol in a child of a family where other members have hypercholesterolaemia, it is unlikely to be caused by multifactorial hypercholesterolaemia.

Heterozygotes can be treated because they have one functioning LDL receptor gene. Two types of drugs are used to lower the plasma cholesterol. One group of bile acid binding resins (cholestyramine and cholestipol) stimulate the production of LDL receptors. On average, they lower the LDL cholesterol by 15–20 per cent. The statins are another group of drugs that may be used alone, or in combination with bile acid binding resins. They reduce intracellular cholesterol synthesis. This stimulates production of LDL receptors and so promotes clearance of LDL from the circulation. One member of this group, simvastatin, can reduce LDL cholesterol by as much as 50 per cent within 6 weeks (Drug and Therapeutics Bulletin 1990). It is also used in patients with severe multifactorial hypercholesterolaemia, particularly if they have cardiovascular disease. Drugs must be combined with a diet low in cholesterol and saturated fats.

Prenatal diagnosis

This is possible by CVS and DNA studies, for couples at risk of homozygous children, as there is a genetic as well as a biochemical marker for the disorder.

Role of the primary care worker

On average, there will be 2–4 patients with familial hypercholesterolaemia on a family practitioner's list. Family doctors will be alerted to the possibility of this disease in offspring of a parent who developed early ischaemic heart disease. Once carriers of the gene are identified, they should be advised not to smoke, and to make strenuous efforts to keep body weight within normal limits (Chapter 8). Most will be regularly monitored in a lipid clinic. Regular exercise is likely to be helpful, and hypertension should be energetically treated, in spite of the lack of confirmed direct relationship between hypertension and death from coronary heart disease.

X-Linked disorders

Haemophilia

Classical haemophilia (haemophilia A) is caused by factor VIII deficiency and accounts for about 85 per cent of inherited bleeding disorders.

Most of the remainder are due to a deficiency of Factor IX (Christmas disease).

Frequency

The birth incidence of haemophilia A in European countries is 1 per 10 000 boys: most probably it is the same world-wide as the mutation rate is quite high. New mutations often seem to have occurred during spermatogenesis in the maternal grandfather, leading to a carrier female, who may then have an affected son without a family history. Organized high quality treatment has led to a near-normal length and quality of life for most haemophiliacs in most developed countries, so in the past half-century the population prevalence has been steadily approaching the birth incidence (Fig. 17.3) (Larsson 1985). Unfortunately, infection of many patients with HIV recently changed this otherwise successful picture in several countries.

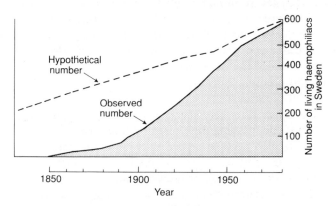

Fig. 17.3 Living haemophiliacs in Sweden in relation to time. The 'hypothetical number' shown assumes a birth incidence of 1/14 000 births (both sexes) and normal survival. With the passage of time, treatment has increased patient survival to the point where prevalence is almost equal to birth incidence. (Based on Larsson 1985.)

Molecular basis

Factor VIIIc (clotting activity) is a key element in the series of reactions that leads to clotting. The Factor VIII gene is located on the X chromosome, and has been identified and sequenced (see Table 2.3). A number of RFLPs linked to the gene have been identified, and can be used for carrier and prenatal diagnosis. Numerous point mutations and deletions cause reduced or absent synthesis of Factor VIII, resulting in a mild bleeding disorder in about 50 per cent of cases, and a severe disorder in the rest. Ten per cent of cases are very severe, i.e. there is practically no clotting activity. Some female carriers also have a bleeding tendency.

Clinical features

Factor VIII does not cross the placenta, so bleeding may occur in affected newborn males, leading to e.g. a cephalohaematoma at birth, prolonged bleeding from the umbilical cord, haematomas after injections or bleeding after circumcision. However, many affected babies have no problem until they start to crawl and later walk, when falls may lead to excessive bruising, or minor cuts lead to prolonged bleeding. Ninety per cent of those with severe disease have shown clear evidence of an excessive bleeding tendency by 3 or 4 years of age, but a smaller group with more than 6 per cent of the normal level of Factor VIII activity have relatively few symptoms, and may not be diagnosed until much later. The commonest severe complication is bleeding into joints such as the elbows, knees, or ankles. This may follow minor injuries, but often seems to be spontaneous. Recurrent bleeding often results in one or more fixed, unusable joints. Internal haemorrhage around the face and neck severe enough to obstruct respiration is a serious but rare emergency. Pressure from accumulated blood on other structures may cause problems such as claw-hand deformity resulting from pressure on a forearm nerve. Haematuria and gastro-intestinal bleeding can also occur. In fact, most of the tissues of the body may be involved. Until recently, the commonest cause of premature death, particularly in severely affected patients, was intracranial bleeding.

Modern management is very successful in preventing and treating these problems. While encouraging their affected sons to lead as normal a life as possible, parents also need to advise them to avoid foolhardy risks: this may involve encouraging intellectual as distinct from manual pursuits, and swimming rather than football or cricket using a hard ball.

Haemophilia is managed by regular injections of fresh plasma or plasma concentrate containing Factor VIII. The cheapest concentrated preparation is cryoprecipitate, prepared in the blood bank from the plasma left by packing red blood cells. This treatment can be given intravenously at home, following instruction from staff at a specialist centre. There are also commercial preparations of nearly pure Factor VIII prepared from pooled plasma samples. These are more expensive than cryoprecipitate but much more convenient for home administration, and had been widely administered to haemophilic patients before the risk of HIV infection became evident.

Ninety per cent of haemophiliac patients in the UK are managed through a national network of specialist haemophilia centres (Rizza *et al.* 1983). Unfortunately, many patients were infected with hepatitis B or C virus by blood products, and more recently and more catastrophically, many have become HIV positive. The virus was transmitted mainly in commercial factor VIII concentrates prepared in the United States from pooled plasma obtained from paid blood donors. In European countries that make their own concentrate, or still use cryoprecipitate, most haemophiliacs are unaffected; but in countries (such as the UK) that bought Factor VIII prepared in the United States, many haemophiliacs are now HIV positive, and AIDS has become the commonest cause of premature death among them. Several wives of HIV-positive haemophiliac husbands also developed AIDS. These risks can now be avoided by heat-treating the concentrate (this will also kill any other viruses present), or by using genetically engineered Factor VIII (p. 157). Patients who are not immune to hepatitis B should be vaccinated.

Carrier diagnosis

Because of the considerable overlap between carrier and normal ranges of Factor VIII, it was formerly only possible to give a woman with a positive family history a probability of being a carrier. As a result, many women who were not carriers underwent prenatal diagnosis in every pregnancy. DNA methods now allow clear identification of women not at risk, and early prenatal diagnosis for those who are.

Prenatal diagnosis

Table 17.5 shows how prenatal diagnosis for haemophilia has developed over the last 20 years or so. First trimester prenatal diagnosis by DNA methods is requested by many women at risk, especially since the complications of maintenance treatment have become apparent. For many other X-linked conditions, we are still only able to provide fetal sexing.

Role of the primary care worker

Multiple bruises of different ages may be caused by physical abuse, but haemophilia is a possible diagnosis if a male baby develops abnormal bleeding, especially if there is a positive family history on the mother's side. If a haemophiliac presents with an unexplained symptom, an undiagnosed bleed may be the cause. For example, abdominal pain and right-sided tenderness may be caused by a haematoma rather than appendicitis. A large bleed into a joint or tissue causes fever, anorexia, and may be accompanied by a leucocytosis — thus mimicking sepsis. The patient can become anaemic and need a blood transfusion. Drugs such as aspirin which interfere with platelet function should be avoided. The families of patients with haemophilia in the UK can be reassured that HIV is now no longer a risk. The benefits of testing female relatives should be discussed, and it is worth checking that the family is in touch with a support organization (p. 373).

Table 17.5 Evolution of methods for prenatal diagnosis of haemophilia

Years	Method	Option available	Uptake
To 1975	Fetal sexing by amniocentesis	Abort all males in mid-trimester. 50% would be unaffected	Relatively small
1975–85	Fetal sexing by amniocentesis. If male, definitive diagnosis by fetal blood sampling	Abort affected males only, in mid-trimester	Considerably increased
1985 onwards	Definitive diagnosis by CVS, DNA and chromosome studies	Abort only affected males, in first trimester	Very substantial interest

Duchenne muscular dystrophy (DMD)

Frequency

Duchenne muscular dystrophy affects one in 4000 boys in European countries. In about two-thirds of cases, the mother can be shown to be a carrier, but about one-third of cases are due to new mutations. Since so many cases are due to new mutations, its incidence is probably much the same world-wide.

Molecular basis

Duchenne muscular dystrophy is due to mutation or deletion in a large gene on the X chromosome (Table 2–3). This gene codes for the protein dystrophin, a membrane protein that is important in calcium transport in muscle cells, and was the first gene product identified by reverse genetics (p. 141). Its absence leads first of all to hypertrophy of the cell, and later to muscle cell death. A number of RFLPs linked to the gene have been identified. When the disease is due to a sizeable deletion, female carriers and affected males can be diagnosed directly; when it is not, they can be diagnosed by family studies using RFLPs.

Clinical features

An affected boy usually seems to be developing normally in the first year of life. Often there is a delay in beginning to walk. Over 90 per cent of normal infants can walk by 18 months compared with about 60 per cent of boys with DMD. The disease is evident in 50 per cent by the age of two and a half, and in nearly all affected boys by 5 years. In the early stages there is usually enlargement of the calf muscles due to excess adipose and connective tissue. Other muscle groups, for example around the shoulder, may become bulky later on. Though hypertrophied, these muscles get gradually weaker and eventually waste. Initially the lower limbs are usually weaker than the upper, and proximal muscles are affected more than distal ones. Weakness of muscles of the buttocks and around the hip leads to a waddling gait. When affected children start having difficulty rising to a standing position, they characteristically using their arms to 'climb up' their legs, and then push themselves upright. When the muscles of the shoulder become involved, attempts to lift the child from behind lead to him sliding through the examiner's arms.

The disease progresses in a relentless manner, though sometimes in stops and starts. About 60 per cent of affected boys are confined to a wheelchair by the age of nine, though at first this may only be needed out of doors.

Usually, the later this stage is reached, the longer the child will live. Fifty per cent will have died by fifteen and a half and 90 per cent by nineteen and a half (Emery 1988). The figures have not changed significantly during this century. Death is usually due to aspiration pneumonia, though heart muscle also appears to be involved and heart failure causes 10 per cent of deaths. The IQ of affected boys tends to be about 20 points lower than normal.

Management

Even though treatment does not prolong life at present, appropriate management will make life more pleasant. The boy should lead as normal a life as possible for as long as possible. If the child is ill, prolonged bed rest may increase the muscular weakness. Once the boy is confined to a wheelchair, immobility tends to lead to overweight, aggravating respiratory complications and making movement and lifting more difficult. In the case of a chest infection, physiotherapy and early treatment with antibiotics are important. Physiotherapy also helps to slow the development of the muscular contractures that develop in the later stages of the disease. Scoliosis, which also impedes respiratory function, is common in teenagers and may be corrected surgically. Psychological problems are aggravated by the progressive immobilization of the patient as his peers are entering the most active time of life.

There are many forms of muscular dystrophy with different modes of inheritance, and not all pursue such a severe or inexorable course: some are of later onset and cause progressive disability in late adult life (p. 287). It is essential to know which type of muscular dystrophy is involved in any particular family, to be able to offer correct advice and support.

Carrier and prenatal diagnosis

There is a substantial demand for definitive carrier diagnosis from female relatives of affected boys, and for prenatal diagnosis from women who are shown to be carriers. As with haemophilia, before the advent of DNA methods it was only possible to give female family members an assessment of risk that they were a carrier, based partly on measurement of serum creatine phosphokinase (CPK). Definitive carrier diagnosis is a major benefits of DNA methods.

The role of the primary care worker

It is important to be sure that a definitive diagnosis has been made in every case, that DNA from an affected boy has been stored, and that female relatives receive counselling and the offer of carrier testing.

The diagnosis of DMD should be considered in male infants whose motor development is significantly delayed, particularly if it initially seemed to be all right. Various other forms of congenital muscular dystrophy and neurological disorders may be mistaken for DMD. DMD rarely causes weakness in the neonatal period. A 'floppy baby' is more likely to have neurological problems e.g. the autosomal recessive condition Werdnig–Hoffman disease, where degeneration of the anterior horn cells of the spinal cord leads to lower motor neurone weakness. The diagnosis of DMD is biochemically confirmed by measurement of CPK which leaks into the serum from damaged muscles. It may reach 50–100 times normal in an affected child, and permits reliable neonatal diagnosis (Skinner *et al.* 1982).

We recommend the book on DMD by Alan E. H. Emery for any general practitioner who has a young patient with the disease.

Glucose-6-phosphate dehydrogenase (G6PD) deficiency

Like the haemoglobin disorders, G6PD deficiency occurs predominantly in areas where malaria is or formerly was endemic, and is common in the same ethnic groups (see Table 15.2). About 7 per cent of the world population carries a gene for G6PD deficiency. It is a globally important cause of neonatal jaundice, which can lead to kernicterus and in turn to death or spastic cerebral palsy. It can also lead to life-threatening haemolytic crises in childhood and at later ages, precipitated by drugs, infection or eating broad beans. These complications can be avoided by simple preventive measures (Luzzatto and Mehta 1989; WHO 1985, 1989).

Molecular basis

G6PD is a 'housekeeping' enzyme, responsible for producing the biological reducing agent NADPH (reduced nicotine adenine diphosphate), which is needed to protect the red cell and its haemoglobin from oxidation. G6PD deficiency has very little clinical effect under ordinary circumstances because normal red cells contain much more G6PD than they need. However, its lack may become dramatically apparent in the presence of oxidizing stresses which normal red cells can easily deal with. G6PD deficiency is fully expressed in hemizygous males and homozygous females, and in a proportion (perhaps 10 per cent) of female heterozygotes.

Clinical features

In affected males there is very mild chronic haemolysis (the mean Hb level being about 1 g dl less than 'normal'). This can cause neonatal jaundice, which occurs spontaneously in some G6PD-deficient infants, usually as a profound form of physiological jaundice. Babies with infections, and premature or breast-fed babies are more likely to develop G6PD-deficiency-associated jaundice. Jaundice may also occur if susceptible babies are exposed to toxic environmental agents, e.g. clothes that have been stored in contact with moth (naphthalene) balls, or certain herbal medicines; or if the umbilical cord stump is dressed with antiseptic 'mentholated' powders. Mildly affected babies can be protected by exposing them to light, but more severely affected infants will require an exchange transfusion.

In older children and adults, an acute haemolytic reaction can occur as a result of eating broad beans. This condition is called favism after the botanical name of the bean, *Vicia faba*. Favism is common in mediterranean countries and southern China where broad beans are an important part of the diet, but is rare in Africans. It occurs most frequently in children below 5 years of age, is relatively uncommon in adults, and can be fatal. Acute haemolysis may also be caused by a limited range of drugs (Table 17.6). The same agent may cause haemolysis in one G6PD-deficient person but not in another, and in the same person at one time but not at another.

Acute haemolysis begins within one or two days of eating fava beans or the administration of the drug. The picture varies from a transient mild anaemia to rapidly progressing anaemia with back and abdominal pain, jaundice and haemoglobinuria, and transient splenomegaly. Heinz bodies (inclusions of precipitated oxidized haemoglobin attached to the inside of the red cell membrane) are found in the peripheral blood. They indicate that abnormal oxidation of haemoglobin has occurred. Anaemia can be so severe that an urgent blood transfusion is essential.

Table 17.6   Common drugs that can be associated with haemolysis in G6PD deficiency

Category	Examples
Antimalarials	Primaquin
	Pamaquin
	Chloroquin?
Sulphonamides	
Sulphones	DDS, dapsone
Antibiotics	Nitrofurantoin
	Nalidixic acid
	Choramphenicol?

(A comprehensive list may be found in WHO 1989).

Diagnosis

Screening for G6PD deficiency is cheap, simple, and available at most haematology laboratories. It should be requested for at-risk neonates with deepening jaundice, or for children or adults with an acute haemolytic reaction. It could easily and usefully be added to routine screening for haemoglobin disorders, since both conditions occur in the same ethnic groups. Prenatal diagnosis is not indicated for the common form of G6PD deficiency, which is not a disease. However, it might be appropriate for some rare forms that cause chronic severe haemolytic anaemia.

Role of the primary care worker

Affected people need to be given correct information about the risks of the condition. The drugs that can cause problems in people with G6PD deficiency do *not* include many common medicines such as penicillin and paracetamol, and there is very little evidence against aspirin (WHO 1990*a*). When a G6PD deficient individual is found, the family should be screened in order to identify female carriers and other affected males, and to give them advice about the use of drugs etc.

Fragile X mental retardation

About 3 in 1000 females and 4 in 1000 males are severely mentally handicapped. The excess of males is attributed to inherited defects carried on the X chromosome, so a quarter to a third of all severe male mental handicap is X-determined.

In about one-third of families transmitting an X-linked form of mental retardation, a characteristic marker — a 'fragile site' — is visible near the tip of the X chromosome of affected males in 5–50 per cent of cells studied. A fragile site is a place where stain is not taken up, making the chromosome appear vulnerable to breakage (which it is not). A fragile site is also found on one X chromosome in some female relatives of affected males. In these families, the presence of the characteristic fragile site is associated with a syndrome of mental retardation, a characteristic long face with large ears, and usually with large testes in the male. In general, but by no means always, the parcentage of cells expressing the fragile X site is proportional to the degree of mental retardation, but the fragile X site is an unsatisfactory marker. It is expressed only under particular cell culture conditions; it is not found in all affected families; it is not found in all affected males; and it may occur in unaffected males. Its presence in females is sometimes associated with mental retardation, presumably because of variable inactivation of the X chromosome (see p. 31).

In the extended family, the penetrance of severe mental retardation in males with a fragile X site is 90 per cent. About 30 per cent of female carriers are severely retarded but less so than their affected male relatives (Figure 17.4), giving an incidence of fragile X mental retardation of about 1 per 1000 in males and 0.6 per 1000 in females (since there are twice as many female

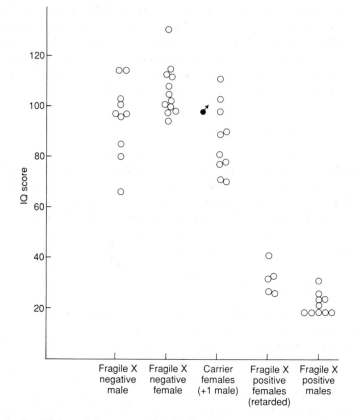

Fig. 17.4 The IQ in one large extended family with members with fragile X mental retardation. (Based on Veenema *et al.* 1987.)

carriers as male hemizygotes). Unaffected carrier males and females have a slightly lower IQ than related non-carrier males. The fact that fragile X mental retardation is almost as common as Down syndrome was confirmed in a recent survey of a group of children with an IQ of 50 or less (Asthama *et al.* 1990). Mental handicap is discussed further in Chapter 20.

Affected individuals rarely reproduce, but unaffected hemizygous males and female carriers may transmit the abnormality. There is also a high rate of spontaneous mutation. Different types of mutation may explain the considerable range of severity of the syndrome in different affected families.

The fragile X gene has recently been identified: this makes accurate carrier diagnosis possible and opens the door to population screening.

Prenatal diagnosis

Prenatal diagnosis for fragile-X mental retardation is possible, but at present is rather unsatisfactory. The limitations mentioned above make it hard to exclude the condition reliably, but combined cytogenetic and DNA tests give more reliable results. It is also impossible to be sure that a male fetus carrying the abnormality will develop into a severely retarded person, or that a female carrier will be unaffected.

Role of the primary care worker

The diagnosis must be considered in mentally retarded males when there is no other obvious cause such as Down syndrome. Female relatives in particular need genetic counselling and carrier testing, because of their high genetic risk.

The condition is so common that when the molecular defect has been better defined, population screening for carrier females could be realistic.

Key references

For most conditions, the relevant chapters in:

Emery, A. E. H. and Rimoin, D. L. (2nd edn) (1990). *Principles and practice of medical genetics*, Volumes 1 and 2. Churchill Livingstone, Edinburgh.

Scriver, C. R., Beaudet, A. L., Sly, W. S., and Valle, D. (ed.) (1989). *The metabolic basis of inherited disease*, (6th edn). McGraw Hill Inc., New York.

For Huntington's chorea: Hayden, M. R. (1981). *Huntington's chorea*. Springer Verlag, Berlin.

For achondroplasia: Murdoch, J. L., Walker, B. A., Hall, J. G., Abbey, H., Smith, K. K., and McKusick, V. A. (1970). Achondroplasia—a genetic and statistical survey. *Annals of Human Genetics*, 33, 227–44.

For haemophilia: Living with haemophilia. Castle House Publications, Tunbridge Wells, Kent.

For Duchenne muscular dystrophy: Emery A. E. H. (1988) *Duchenne muscular dystrophy*, (revised edn) Oxford University Press.

For Glucose-6-phosphate dehydrogenase deficiency: WHO (1989). Glucose-6-phosphate dehydrogenase deficiency. Report of a WHO Working Group. Bulletin of the World Health Organization, 67, 601–11.

18.   Chronic disorders of childhood

Congenital malformations

In general, a 'major' malformation means one which is life threatening, or has severe implications for form, function, and psychological well-being if uncorrected. Some disorders inevitably kill the newborn, e.g. when an organ or tissue is missing as in anencephaly or agenesis of the kidney. Others, such as congenital heart disease or unilateral disorders of the renal tract can cause severe chronic illness and premature death. Yet others, such as cleft lip and palate, talipes, congenital dislocation of the hip, blindness or deafness, and malformation or ambiguity of the genitalia, are not life threatening but can have devastating social effects.

Many congenital malformations can be repaired by paediatric surgery, but severe malformations that cannot be repaired are reasonable targets for prevention by prenatal diagnosis and selective abortion. These include multiple malformation syndromes, disorders of the central nervous system, severe malformations of the heart, bilateral abnormalities of the kidney, reduction deformities of the limbs, and major defects of the abdominal wall. In this chapter we discuss selected common disorders to demonstrate points that are particularly important in primary care.

Many malformations in babies are detected at birth (or even by ultrasound examination before birth), and some have been operated on even before the baby is brought home. Parents of a child with a congenital malformation are likely to be concerned about

(1) whether the child could have other abnormalities or be mentally retarded;
(2) whether the malformation is fully correctable (or fully corrected), and how significant any handicap will be;
(3) what support is available to help them cope;
(4) whether the problem was caused by something the mother did or ate; and
(5) whether there is a risk of recurrence in further children.

These parents and those with a mentally retarded child usually need a consultation with a clinical geneticist. The commonest outcome is reassurance that the risk of further children being affected is relatively low.

With earlier discharge from hospital after delivery, infants with less obvious problems are increasingly likely to be missed (see p. 340). Milder congenital disorders or those with later onset tend to present to primary care workers, who also often see children who look a bit odd. The question is which children should be referred to a clinical geneticist or a paediatrician for an opinion on the possibility of a serious underlying condition? In some conditions such as Down syndrome (p. 287) and fragile X mental retardation (p. 251), there is a characteristic facial appearance. Other children should be referred if there is more than one seemingly minor type of malformation (not skin blemishes or over-riding toes), or additional cause for concern such as slowness in passing developmental milestones, poor growth, feeding problems, an unusually shaped or big head (though the commonest cause for this is a familial big head), or a history of other similar cases in the family.

The study of children with structural defects is called dysmorphology. Physical abnormalities are often part of a specific syndrome. Precise diagnosis requires a detailed physical examination and a full family history, and can be very useful in giving a prognosis and recurrence risk. Infants with single malformations have a much better prognosis than those with multiple malformations. The distinction between problems of prenatal onset (due to disturbed embryonic or fetal development) and problems of postnatal onset is not merely academic. The latter are often attributed to difficulties at delivery, with implications in terms of guilt and responsibility, an issue that is particularly relevant to cerebral palsy (see below).

It is difficult to obtain reliable figures for the true incidence of congenital malformations, given the uncertainties in defining them and the fact that there is no clear dividing line between the normal range of physical variation and malformation (e.g. is squint a malformation?). Many are internal or can be detected only by thorough physical examination, and even borderline abnormalities can be important. For example, ureteric reflux, a functional result of a common and minor malformation of the vesico-ureteric valve, can lead to chronic kidney disease and hypertension. In addition, the incidence of some malformations varies from place to place (e.g. neural tube defects) and of others varies with time e.g. with rubella epidemics. Reliable figures require long-standing and comprehensive recording systems.

Table 18.1 gives average figures for twelve European centres collected in the 'Eurocat' study (de Wals *et al.* 1985; Eurocat working group 1989). A representative figure is about 30 malformations per 1000 births. However, from 20–30 per cent of malformed babies have multiple malformations, so the incidence of malformed babies (about 20/1000 births, i.e. 2 per cent) is less than the incidence of malformations. The table also shows how different groups of malformations can have different effects in terms of mortality and morbidity, and includes estimates of the success of surgical repair (Czeizel and Sankanarayanan 1984).

Less than half of congenital malformations have a clear cause (Trimble and Doughty 1974) (Table 18.2). Many follow the 'rules' of multifactorial inheritance (p. 106) with a recurrence risk of the order of 1–6 per cent. A small proportion of most kinds of congenital malformation is due to inherited

Table 18.1 Congenital malformations: birth prevalence, effects, and effectiveness of treatment

System involved	Incidence/10 000 births (EUROCAT study)	Average of 70 potential years			Prenatal diagnosis (Detectable by ultrasound before 24 weeks gestation)
		% lost	% chronic disability	% normal with treatment	
Central nervous system	18–50	79	18.5	2.5	Most severe
Eye	3–12	19	46	36	
Ear, face, and neck	7	0	30	70	
Cardiovascular system	40–96	34	17	49	Most severe
Respiratory system	4	35	36	29	
Palate ± lip	14	3	22	75	
Digestive system	12–38	23	4	73	
Genitalia	11–24	0	24	76	
Urinary tract	9–16	19	36	45	Most severe
Musculoskeletal	43–89	0	1	99	
Skeletal	21	19	18	62	Many
Other (including multiple)	50–60	23	20	57	Many, especially multiple
TOTAL (approximate)	290	22	24	54	

(Based on de Wals *et al*. 1985 and Czeizel and Sankanarayanan 1984)
The estimated totals were calculated from the original tables of Czeizel and Sankanarayanan.

Table 18.2 Estimated proportion of congenital malformations due to known causes (assuming a basic incidence of 30/1000)

Cause of congenital malformation	Incidence/ 1000 births	% of all malformations
Single gene mutation	2.25	7.5
Chromosomal (*ca.* 30% of aneuploid infants are malformed)	1.8	6.0
Maternal infections (rubella, CMV, *Toxoplasma*) when no epidemic	0.6	2.0
Diabetes (9% of 4.7/1000 diabetic mothers have malformed infants)	0.43	1.44
Poor diet (may be responsible for up to 4/1000 NTDs[a] in genetically susceptible populations)	0–3	0–10
Sub-total	8.1	27
When there is a rubella epidemic in a non-immune population	$\leqslant 7$	
Total	$\leqslant 14$	40% of total

(Based on Trimble and Doughty 1974)
[a] NTD = neural tube defect

single gene defects that carry a high recurrence risk, but these are hard to distinguish unless there is a strong family history.

Neural tube defects: spina bifida and anencephaly

Neural tube defects (NTDs) are the target of one of the principle prevention programmes based on screening of pregnant women in the UK. We describe them in some detail because most pregnant women undergo testing, and some will ask questions or need advice.

The incidence of neural tube defects varies with geographical area, being relatively low (1–2 in 1000) in much of Europe and the USA, but higher in the UK, Hungary, and parts of Egypt, India, and China. Before maternal serum AFP screening was introduced, the average birth prevalance in the UK was about 4–5 per 1000, ranging from 5–7 per 1000 in Wales, Ireland, and Western Scotland to 1–2 per 1000 in parts of South-East England. As a result of maternal serum AFP (p. 331) and ultrasound screening in pregnancy the UK is now a low incidence area (Cuckle *et al.* 1989).

The steady fall in incidence of neural tube defects in the UK in the past 15 years indicates the importance of environmental factors, while the fact that Americans of Irish extraction have a higher risk than other Americans underlines the importance of genetic factors. Siblings of an affected person have a risk about 10 times the local general risk, and in a few families the condition appears to be inherited as a Mendelian recessive gene. There is also evidence that dietary deficiency of folic acid is involved (MRC Vitamin Study Research Group 1991): the genetic predisposition could be concerned with metabolism of this vitamin. The abnormalities in embryological development that lead to neural tube defects are described in Chapter 4. Neural tube defects are also more common in twins (Vogel and Motulsky 1986).

Clinical features

There are several types of neural tube defects, ranging from anencephaly to spina bifida occulta (Table 18.3). In addition, about 20 per cent of spinal X-rays of healthy people show a minor defect, usually in the lower lumbosacral vertebrae. A few of these individuals have a significant bony deficit, and rarely may have a neurological defect in the lower limbs and/or a disturbance of bowel or bladder function (Huttenlocher 1983).

Infants with anencephaly and about one-third of those with spina bifida die in the later stages of pregnancy, at birth or in the neonatal period. Liveborns with anencephaly are incapable of surviving for more than a few days. Open spina bifida (a mid-line defect in the skin, vertebral arches and spinal cord) presents much more complex problems. The clinical picture depends on the level of the lesion, and whether a meningomyelocele is present. It includes children with a minor neurological abnormality of their lower limbs and children with severe hydrocephalus, incontinence of faeces and urine, and paralysis with loss of sensation below the level of the lesion.

The severely affected newborn baby will have a transparent membrane (a meningomyelocele) covering the defect, which may contain neural elements, and initially leaks cerebrospinal fluid (CSF). Leakage lessens as the membrane dries, and the membrane gradually bulges as CSF accumulates beneath it. The defect can cause problems lower down the body through interrupting the nerve supply to various organs, and higher up through interfering with circulation of the CSF.

In the embryo the spinal cord is as long as the vertebral column. During fetal life and childhood, the vertebral column elongates faster than the cord, which slips relatively upward, leaving a 'mare's tail' of lumbar and sciatic nerves inside the lower part of the spinal column. However, in spina bifida, the spinal cord is tethered to the vertebral column and skin and cannot slip upwards, so it often pulls the medulla downwards into the foramen magnum, creating the 'Arnold–Chiari malformation'. This compression may occlude the opening through which the CSF normally escapes from within the cerebral ventricles, and can create an internal hydrocephalus (Fig. 18.1).

Table 18.3 Classification of neural tube defects

Disease	Abnormality		Outcome
Anencephaly (*ca.* 50% of total)		Brain development very defective	Late intrauterine or neonatal death
Spina bifida (*ca.* 50% of total)	(open)	Open defect in back with meningomyelocele: problems depend on the level of the lesion: the higher the worse	Ranges from intrauterine death, or death in childhood, to active adult life
	(closed)	Defect covered by skin. Ranges from severe to normal	Can be late presentation with problems in walking, continence etc.

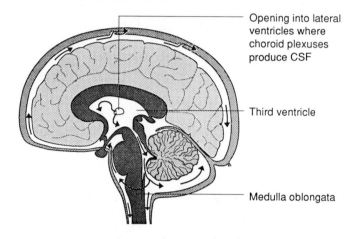

Opening into lateral
ventricles where
choroid plexuses
produce CSF

Third ventricle

Medulla oblongata

Fig. 18.1 Circulation of the cerebro-spinal fluid (CSF). CSF is generated by the choroid plexuses in the lateral ventricles and the third ventrical of the brain. It flows from the inside of the brain through apertures in the walls and roof of the 4th ventricle, below the cerebellum. Thence it circulates over the surface of the brain, and is reabsorbed through the arachnoid granulations in the main venous sinuses of the skull.

Isolated hydrocephalus may also occur in connection with congenital stenosis of channels for CSF within the brain, some brain tumours, inherited diseases, congenital infections, and meningitis acquired in childhood.

The head of an untreated infant with hydrocephalus grows at an abnormally rapid rate, the circumference crossing the centile lines on serial measurement. Eventually there is separation of the sutures and bulging of the anterior fontanelle and forehead. In severe cases, the scalp skin becomes thin and shiny, the veins are flattened, and the eyes deviate downwards (the 'setting sun' sign). Two-thirds of children with hydrocephalus need a shunt to drain the excess CSF elsewhere in the body (usually into the peritoneum) to prevent these developments. However, parents need to be able to diagnose when a shunt has become blocked or infected, producing drowsiness, vomiting, or unsteadiness, and urgent surgery may be needed if blockage persists. Shunts may also cause a low grade infection presenting with indolent fever and splenomegaly, or low grade peritonitis. Fortunately many children stop needing a shunt with the passage of time.

Complications below the site of the lesion disturb the nerve supply to bladder and bowel and can lead to paralysis and incontinence of urine and faeces. Urinary tract infections are very common. Absence of sensation below the lesion leads to pressure sores and unnoticed leg injuries. An acutely unwell child with spina bifida may thus have a blocked shunt, a fever from a urinary tract or chest infection, or shock from an undetected fractured femur. Many children will be confined to a wheelchair with complications as in Duchenne muscular dystrophy (p. 247) — obesity, kyphoscoliosis, and respiratory insufficiency.

Treatment

Difficult decisions often have to be made about infants with neural tube defects. The decision whether or not to operate on an affected newborn is now usually 'selective', taking into account the findings on examining the baby (Table 18.4). The decision whether or not to terminate a pregnancy with an affected fetus cannot be so selective because clinical examination of the fetus is not possible. Both types of decision call for knowledge of the natural history of the disease when all infants are treated. Fortunately this information is available.

Hunt (1990) reported a long-term follow-up study of a group of 117 babies with open spina bifida born between 1963 and 1971, operated on within 48 hours of birth, and reviewed 16–20 years later. The group's survival curve is shown in Fig. 18.2. Forty-six (41 per cent) had died before the age of 16, half of these within the first year. Causes of death were: renal failure (12), hydrocephalus (7), CNS infection (8), cardiorespiratory failure (14), others (5). Table 18.4 shows the percentage of the surviving patients with severe, moderate, or minimal disability at 16 years of age. The spinal level at which the infants could perceive sensation had perhaps the most predictive value for their long-term prognosis: young people with sensation below L3 had least disability, while those with sensation only above T11 had most disability. The table also illustrates the point that when the most severely affected patients die early, the condition of the survivors can give an over-optimistic picture of a disease.

Pregnancy screening and prenatal diagnosis for neural tube defects are outlined in Chapters 22 and 13. A family doctor is likely on occasion to be

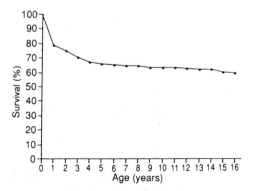

Fig. 18.2   Survival to the 16th birthday in 117 consecutive cases of open spina bifida, all treated (unselectively) soon after birth. (Reproduced from Hunt 1990.)

Table 18.4   Summary of history in 112 patients with spina bifida aperta, treated non-selectively at birth[1]

	Sensory level				Total
	Above T 11 (at or above umbilicus) No. (%)	T 11–L3 (umbilicus to knee) No. (%)	Below L 3 (below knee) No. (%)		No. (%)
Total	42	32	38		112
Died before 16 years	22 (52)	14 (44)	10 (26)		46 (41)
Survivors	20	18	28		66
Condition of survivors					
Can walk >50 yards	0 (0)	7 (40)	25 (90)		32 (48)
Continent	0 (0)	3 (15)	13 (45)		16 (11)
Able to live without supervision in adapted housing	2 (10)	8 (45)	24 (85)		34 (52)
Employable, sometimes in sheltered work	4 (25)	14 (80)	27 (95)		45 (68)
IQ > 80	10 (50)	13 (70)	22 (80)		45 (68)

(Based on Hunt 1990)

[1] Excludes five patients with very asymmetrical sensory levels

consulted by a pregnant woman who has been told she is carrying a fetus with open spina bifida; counselling needs to be based on the whole picture.

Prevention: vitamin supplementation for the mother

Initial studies in women who had already had a child or fetus with a neural tube defect indicated that supplementing their diet with a multi-vitamin preparation or folic acid before conception and in the very early stages of pregnancy considerably reduced the recurrence risk (Smithells *et al.* 1981; Wild *et al.* 1986; Laurence 1983). These studies raised three questions.

1. Were the results right? There could be some doubts, as the background incidence of NTDs seemed to be falling quite fast at the time, and the initial studies did not include randomized controls.
2. Would vitamin supplementation for all women reduce the incidence of NTDs in both high risk and low risk populations?
3. Could vitamin supplementation itself cause fetal abnormalities?

Three different kinds of study were set up aiming to answer these questions:

1. A collaborative international study of the safety and efficacy of vitamin supplementation for women at high risk. This study proved difficult to establish. Many doctors felt that an untreated control group was unethical in view of (a) the existing evidence in favour of vitamin supplementation, (b) the fact that all the women in the trial would already have had one abnormal child or fetus, and unsupplemented controls would be exposed to a possibly avoidable risk of recurrence, and (c) the lack of evidence of risk associated with multivitamin preparations in the doses given. However, the results have now been published (MRC Vitamin Study Research Group 1991). The prevalence of neural tube defects was 6 in the 593 completed pregnancies of the women who took 4 mg of folic acid from the time of conception to the 12th week of pregnancy, and 21 in the 602 pregnancies in the control groups. In other words it appears that folic acid prevented 72 per cent of neural tube defects. The other vitamins did not offer significant protection.
2. A second, uncontrolled study suggests that vitamin supplementation may be more effective in high incidence than in low incidence areas — i.e. there may be a background incidence of neural tube defects due to some other cause (Sellar and Nevin 1984; Smithells *et al.* 1989).
3. In Hungary, a large scale, randomized study is being made of periconceptional vitamin supplementation in young women without a previous affected child (Czeizel *et al.* 1990).

Congenital heart disease

With an incidence of about 8 per 1000 births, congenital heart disease is the commonest kind of congenital malformation. It is particularly common in infants with other problems such as chromosomal abnormalities or an intrauterine infection, and is a common malformation in children of diabetic mothers (p. 318). A few specific syndromes are recessively inherited.

About half of congenital cardiac abnormalities are mild. Some ventricular septal defects correct themselves with time and other abnormalities may be repaired by relatively simple surgery. However, half are very severe, and congenital heart disease accounts for over half the deaths from congenital abnormalities in childhood. Signs of congenital heart disease vary according to whether there is a shunt. A left to right shunt produces cyanosis, as in Fallot's tetralogy. A right to left shunt produces increased pulmonary blood flow as in the common forms of patent ductus arteriosus, atrial septal defect, and ventricular septal defect. The age at presentation depends on the type of the lesion. Neonatal diagnosis of congenital heart disease is discussed on p. 341. Some severely affected infants present in the neonatal or post natal period with heart failure, but quite often a congenital heart defect is detected only later in childhood when an abnormal cardiac murmur is found. Many initial investigations, e.g. echocardiography or ultrasound are now relatively non-invasive. Some affected infants need operation in the neonatal period, for instance for a patent ductus arteriosus. Many get worse if left undiagnosed, especially those in whom pulmonary hypertension may develop.

Prenatal diagnosis

Most severe forms of congenital heart disease can now be detected in early pregnancy. A basic '4-chamber scan' after 16 weeks included in routine second trimester fetal anomaly scanning constitutes a screen for congenital heart disease. It does not detect all disease, but identifies major defects in about 2 per 1000 pregnancies. Table 18.5 shows the outcome in the first 110 cases diagnosed prenatally at Guy's hospital in London (Allen *et al.* 1986). In 47 out of the 57 severely affected fetuses detected before 24 weeks' gestation, the parents chose abortion (because of multiple anomalies or severe cardiac disease). Only 10 per cent of the 63 continuing pregnancies led to a living, chromosomally normal child.

Recurrence risk

This is generally low. There is on average a risk of about 5 per cent of recurrence in a sibling. It seems that parents who themselves have a corrected CHD also have a risk of about 5 per cent in each pregnancy of having an affected child. Expert ultrasound scanning in pregnancy should be offered

Table 18.5 Outcome in 110 cases where severe congenital heart disease was diagnosed prenatally (1980–86)

	Termination of pregnancy	Death in utero	Neonatal death	Alive	Total
All cases	47	23[1]	30	10	110
Number with multiple anomalies	25	8	7	3[2]	43

(Based on Allen *et al.* 1986)

[1] Pregnancy not terminated because diagnosis was made too late.
[2] Two babies had Down Syndrome

to all such mothers and those who have already had one affected child, and to their first-degree relatives.

Diagnosis in primary care

In primary care, the commonest problem is to decide whether a murmur is, or is not, a sign of heart disease. Most murmurs are not. Benign murmurs are systolic only, are relatively short, and of mild intensity, are not harsh or accompanied by a thrill. They are usually best heard to the left of the lower sternum and start after a clear first sound. The child will be growing normally with no abnormal breathlessness on exertion, and is unlikely to have more than the usual amount of respiratory infections. The murmur may lessen or disappear when the child breathes in. A venous hum is another common harmless sound, usually heard best in the neck and below the collar bones. It extends through systole and diastole, and may disappear when the child lies down, or turns his head. Compressing the jugular vein also makes the murmur disappear.

Abnormalities of the urinary tract

These are common, being found in 9–16 per 1000 babies. In newborns, a renal abnormality may be diagnosed because the baby develops non-specific signs of a neonatal infection, failure to feed, apathy, floppiness, and listlessness, sometimes with vomiting. The standard management of any neonate or young infant that is failing to thrive includes a urine culture. A child with a renal tract abnormality may have a palpable mass in the abdomen if there is a hydronephrotic kidney, or a palpable bladder if there are posterior urethral valves, or there may be some abnormality of the external genitalia. Children with less severe (often unilateral) renal abnormalities often present later in life, usually with a urinary tract infection. Renal abnormalities are commoner in males, while urinary tract infections unassociated with abnormality are commoner in females. People who had a urinary tract infection in childhood and consequent renal scarring have a 20 per cent risk of hypertension and a 10 per cent risk of chronic renal failure (Guidelines for the management of acute urinary tract infections in childhood 1991).

About 30 per cent of children with infections have vesico-ureteric reflux. This is by far the commonest abnormality of the renal tract, and the commonest cause of chronic kidney damage.

Prenatal diagnosis

Urinary tract abnormalities have been detected by ultrasound scanning, at different centres and different stages of pregnancy in from 1–7.6 per 1000 pregnancies: perhaps 1 in a 1000 fetuses have a severe abnormality (Livera *et al.* 1989). Abortion is a realistic option for severe abnormalities, while early treatment and surveillance is the right approach for fetuses with treatable or unilateral problems.

Renal malformation may be associated with other abnormalities. For instance, 9 of 39 fetuses with obstructive uropathy identified before 24 weeks' gestation by Nicolaides and co-workers (1986) were chromosomally abnormal. Only severe abnormalities can be detected before 24 weeks' gestation, and it seems that a reasonably reliable prognosis can be given, e.g. bilateral disorder is associated with a poor prognosis. In the same study, of 30 pregnancies (75 per cent) considered to have a poor prognosis, 22 were terminated, and the remaining eight ended in intrauterine or neonatal death. The nine pregnancies given a good prognosis continued to term, and eight of the babies were treated successfully.

Many abnormalities detected later in pregnancy are milder, and can be treated successfully after birth by early operation or treatment with prophylactic antibiotics (Thomas *et al.* 1985). The commonest problems are obstructive, and the usual sign is dilatation of part of the renal tract. Paradoxically, severe abnormalities such as renal hypoplasia or agenesis can be harder to see in later pregnancy because they cause oligohydramnios.

Diagnosis in primary care

These are the congenital malformations that family doctors are most likely to diagnose themselves. A high index of suspicion for urinary tract infections and underlying pathology is required. A urine culture should be requested, particularly in infants who present with a fever that has no obvious specific cause and does not resolve spontaneously in 24–48 hours. Bacterial growth is significant when a midstream urine specimen contains more than about 10^5 organisms mm^{-2}. Many children with a renal abnormality will need long-term low-dose maintenance antibiotics, and some need an operation to correct the underlying problem.

Cerebral palsy

Cerebral palsy, which occurs in 2–4 per 1000 babies, is one of the commonest causes of lifelong disability. It is defined as 'a disorder of posture or movement which is persistent but not necessarily unchanging, and is caused by a non-progressive lesion of the brain acquired at a time of rapid brain development' (Hall 1989). It is rarely familial. It used to be thought to be due to asphyxia during delivery, but its incidence is not falling with improved obstetric care (Anonymous 1989) and frequently there is no convincing indication of cause. It may be due to chronic or recurrent hypoxia during fetal life (for instance associated with intra-uterine growth retardation) or to accidents during development of the cerebral vascular tree. Vascular accidents may occur in any developing organ and may account for some congenital malformations, but the developing brain may be particularly vulnerable.

Cerebral palsy may also be caused by vascular or haemorrhagic lesions of the brain around the time of birth, but it is now thought that only about 8 per cent of cases are due to events at birth (Hall 1989). Most babies with cerebral palsy did not have birth asphyxia, and most babies with birth asphyxia do not develop cerebral palsy. More babies with birth asphyxia develop cerebral palsy than those without it, but both cerebral palsy and difficulty in resuscitation can be due to pre-existing brain damage.

Nevertheless, the fact that cerebral palsy is still often called a 'birth injury' promotes the idea that it is due to negligent management of the birth, and encourages parents to sue their obstetrician, and judges to convict. Very large sums can be awarded in compensation. This encourages defensive medical practice and has dramatically increased the cost of medical insurance, particularly for obstetricians. The problem illustrates the high social cost of our ignorance of the causes of congenital disorders, and that we can do little to prevent them unless we know their true causes.

Diagnosis

The average family doctor will only look after one or two children with cerebral palsy. The diagnosis is most often made during the first year of life, because of abnormalities in motor development. Cerebral palsy may affect any combination of head, limbs, or trunk, and may be spastic, athetoid, mixed, or hypotonic in type. In some cases it is accompanied by mental retardation, defective vision, or epilepsy, but severely physically handicapped people with cerebral palsy may be mentally normal. Spastic cerebral palsy, the commonest form, usually presents with delay in motor and often in social development, or with failure to use an arm, or dragging of a leg. There may be increased extensor tone, so that an infant held suspended face-down may arch his back with extension of his neck, giving a false impression of good motor power. However, on pulling the baby up from a lying to a sitting position there will be gross head lag, as flexion is weaker than extension. Increased muscle tone reveals itself at 4–6 months when the baby is held vertically. Normally the hips and knees are flexed, but in diplegic spastic cerebral palsy they are extended and crossed ('scissored'), because of adductor spasm. Scissoring of the legs is particularly striking in spastic diplegia, and can lead to secondary dislocation of the hip. Tendon reflexes are brisk (a difficult sign to evaluate) and primitive reflexes like the Moro persist too long. If spasticity is one-sided, the hand is more severely affected than the foot. It may be kept closed at a time when small bricks are picked up with a pincer grip with the 'good' hand. Alternatively objects may be approached by the whole hemiplegic hand, rather than with the fingers.

In hypotonic cerebral palsy the baby is generally floppy, the striking physical sign being a tendency for the infant to slip through the hands when held. Joints are also abnormally mobile. Ataxia or athetosis usually develop subsequently. The child with athetoid cerebral palsy makes characteristic writhing athetoid movements. Hypotonia is a feature of many other conditions besides mental retardation and cerebral palsy, e.g. neuromuscular disorders, biochemical disturbances such as hypercalcaemia, and vitamin D deficiency.

Accurate diagnosis is important for genetic counselling as well as for treatment. The differential diagnosis includes some conditions with a poor prognosis and a high recurrence risk, such as Duchenne muscular dystrophy (p. 247). If cerebral palsy is confirmed, the family should be referred to a unit that can provide the comprehensive care the children need.

Chromosomal disorders

Table 18.6 summarizes the types, incidence, and clinical features of the commonest chromosomal disorders. They fall into two groups in terms of severity: autosomal aneuploidies and sex chromosome aneuploidies. Their causes, and changes in incidence with changes in parental age distribution, are discussed in Chapters 4 and 8.

Down syndrome

Most children with Down syndrome are identified at or soon after birth by their characteristic physical appearance, including a flat occiput, prominent

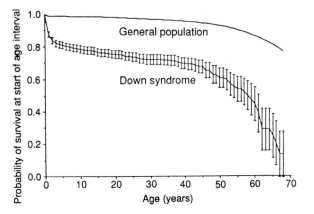

Fig. 18.3 Survival curves for people with Down syndrome, and for the general population, in British Columbia. (Reproduced from Baird and Sadovnick 1988, with permission.)

Table 18.6 Common chromosomal disorders

Chromosomal constitution	Name of syndrome	Birth incidence	Effects
Autosomes			
47 +21	Down syndrome	1/600–1/1000	Severely mentally retarded Reduced survival Malformations common
47 +18	Edward's syndrome	c. 1/3000	Very severely mentally retarded Growth retarded
47 +13	Patau syndrome	c. 1/5000	Multiple malformations Survival usually <1 year
Mosaics	–		Often severely affected
Sex chromosomes			
47 XXY	Klinefelter syndrome	c. 1/700 males	Mild educational handicap; tall, slightly feminized; usually infertile
45 XO	Turner's syndrome	c. 1/2500 females	Short stature; infertile; some congenital malformations
47 XXX		c. 1/8000 females	Mild educational handicap; some are infertile; occasional psychiatric problems
47 XYY			Usually little effect
Mosaics			Partial effects

(Based on Royal College of Physicians 1989)

epicanthic folds, flat bridge of the nose, and malformed ears coupled with hypotonia. About 40 per cent of affected babies have a congenital heart defect, and intestinal obstruction also occurs. But in a few affected neonates the signs are not obvious enough to ring warning bells.

Survival is shown in Fig. 18.3. In the past, around 20 per cent of the children died in the first few years, usually as a result of multiple congenital malformations or pneumonia. There was relatively little help and the stigma associated with mental subnormality in the family was far greater than today, and most severely mentally retarded children were placed in institutions. The attitude being generally pessimistic, care was custodial rather than aimed at the best quality of life. The majority of affected children died of infections before the age of 10, and older patients lived out segregated lives in single-sex institutions. However, survival began to improve in the 1930s, and Baird and Sadovnick (1988) calculate that today 72 per cent of Down syndrome individuals should survive until the age of 33, and 44 per cent will live until they are 60 (compared with 86 per cent of the general population). Two-thirds of these will die between 60 and 68 years of age.

The most severe physical problems are likely to occur in the most profoundly mentally handicapped children. The children often have additional problems such as defects of sight and hearing which may, unless corrected, aggravate their handicap. Hypothyroidism is relatively common, and there

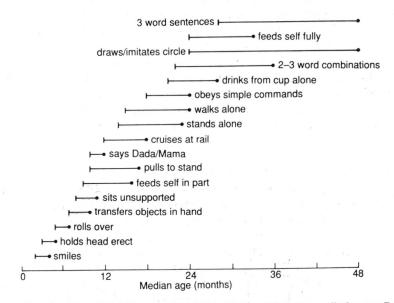

Fig. 18.4 Developmental milestones in Down syndrome. The 50th centile for non-Down syndrome children is shown at the left-hand end of each bar, and the mean for Down syndrome children is shown as a circle at the right hand end of each bar. (Reproduced from Epstein 1989, with permission.)

is an increased risk of acute lymphatic leukaemia. Many Down syndrome individuals develop a form of Alzheimer's dementia at a relatively early age, confirmed by autopsy studies of individuals who have lived to over 40. This observation helped in localizing a gene associated with Alzheimer's disease to chromosome 21 (McKusick 1988).

It is only possible to give a range of likely achievements to the parents of a Down syndrome baby. The IQ is very variable, ranging from below 20 to about 80 with a mean of about 40, and tends to fall further behind with age (Epstein 1989). Figure 18.4 shows a comparison of Down syndrome and normal children in passing developmental milestones. A few Down syndrome individuals are only mildly retarded, some are profoundly handicapped, and most are intermediate. Most have an open and sunny personality. There is some evidence that girls may be a little more advanced in some fields than boys.

General aspects of mental retardation in children are discussed below, and in adolescents in Chapter 19.

Sex chromosome aneuploidies

A sex chromosome abnormality is present in about 0.26 per cent of live-born babies, and they constitute a quarter of the chromosomal abnormalities found at amniocentesis. In general they have much less severe effects than autosomal aneuploidies. Congenital malformations are less common and less severe. The detrimental effect on the development of the brain is not marked though the IQ may be below average — in general, the greater the number of X chromosomes, the more severe the mental retardation. The main effects are on development of the ovaries or testes, and on the development of secondary sexual characteristics. Klinefelter syndrome (XXY) is commoner than Turner's syndrome (45 X) because fetuses with 45 X tend to miscarry. In fact, 5–10 per cent of all spontaneously aborted fetuses are 45 X.

Turner's syndrome (45 X female)   The incidence is about 1 in 2500 at birth. About two-thirds of people with Turner's syndrome are 45 X, the remainder being mosaics (p. 64) who are often less severely affected. The syndrome may be suspected *in utero* because of fetal oedema, particularly around the neck, which can progress to intrauterine death but sometimes resolves spontaneously. It may be suspected at birth if the baby develops oedema of the dorsum of the hands and feet and loose skin folds around the nape of the neck. The birth weight is lower than normal and the height is below the third percentile, the final height usually being between 140 and 150 cm. The clinical features are very variable. They can include a short webbed neck with a low hairline, broad chest with apparently widely spaced nipples, cubitus valgus, short metacarpals or metatarsals, epicanthic folds, high arched palate, and squint. As the girl gets older, pigmented naevi become more prominent. There may be associated internal congenital anomalies such as coarctation of the aorta, ventricular septal defects, and renal anomalies such as a horseshoe kidney. However, some people with Turner's syndrome have few of these features and just present as short adolescents or adults in whom menstruation has failed to occur.

Sexual maturation and puberty fail because the ovaries have been replaced by yellowish-white streaks. Plasma oestrogens are low, the secondary sexual characteristics remain infantile, there is no menarche and the patient is sterile. The ovarian failure is primary and the plasma levels of gonadotrophins are usually markedly raised. Typically the intelligence is within the normal range, though there are reports that a slightly lowered IQ is more common than usual.

Treatment with replacement oestrogen/progesterone hormones allows secondary sexual characteristics including the breasts to develop and the vagina and uterus to attain adult size. Hormone replacement therapy may be delayed until the early or mid teens to minimize the chance of early closure of the epiphyses. Though menstrual withdrawal bleeds will occur, the woman with Turner's syndrome remains sterile. Some may be candidates for ovum donation.

Klinefelter syndrome (XXY male) This condition is surprisingly common, occurring in about 1 in 700 males. There will be one or two such patients on the list of the average family doctor, but some may be undiagnosed.

The condition is difficult to diagnose before puberty as affected boys usually appear normal. Adult males with Klinefelter syndrome are tall, their final average height being 2–5 cm more than normal. The cardinal abnormal physical sign is that the testes are very small (1–2 cm post-pubertal compared with the normal 3.5–4.5 cm). The scrotum may appear normal. Nearly all patients are sterile, with azoospermia, though fertility has been proved in rare instances. The testicular Leydig cells are defective and the plasma testosterone is reduced, with raised gonadotrophins. The degree of development of the secondary sexual characteristics is very variable. Typical features, such as a female distribution of body fat, high-pitched voice and scanty body hair, may not be present, but most affected men have a poor growth of facial hair and need to shave only once or twice a week. Gynaecomastia occurs in about half of them and may be very upsetting. Klinefelter males may be of average intelligence but often have a mild or moderate degree of retardation, and there is a higher than average incidence of disturbed behaviour. Treatment is by replacement therapy with a long acting testosterone. In rare instances, breast surgery may be needed for cosmetic reasons to correct the gynaecomastia. There is an increased incidence of breast cancer in patients with Klinefelter syndrome.

Other disorders   Other sex chromosome disorders include XYY (male) and XXX (female). XYY males are tall and may have severe acne, but are otherwise usually physically and intellectually normal, and fertile, with chromosomally normal children. At one time they were thought to be over-represented in institutions for violent offenders, but this has been disproved. XXX females are often physically normal and fertile, but on average have reduced intelligence. Table 18.7 summarizes the results of one long-term study of mental ability in children with sex chromosome disorders (Jones and Ratcliffe 1990).

When a sex chromosome abnormality is diagnosed prenatally, the parents are presented with a difficult choice. Counselling requires detailed knowledge of the likely range of clinical problems, and in such intermediate conditions is particularly likely to be influenced by the counsellor's personal views. Holmes-Siedle and co-workers (1987) reported on Scandinavian parents' decisions following prenatal detection of a sex chromosome abnormality in 40 out of 7299 amniocenteses, in Scandinavia. About two-thirds of the fetuses were aborted, including all six with 45 X, and five of eight that were 45 X mosaics. However, the rate of abortion for sex chromosome abnormalities is lower in some other countries.

Mental retardation

The IQ, a measure of mental as opposed to physical age, is a rough indicator of a child's capability. It is calculated by dividing the mental age by the real age and multiplying by 100. Thus the general development of a 10 year old with an IQ of 40 is equivalent to that of an average child of 4 years old. Knowledge of the IQ can give parents an approximate assessment of what their child will, or will not, be able to do in the future, but IQ is only one of several factors that influence achievement.

About 0.3–0.4 per cent of the population are severely mentally handicapped and are described as having 'severe learning difficulties'. About one-third of these have Down syndrome and a slightly smaller proportion have fragile X mental retardation (p. 251), but in many cases no obvious cause can be found. Children with an IQ below 20 need nursing care all their lives and have only a minimal development of language. Those with an IQ between 20 and 35 will need full supervision even as adults, but should be able to feed themselves and to keep themselves clean and dressed. Children with an IQ between 35 and 50 should be able to talk and look after their basic needs. With supervision and guidance they may be expected to work in a sheltered environment. Their problem is usually detected in the pre-school years because of markedly slow general development.

About 2.5 per cent of the population are mildly mentally handicapped. Social and environmental factors seem to be particularly important causes of milder mental retardation (Lamont and Dennis 1988). Children with an

Table 18.7 Results of a study of IQ and educational attainment in children with sex chromosome aneuploides

	Chromosomal anomaly			Normal controls	
	XXY	XYY	XXX	Male	Female
Number tested	19	19	16	149	
IQ score (Wechsler) mean	94.5	102.3	85.2		
range	67–131	74–121	64–110		
% requiring speech therapy	47	44	50	11.2	5.4
% requiring remedial reading	77	54	81	18	12
Number skills at 13 years	Reduced	Normal	Reduced		
No. in further education (26 school-leavers)	3	1	2		
Potentially fertile	No	Yes	Yes		

IQ between 50 and 75 are sometimes not noted to be backward until their early school years, although they may have been behind at their developmental checks. They will usually be slow to get clean and dry, are likely to have speech delay and perhaps be late in walking. Such children will usually learn basic reading and writing skills. As adults, some are self-sufficient with modest family or community support, particularly if they have stable personalities. They may get simple jobs in areas where there is little unemployment.

Some metabolic causes of severe mental retardation such as phenyl-ketonuria, congenital hypothyroidism, and neonatal jaundice can be detected in the newborn and successfully treated, and some environmental causes such as maternal infections or alcohol addiction could, in theory, be avoided (Chapter 21). However, genetic causes can be neither prevented nor corrected: at present the only way to avoid them is by prenatal diagnosis and selective abortion.

Diagnosis

Suspicion may first be raised by an abnormal appearance of the newborn, or if an infant is born into a family known to be at risk (e.g for fetal alcohol syndrome, p. 315) or there is evidence of associated problems such as blindness or deafness. Whenever mental retardation is suspected in the neonatal period, it is essential to ask for karyotyping in order to diagnose chromosomal abnormality.

A mentally handicapped baby will be retarded in all developmental areas, but the picture is usually patchy, slowness in social responsiveness being most marked. For example, children with Down syndrome are at their best in the first 6–9 months when motor development dominates the scene. Children who are slow in physical rather than social development may have a neuro-muscular disorder.

The degree of interest and attention the baby pays to his surroundings is the most telling. Thus, a 9 month old retarded baby may sit on his mother's knee inertly, paying only fleeting attention to any toy that he is presented with and making very little spontaneous attempt to pick up the toy himself. He may make a poor range of sounds and the examiner may be concerned that he is deaf because he fails to turn to sounds, or only turns very slowly, when in fact his hearing is normal. His mother may complain that he sleeps excessively or that he is a very 'good' baby, causing no bother because he does not cry as normal babies do to attract attention. All these can be very sinister signals. So can the persistence of a pattern of behaviour beyond an age at which it is considered appropriate. Whereas a normal 9 month old is usually able to chew, the retarded child may have great difficulty with lumpy foods and may still be eating strained products. In the second half of the first year, infants put all objects into their mouths. Mouthing usually stops during the second year but may persist in retarded children. In contrast to the globally retarded child, an intelligent deaf child, for example, may show a lack of responsiveness to speech but normal or increased visual alertness and ability to manipulate.

It is important not to miss the diagnosis when early treatment could make a difference, but as most retardation is mild or moderate, there is no urgency in referring children who appear to be healthy but are slow in passing their developmental milestones, for specialist advice. It is wise to refer the baby for assessment if the parents are concerned, but if the anxiety arises from the doctor alone, reassessment in a few months may be the best plan. It is important to get the health visitor to study family circumstances to see if the child's slowness could be due to poverty of the environment and lack of stimulation.

Key references

Goodman, R. M. and Gorlin, R. J. (1983). *The malformed infant and child*. Oxford University Press.

Modell, M. and Boyd, R. (1989). *Paediatric problems in general practice*, (2nd edn). Oxford University Press.

Hunt, G. M. (1990). Open spina bifida: outcome for a complete cohort treated unselectively and followed into adulthood. *Developmental Medicine and Child Neurology*, **32**, 108–18.

Czeizel, A., Dudas I., Fritz, G., Tecsoi, A., and Bod, M. 1990. Methods and results of the optimal family planning program in Hungary. (In press)

19. The impact of chronic disorders on patient and family

The disorders we are concerned with range from illnesses where a well-managed child or adult is healthy for most of the time (a 'sick/healthy' person) to disorders involving severe physical and/or mental handicap in which affected individuals need constant attention. Despite this range, the patients' families have much in common, and share many of their problems with families of children with acquired disabilities. Table 19.1 summarizes some factors that can affect the impact of a disorder and its management on the family. In this chapter we also consider some relevant ethical issues. In many cases there is no 'right' or 'wrong' approach, and the views we express are our own.

Table 19.1 Factors affecting the impact of chronic disease on the family

Problems

- Age of onset
- Danger of premature death
- Physical handicap
- Mental handicap
- Pain
- Unusual appearance
- Subsidiary handicaps such as of speech, hearing, endocrine disorders etc.
- Problems of pubertal development, sexuality, and reproduction

Problems of treatment

- Type of treatment: drugs, other substances, physiotherapy
- How effective?
- Required daily or periodically?
- How time consuming?
- How burdensome, uncomfortable or painful?
- Done at home, or requires travelling?

Social and psychological factors

- How the news was first given to the parents
- Is prenatal diagnosis available? If so, was it offered to them?
- Social circumstances of the family
- Availability of social and financial support

Facing the diagnosis

The decision to treat

Many congenital disorders are diagnosed at birth, especially when a baby has obvious malformations. Sometimes difficult decisions have to be made rather rapidly about how to manage an infant with a severely handicapping disease, especially whether or not to attempt to prolong life. This is a very difficult area. The parents of a baby with, for example, severe spina bifida must be fully informed about the likely prognosis with or without surgery (p. 261), but the team in charge of the baby's care must be prepared to make the final decision about management if the parents are undecided. This decision must take into account both the interests of the child and those of the rest of the family. It is sometimes acceptable not to try actively to prolong life. It is not acceptable to try actively to shorten life, though on occasions children will die earlier than they would otherwise have done because of side-effects of drugs given to relieve unnecessary suffering.

The parents' decisions are not over-riding. The right of a child to treatment may transcend those of parents to refuse consent. For example, in the UK the child of a Jehovah's witness may be made a Ward of Court if a transfusion is essential and the parents refuse permission, and in an emergency, treatment can be given if the responsible doctor has the written support of a colleague. Occasionally the parents disagree between themselves about whether an affected child should be treated or not: in such cases, in general the mother's views should prevail.

The parents' reactions

Once it is clear that a child definitely has a serious congenital disorder, the news should, if possible, be given to both parents together, emphasizing at the outset that this is a burden that they will have to carry jointly. This reduces the possibilities for misunderstanding between the parents and between them and the doctor because, if the mother does not ask an important question the father may do so, and vice versa. In the case of a single parent, it is helpful for a close relative or friend to come to the early consultations.

The parents' reactions classically undergo a development with time (Lansdown 1980). The usual initial reactions are *shock*, *confusion*, and *disbelief*, and several interviews may be needed before the parents can take in anything more than the most basic information. Initial disbelief may be especially long-lasting when the child has appeared normal in the early months or years, and parents may accept that their child has a severe problem only after they have been embroiled in several cycles of illness and recovery — for example, antibiotics and physiotherapy for cystic fibrosis, or regular blood transfusions for thalassaemia.

Parents need the opportunity to express their *anger*, which is often more marked when the mother had sensed for weeks or months that something was wrong with her child, but was unable to get professionals to accept it. It is difficult for parents to understand that a delayed diagnosis may be inevitable in the absence of a family history. Recurrent respiratory infections are common in small children and in the vast majority of cases are not a sign that the child has cystic fibrosis. The child with Duchenne muscular dystrophy is healthy for the first year or two of life, though there may be a delay in learning to walk: the fact that these children's average IQ is around 80 may add to delay in diagnosis because people adjust to the fact that the child is a bit slow, and the child himself takes longer to perceive and complain that there is something wrong. If the parents have suspected for a long time that something was wrong there may be an element of *relief* when the diagnosis is finally made.

Most parents will naturally become *depressed*, feeling pessimistic about the future and lacking interest in and enjoyment of their usual activities. Though most will emerge from this stage, it is likely to recur on occasion over the subsequent years.

Parents feel a varying amount of *guilt*, which very occasionally has a partial basis in reality. The child may have been born handicapped following a pregnancy in which the mother drank or smoked heavily or was on drugs, or requested or attempted an abortion. The parents may feel they delayed in seeking help for a baby who was failing to thrive because of an occult urinary tract infection, or feel guilty because of the inevitable negative emotions mixed with positive feelings that are aroused by a deformed or sickly infant. They may be ashamed to talk about their anger or hostility towards the child and it can be helpful for the doctor or other counsellor to assume these emotions are there, and to indicate spontaneously that they are normal in the circumstances. The parents' anger and guilt is sometimes directed at health workers, who should accept that it is better for the parents to direct these feelings outwards than to tear themselves to pieces.

Guilt may be complicated by feelings of *shame*, especially if the parents fear, or experience, negative reactions within their family or from society. Improved community education about disability helps to adapt society to the handicapped person, as well as the person to society.

Denial is an important, sometimes helpful but sometimes maladapted defence mechanism, that is often adopted by parents and older adolescent patients, particularly when a condition is intermittent or treatment is demanding. The mother of a child with a congenital disorder who tries to make her child normal by pretending that it is normal, can be devastated when the birth of a healthy sibling unmasks her fantasy, while a deterioration in the child's condition can release overwhelming anxiety. Adolescents and adults with sickle cell disease often lead an essentially normal life for extended periods, and may deny their disease and default from clinic attendance, only to be demoralized when laid low by a painful crisis. Complete denial is less common, because less feasible, among patients whose disease makes regular demands on them, like thalassaemia or cystic fibrosis. But adolescents with any kind of chronic disorder are quite likely to try to dispense with treatment at some stage.

In recent years, the availability of prenatal diagnosis has added an additional factor. Parents who have a child with an avoidable condition, without being forewarned and offered the option of prenatal diagnosis, often find it especially difficult to accept the child's disorder. What would previously have been perceived as a misfortune has been converted into an injustice, and their confidence in the medical system upon which they are now so heavily dependent is seriously shaken. It is extremely important for the future well-being of patients and their families that all couples be informed about such avoidable risks before or during pregnancy. An informed choice against prenatal diagnosis or abortion means that an affected child will be accepted from the very beginning, and the scene is set for a confident working relationship between health workers and the family. Couples who have decided against prenatal diagnosis should be offered neonatal diagnosis, which will relieve their minds quickly if the child is healthy, and save them the distress of delays in diagnosis if it is not.

In most cases, after the initial reactions the parents come to accept the situation with varying degrees of equanimity. They build up psychological defences and adjust their expectations and style of living, though there are likely to be periods of anger and despair. 'The family life of sick children' by Lindy Burton (1975) is an excellent general description of the impact of cystic fibrosis on the family.

Childhood

The personal doctor

Many professionals are involved in the care of children with chronic problems: doctors, social workers, health visitors, physiotherapists, and teachers. It is particularly important for one doctor to be the child's personal physician — the individual who works most closely with the family, who is readily available, and with whom they can build up an open long-term relationship. This may be the family doctor or the hospital specialist, as long as the latter sees the child regularly and does not delegate care to a succession of junior staff. In some instances, the family's main point of reference will be a senior nurse. With children whose chronic illness is subject to acute or life-threatening complications, the family is particularly apt to consider the hospital doctor or the sister on a specialized children's unit as the best source of immediate advice in a crisis. This is reasonable, especially if the family doctor knows little about the child's illness, or feels insecure about dealing with the child when sick because of a complicated management regime.

Nevertheless, the family doctor may be able to act as a bridge between family and hospital staff, especially if doctor and parents have built an easy relationship. This may allow the parents to raise topics that they thought too trivial or embarrassing to discuss on a visit to the out-patient department. The family doctor may explain the family's fears and discontents to the consultant, or encourage them to comply with a difficult but essential treatment regime. Consultants in turn can encourage a practice to take a more active role in managing the child's condition by sending full reports, and by inviting the family doctor or health visitor to attend case conferences. All this can take quite a lot of time and effort. It is easy for family doctors to lose their grip on the care of children with rare diseases and complicated problems who are managed completely by the hospital. Adolescence may also be a difficult time administratively, when it becomes uncertain whether the patient is under the care of a paediatrician or an adult physician. The family doctor can help ensure that patients have a smooth transition from paediatric to adult care.

In the presence of a chronic disorder, other common problems can have a more serious impact on quality of life than they do in healthy children. It is specially important to ensure that these children have the full range of immunizations, and to diagnose e.g. deafness or problems with vision and ensure appropriate treatment as early as possible. On occasion, if the family doctor is seeing the child for, for example, an intercurrent infection, it may be helpful to discuss whether the family is receiving the necessary help. Have services been promised but not actually provided? The types of help available vary considerably between countries. In the UK, with the assistance of a social worker, families can obtain practical help in several areas, though criteria are quite strict and some benefits may be refused for children the doctor considers substantially handicapped. Grants available include Income Support, Family Credit, or payments from the Social Fund for some needy families, and funds from local authorities and voluntary organizations to alter houses and flats so that they are more suitable for the handicapped.

Families need consistent support from their doctor, who should not necessarily feel expected to change the course of the disease. Even when there seems nothing specific to offer, a positive comment on the child's progress or good caring by the parents may be of great value in raising morale.

Advice on managing children with a chronic disorder

(The following remarks apply particularly to children with physical rather than mental disability.)

The key to successful management is continual personal support and the two-way flow of information. The parents, and ultimately the child, should be well-informed about the disease, its management and complications. Children are best able to accept their illness when the parents also accept it without shame or evasion, and honestly explain the situation to them. Patients are less likely to comply with treatment if there is denial on the part of the parents.

Children's questions should be answered honestly at a level appropriate for their age and understanding. This may be more a matter of the doctor's tactful and appropriate response to comments by the child, or to parental remarks made in front of the child, than of a formal educative session. Healthy siblings should not be kept in the dark, but unless they are several years older than the sick child and mature enough to cope, they should not be given more information about the illness than the patient has. Any other course is likely to hinder easy communication between siblings. When it is necessary to have a private discussion with the parents in the consulting situation, it is usually better not to send the child out of the room, because children's fantasies about plans being made in their absence are likely to be worse than the reality. However, if there is a separate examination room it is easy to see the parents alone for a moment whilst an older child is getting undressed; at least this provides an opportunity to arrange a separate consultation with the parents at a later date, if necessary.

If the question arises, it is usually a mistake to deny inaccurately to the child that his life may be shortened. In these days of open discussion, frequent medical television programmes, and self-help groups including families whose children have died of the disease, any other policy would undermine the patient's confidence. However, survival is steadily increasing in many conditions, and it is usually possible to discuss the matter in an optimistic and open way.

While the doctor must not evade helping the patient to accept tragic and unpalatable truths, such information should not be forced on the child (and family). It is better to wait until questions arise. A common mistake is to give untruthful off-the-cuff reassurance when caught suddenly unaware by a remark from the parent or child. If the doctor can think quickly enough, an appropriate response to the question, 'He isn't going to die is he, doctor?' could be, 'Is that what you are afraid of?'

Even if life expectancy is seriously curtailed, the child should be encouraged to lead as normal a life as possible for as long as possible. This includes attending school, doing homework, and sitting examinations: any lack of interest in this part of the child's life conveys the message that there is no future and no hope. Parents need to be encouraged to treat sick children as normally as possible: over-indulgent or over-permissive attitudes can make the child uneasy and insecure. Normal management means normal discipline, even punishment, if appropriate. However, the unpredictable course of many congenital conditions can make it very difficult to keep to rules. How long will a child with muscular dystrophy be able to walk, or stay at school? Will the next chest infection in the child with cystic fibrosis be fatal?

A child will need to be given clear information (usually reassurance) about future sexual function, the possibility of having a family of their own, and the chance (usually small) that their own children will be affected.

Parents should be encouraged to bring new therapeutic suggestions inevitably raised by the media, friends or relatives to the doctor for discussion. In conjunction with the specialist it may be useful to offer them a second opinion, to back up their confidence that the diagnosis is correct and the right course is being followed. Parents often get support and practical advice from meeting other couples with similar problems, so it is usually helpful to put them in touch with the appropriate self-help organization (see Apendix) at an early stage, unless the family does not like the idea. Families who don't attend meetings can still get a lot of help and encouragement from newsletters etc. Many support associations provide clear and well-written leaflets and books not only for parents and health workers, but also for affected children, adolescents and adults, and sometimes for their siblings.

Many affected children require regular medication, special diets, or physiotherapy, and paradoxically as treatment becomes more effective, life often gets harder for everybody concerned. However, if improved treatment offers increased hope of a normal life or longer worthwhile survival, it will be possible to maintain the family's morale. Twenty-five years ago it was technically easy to manage children with cystic fibrosis or thalassaemia, but the disease was fatal within a few years. Though management varies with disease, all regimens have the effect of limiting the family's freedom and may lead to conflict between parents and child, which is often heightened at adolescence. However, sometimes the doctor can hold out hope of lessening dependence on gruelling treatment as skill in managing the disease progresses. For example, if an oral, instead of an injectable iron-chelating agent for treatment of thalassaemia becomes available in the next few years, compliance with treatment is likely to increase.

Adolescence and adulthood

People with handicaps should be encouraged to develop and expand those skills that are not stunted by their disability, in the same way that the child with severe myopia may follow intellectual rather than sporting pursuits. This aim may need explicit discussion with the adolescent. It is also important to update young peoples' knowledge as they get older. The doctor closely involved with the child and the mother over many years can sometimes find it difficult to relate more directly to the teenager as a maturing individual; a problem compounded by the small size and delayed puberty of many adolescents with a chronic disease. Physical problems may also be more complex in this age group; e.g. severe lung and liver disease in CF, or liver, heart, and endocrine damage from iron overload in thalassaemia.

At this stage the 'personal' doctor becomes specially important. Adolescents need help and support in moving from the parents' protection towards independence. Some, when old enough to decide for themselves, even try rejecting life-supporting therapy, and milder problems with compliance are common as the young person tests out the reality of their predicament. Problems of puberty and sexual development and functioning also come to a head, and an endocrinologist's help may be needed to bring about normal sexual development. For severely physically disabled young people, adolescent sexual counselling may need to include a discussion on methods of masturbating, as well as giving an opportunity to express fears such as lack of physical attractiveness.

Adolescent and young adult patients nearly always require help and support from a specialist in their disease. If they are not already attending a specialist centre (see below), one should be identified and the young person referred for an assessment. This will also help them to establish for themselves that they are receiving the most up-to-date management and that their local doctor is in touch. Many adolescents benefit particularly from contact with a peer group of similarly affected people. 'I used to think there was no place in the world for me, but now I feel that I belong'.

Teenagers and adults should be given as much control of their disease as possible. For example, patient-held notes are now available for several conditions. These make it possible for affected people to travel with the confidence that local doctors will be fully informed about their case by referring to their notes, and their treatment protocol will be followed. They also need detailed books specially written for them, including objective descriptions of their disorder, its treatment and possible complications, and the psychological challenges they may face. (Such books incidentally, can also be very useful for health staff working with the patients).

Many problems that have been contained or suppressed within the family are likely to come to a head around adolescence. Many patients develop an impressive and attractive maturity, but they can also show depressive, aggressive, excited, or even bizarre behaviour. Some may be tempted to make a hasty marriage and have children early, in an effort to cram as much life as possible into the limited time they feel is available. The more they feel in control of their disease, the more they can relax and allow themselves to mature, and the more considered their marriage and family decisions are likely to be. However, ethical problems can arise. For example, a sick mother may insist on becoming pregnant even though she knows that the pregnancy may make her illness worse and she could die soon after the birth of her child, who then may have inadequate care in the early years of life.

In theory, well-planned psychological support for the family, started early, should help towards normal psychosocial development and avert many of the crises so often seen in the families of adolescent 'survivors of chronic disease of childhood'. Unfortunately, adequate counselling is rarely available, even at expert centres.

Impact on the family

A chronically sick child will have considerable implications for family stability. According to Lansdown (1980), parents who have been married for five years or more are better able to cope with the situation, and studies have not shown divorce to be commoner amongst parents of handicapped children. It is likely that marriages that are fundamentally sound will adjust satisfactorily, whilst the additional stresses will weaken those that are not working well. General social factors associated with an increased rate of divorce include teenage marriage, lower socio–economic group, over-crowded cramped housing, and marriage forced by pregnancy. Marriages appear more stable if both partners have strong religious convictions. The same factors may be expected to influence a family's ability to cope with chronic congenital disease.

Even the birth of a normal child often causes jealous and regressed behaviour in an older brother or sister. When parents are preoccupied with a handicapped child, they can fail to perceive and respond to the physical and emotional needs of their other children. This is even more likely to happen if there is more than one affected child. Parents who have already lost one child are especially liable to be over-protective, and are naturally more pessimistic about the future of a second child with the same disease. Children younger than the patient may be more vulnerable to such problems, presumably because it is more difficult for them to get the intense parenting they need in their early years. The most marked effect seems to be on children whose siblings have inherited a life threatening condition rather than those whose siblings have a chronic non-fatal disease. In the former case, healthy siblings may need to be firmly reassured that they will not get the same illness as the sick child. Despite the inevitable problems, older brothers and sisters are often very protective, defending a younger affected child if he or she is attacked verbally or physically by others.

Occasionally, family stress becomes intolerable, and the health visitor specially needs to sense when extra help is needed. The birth of a sibling, change of job, or marital tension may all highlight the need to deploy more resources, such as a place in a day nursery, or even temporary fostering or a brief admission to hospital.

There are many material effects on a family. In countries with comprehensive National Health and Social Services, treatment of the disease as such is not likely to cause serious financial hardship, but nevertheless a chronically sick child is likely to limit the family's earning power. The father may be reluctant to move to another part of the country in search of a better job, if that means leaving a trusted and familiar paediatric unit or health centre. The mother may stay at home to look after the sick child instead of finding even part-time work. The family's social life is also likely to be curtailed. Not only may parents be apprehensive about leaving a vulnerable child with someone else, but potential baby-sitters may be reluctant to take on the responsibility. When a disease is progressive, the family's mobility can become increasingly restricted.

In many less developed parts of the world, congenital disorders place a much heavier burden on the family. The family often cannot afford to travel to a hospital even when treatment is free, or take time off from the arduous daily task of earning their living, while genetic advice and possibilities for family planning are also limited (Sangani *et al.* 1990).

Disorders starting in adult life

Pullen (1984) gives a concise account of some of the problems of patients with inherited physical problems starting later in life, using Becker muscular dystrophy as an example. In this X-linked condition, muscular weakness and wasting develop in later childhood or adult life. The main points he makes are that, especially in the absence of a family history, there may be a long period of uncertainty before a diagnosis is made. People outside the family may treat the handicapped person in a patronizing manner, and assume, incorrectly, that he has become mentally, as well as physically, impaired. If the individual only gets the disease as an adult he may eventually be forced to stop work and must often accept that he is no longer the main bread-winner of the family. Social isolation may increase, and further adaptations must be made to the house or flat. An insidious delay in diagnosis applies to other late-onset diseases such as Huntington's disease, in which patients eventually become completely dependent on other people. The spouse gradually recognizes that the dependent affected person is very different from the healthy independent being of the early years of the marriage, and may have difficulty in adjusting to the change. The wife may have to both provide increased support and be the main bread-winner, and may need help to cope with the anger and frustration of the new situation with the accompanying loss of financial security and social contacts.

The evolution of mental handicap: Down syndrome

The main clinical features of Down syndrome have been described in Chapter 19.

Apart from their physical problems, babies with Down syndrome do not differ greatly from others during infancy, as all small children are highly dependent. In childhood their quality of life and the family's ability to cope depend very much on the help available. By the age of five, most Down syndrome children are walking, running, climbing, and feeding themselves with some help in picking up food. They need help, especially with buttons, when dressing and undressing. Though clean and dry during the day they are less reliable at night. Language achievement is variable, but most children will be talking in two- or three-word phrases. Difficulties in communication continue into the teenage years. The current provision of special schooling for children, and of sheltered work or day centres for adults, makes it increasingly possible to keep mentally handicapped family members at home. This in turn increases public exposure to the handicapped, and helps to develop more understanding social attitudes.

In their book *The Adolescent with Down's Syndrome*, Buckley and Sacks (1987) describe ninety Down syndrome teenagers. Many points they make apply equally to teenagers mentally handicapped for other reasons. Only 11 out of the 90 Down syndrome babies had been placed in residential care. Another five were adopted or living in permanent foster homes, mostly because of social problems that prevented parents coping — especially if the mother was unsupported. The general health of the older teenagers seemed not very different from other young people of their age. Whilst over half had to wear glasses, only 12 per cent were reported to have a hearing defect. Forty per cent were below the third centile of height and weight, but a quarter were overweight. A third had a congenital heart defect.

Three of the 90 teenagers had no speech, 10 per cent used phrases of three words or less, and about half used only 'key-word' phrases such as 'me go school'. Less then half the girls and only a fifth of the boys could communicate successfully with strangers. Reading and writing skills were very limited; only a quarter could write more than their own name. Except for some older boys, less than a quarter could go alone to a nearby shop and only slightly more were able to cross the road alone. Those who did go shopping relied on shop assistants to take the correct money.

As they get older, by contrast with normal adolescents they spend more time at home, their interaction with their siblings decreasing. Most contacts are with their parents, and they spend an increasing amount of time watching television and in other passive activities. Clearly, the day centre becomes an increasingly important source of stimulus and social contact.

Few of the teenagers studied had emotional disturbances or markedly antisocial behaviour. Those with most difficulty in communicating tended also to have the most difficult behaviour, probably because of frustration. Families tended to feel most positive about the teenagers who were achieving most, but it is uncertain whether the high achievers produced a positive family response or vice versa. There was a general feeling that these young people had a beneficial effect on the families including their siblings, though a third of the families felt that their social life had been more restricted since the birth of the Down syndrome child.

Finally, on the whole, the young peoples' sexual development and behaviour did not present a serious problem, and sexual activity was limited to masturbation. Pregnancy is rare. Bovicelli and co-workers (1982) collected reports on 30 pregnancies in 26 mothers with Down Syndrome. There were three spontaneous abortions, ten pregnancies resulted in babies with Down syndrome, 11 babies were apparently normal, and in addition to the Down syndrome babies, six others were mentally or physically abnormal. It was thought this could be due to the high incidence of incest in these cases, or to difficulties at birth because of the mothers' small pelvis.

Does society have a responsibility to stop certain couples from having children? This question arises when considering the sterilization of a severely mentally handicapped young woman, as it does when deciding whether to insert an intra-uterine contraceptive device into a schizophrenic woman who is acutely psychotic. The doctor may advise termination of a pregnancy if the pregnant woman is judged incapable of bringing up a child because of severe mental disability. However, it is necessary to maintain a balance between the freedom of the individual and the welfare of the community, and at the moment, in the UK at least, it is legally unclear who can decide in these cases (Bicknell 1989).

Death

Many congenital disorders are compatible with a normal or near normal lifespan, but some are not. In this section we discuss the premature death of a child or young person with a genetic or congenital disease. However, even in conditions where death usually occurs in the second or third decade, advances in treatment often increase life expectancy, justifying an optimistic approach to young people with an inherited fatal condition.

In Western societies, the death of a child or adolescent is more harrowing because it is rare. Friends and neighbours have no experience of similar events, have difficulty in identifying with the bereaved family and are uncertain how to react. In the normal progression of life, parents die before their children; the reverse is considered grossly unjust. When parents discover that their baby has a severe chronic disease, they go through a period of mourning for the loss of a healthy child. If the illness is fatal they have to go through a second bereavement when their son or daughter actually dies. It is distressing to become aware of parents distancing themselves emotionally from a dying child as if they were trying to shield themselves from the pain of the impending loss.

Very young children do not have a mature concept of death. They see it as a separation and are not able to grasp that living or dying are not under the control of the all-powerful adult. However, even children of five or younger are probably able to understand that death is irreversible. They have seen dead flowers or animals and begun to witness second-hand the death and destruction shown nightly on television. A young person's reaction to their own impending premature death will be very complex depending on age, previous quality of life, ability to shield themselves from the reality of their predicament, and the amount of anger and resentment they feel for being cheated out of most of life. Though children with diseases such as cystic fibrosis or Duchenne muscular dystrophy are brought up knowing that their life is likely to be curtailed, the final illness can be very short — an over-whelming infection or acute heart failure. In these cases the family may not have time to be reconciled to the situation.

There are many ways of helping the family. Several issues need to be discussed early on in the terminal phase, including the possibility of hospital admission. This should not be seen as a failure. Sometimes the situation gets out of hand and it becomes impracticable to look after the patient at home. Here we discuss the *home* management of a terminal illness, though many of the same principles will apply if the young person's final weeks or days are spent in hospital. Communication between the primary care team and the family should be made as easy as possible. One doctor should have overall responsibility, and this may be one of the occasions when the family is given the home phone number of their family doctor as a back-up to whatever arrangements are made for night or weekend cover. Other members of the team who might be called in an emergency need to be fully briefed, so that the family does not have to go through the whole distressing story when a different person visits. As the illness draws to its inevitable conclusion, daily or twice daily visiting by the doctor or district nurse will be necessary.

Visits to the home should not be hurried. The doctor may sit by the patient often without saying very much, but allowing time for any member of the family to raise any topic they wish. He can facilitate communication between the family members: the mother may express an anxiety to the doctor as she shows him out of the door saying 'don't tell my husband', but the father may have talked confidentially about the same worry on a previous occasion. Certain subjects may need to be explicity discussed with parents, including reassurance that maximum effort will be made to relieve distressing symptoms, the likely way in which the patient will die, and what should be said to the other children: they should be told the truth, but the amount will depend on the age and maturity of the sibling.

The decision that further active treatment is inappropriate is often very difficult. It may be in the best interests of the child to stop before the family is entirely ready to accept this. On the other hand it may be unavoidable to continue treatment beyond this stage so that parents feel reasonably satisfied that everything possible was done for the child. Death cannot be predicted with accuracy, but an experienced doctor may be able to say that the patient has entered their final few days of life. At the time of the death it may be important to discuss with parents the desirability of a post mortem, par-ticularly if the disease has a likely genetic component.

Contact needs to be continued after the funeral of the young person, so that the family does not feel abandoned, and they have the opportunity to discuss their inevitable questions about various aspects of the management of the illness. During subsequent consultations, perhaps several years later, some parents find it helpful if the doctor mentions the child by name and recapitulates some of their shared experiences.

The role of special treatment centres

Information on whether a particular hospital is able to provide 'dedicated' care for a child or adult with a chronic disorder should be an important consideration for the family doctor in choosing where to refer the patient. The child with a chronic problem, who must use medical services regularly and whose life is therefore particularly susceptible to disruption, needs a specially efficient service but all too often ends up at the back of the queue. For example, thalassaemic patients often come to the ward at 9 a.m. for a blood transfusion which may not actually be started till 6 p.m. because the needs of the other, acutely ill, children take precedence. Such families need to be attended to as a priority. This is one argument in favour of specialist centres.

A critical number of such patients need to be attending a centre to make it worthwhile and efficient to make special provision for them. For instance, when more than 40 thalassaemic patients attend a centre regularly, it is inefficient *not* to set up a special out-patient and day transfusion service. The collective needs of a number of patients with different chronic conditions requiring regular treatment may also be met through a day treatment unit. Some of the pros and cons of care at the local hospital versus at a dedicated centre are summarized in Table 19.2. Treatment close to home has the advantage that help and known staff are easily accessible, with fewer strains on the family. If the nearest expert centre is some distance away, a 'shared care' solution may be arranged in which the patient attends the local hospital for basic care, but visits the expert centre for an annual assessment and advice, in the same way that the care of a patient with complex chronic disorders is often shared between a practice and a hospital unit.

However, even when initial diagnosis and management have been correct and a patient has been appropriately referred, many children with less severe chronic problems such as squint or a urinary tract infection, who should be followed up regularly, are lost track of at some point. The commonest reasons are that they fail to attend as out-patients, are not given a follow-up appointment when they should be, or move house. The practice needs to be alert to this possibility.

Key references

Buckley, S. and Sacks, B. (1987). *The adolescent with Down's syndrome. Life for the teenager and the family*. Published by Portsmouth Down Syndrome Trust,

Table 19.2 Relative merits of general hospitals and special treatment centres for chronic disease

Place of treatment	Advantages	Common disadvantages
Expert centres	• Dedicated efficient service • Continuity of care (may include adolescents and adults) • Most up-to-date practice • Social support (meet other families) • Contact with Support Association • Psychological support • 24 hour direct telephone contact	• Distance: cost of travelling in time and money to the family • Difficulty in dealing with emergencies
General hospital	• Closeness and convenience for family • Rapid access in emergency • More medical and nursing staff gain experience with the condition	• Isolation of family • 'End of queue' mentality • Care often delegated to junior staff • Late introduction of new treatments • Inexperience with complications, especially in adolescent patients • Transfer to adult care may cause problems • Lack of psychological and social support

Psychology Department, Portsmouth Polytechnic, King Charles St, Portsmouth PO1 2ER, UK.

Burton, L. (1975). *The family life of sick children: a study of families coping with chronic childhood disease*. Routledge and Kegan Paul, London. (A detailed study of the family life of children with cystic fibrosis.)

Lansdown, R. (1980). *More than sympathy. The everyday needs of sick and handicapped children and their families*. Tavistock, London.

Part 4

Healthy pregnancy

20. Pregnancy (I): environmental hazards

There are now, as there always have been, popular and often exaggerated ideas about things which benefit or damage the fetus, and unverified 'scientific' concepts can take over where folklore leaves off, leading to fads and fashions; for instance 20 years ago 'uterine decompression' was proposed to enhance a baby's intelligence.

Of course, maternal behaviour does affect the well-being of the fetus. Drugs of addiction, alcoholism, and smoking are all harmful, and attempted suicide e.g. by carbon monoxide or cyanide poisoning can lead to severe mental retardation in the offspring. Misconceptions can also lead to parental guilt and to misguided actions. For example, thousands of pregnancies were terminated unnecessarily in quite distant countries following the nuclear accident at Chernobyl (Trichopoulos *et al*. 1987). Therefore before addressing the factors that we know to be harmful, it is important to point out factors that do *not* seem to damage the fetus.

1. *Maternal nutrition*. In principle, if the mother looks after her own health, the fetus will be all right. Mother and fetus are both resilient, and the placenta tends to protect the fetus at the expense of the mother when necessary. After all, many mothers lose weight because of nausea and vomiting in early pregnancy (though this could contribute to the need for adequate vitamin intake both before and soon after conception). Only severe maternal protein-calorie malnutrition or iron deficiency of an order rarely encountered in developed countries, is known to affect the fetus measurably. In developing countries, malnutrition and frequent infections retard many children's growth, leading to smaller adults. Smaller mothers tend to have smaller babies, so malnutrition can exert effects over several generations. This may be one reason for the 'secular trend' of slowly-increasing stature generally observed as infections are prevented and nutrition improves.
2. *Threatened abortion* may be due to a fetal abnormality (p. 325) and can be associated with intrauterine growth retardation. However, there is still no evidence that threatened abortion does a normal fetus any harm.
3. There appears to be no evidence that failed *induced abortion* harms a surviving fetus, whether the attempt was medical or domestic, and whether surgical or medical means (progestogens, aspirin etc.) were used.

However, a mother who has attempted an abortion is likely to feel guilty towards the baby.

4. There is as yet no evidence of harm from normal background levels of *radiation* (but see p. 321), or radiation from radios, television sets, computers, video screens, or microwave ovens.
5. Most *medicines* taken in normal doses are probably harmless (Czeizel and Racz 1990), but it can be hard to reassure an anxious woman, especially as, simply as a general precaution, most drug sheets advise against taking the drug during pregnancy.
6. There is no evidence that *maternal infections* during pregnancy, other than those discussed in this chapter and listed in Tables 20.1 and 20.2, cause a significant increase in congenital abnormalities.

Anxious women without any obvious additional risk may ask for ultrasound scanning or amniocentesis to exclude abnormality due to one of the above. They need clear information on the known risks in pregnancy, and counselling should include a reminder that the prenatal investigations available exclude only a specific, limited range of abnormalities. However, a normal ultrasound scan can be very reassuring. Cytogeneticists often argue whether maternal anxiety is an acceptable indication for amniocentesis, and some studies have shown an unexpectedly high prevalence of chromosomal abnormalities in the fetuses of anxious mothers, possibly because they were subconsciously aware of some risk factor. In the UK, where fetal karyotyping is rather strictly rationed, unless such women are in a recognized risk group, the test has to be done privately (Ferguson-Smith and Ferguson-Smith 1990). Women who have had a cold or influenza, or taken aspirin, paracetamol, or antibiotics in early pregnancy because of an infection, can become quite anxious: they should be reassured.

Infections

Congenital infection must be considered in the differential diagnosis of a sickly infant who fails to thrive, or has a neonatal infection or evidence of a congenital heart defect, even if the mother was healthy throughout the pregnancy. Congenital infections can include three components, depending on the organism and the stage of pregnancy at which maternal infection occurs.

1. Certain malformations are particularly associated with intrauterine infection in the first half of pregnancy. These grade into more subtle abnormalities often detected only when the child is older.
2. The typical picture of severe neonatal infection includes enlargement of liver and spleen, jaundice, and a petechial or purpuric rash associated with thrombocytopenia. Lesions characteristic for specific infections may also occur.

Table 20.1 Common maternal infections that can cause congenital defects: incidence and preventability

Organism	Route of infection	Incidence of maternal infection (%)	% of infections sub-clinical	% transmitted to fetus	% infected fetuses abnormal	Prevention method	PND possible
Rubella	Respiratory	Varies	40–50	0–90	50–10	Universal immunization	At >21 weeks (may be possible earlier)
CMV	Contact	1	Most	30–50	*ca.* 10	No method	At >21 weeks
Toxoplasmosis	Cat excreta, raw meat	0.5–1	Most	30–50	*ca.* 15	Screen and treat in pregnancy	At >21 weeks

CMV: *ca.* 0.3/1000
Toxoplasmosis: *ca.* 1.5/1000 in France, 0.75/1000 in the UK

Table 20.2 Congenital and neonatal abnormalities following viral infections in pregnancy

Virus	Effect on fetus	Clinical features in the newborn	Comments
Varicella (chicken pox) and herpes zoster (shingles)	Severe neonatal illness is most likely to occur if maternal infection occurs in the 4 days before delivery. Chance of congenital defect following maternal varicella very low, but can occur, causing multiple severe abnormalities	May be no infection in the newborn, or typical chickenpox, shingles or disseminated disease	Severity of the neonatal infection does not appear to be related to the severity of the maternal disease. Sensible to give zoster immune globulin to newborn exposed to varicella
Hepatitis A	Unlikely to cause congenital defects.		
Hepatitis B (HBV)	No evidence that the virus causes congenital abnormalities. If the mother is a chronic carrier or develops the infection especially in later pregnancy, there is a high risk of hepatitis developing in the baby during the first few months of life	May be asymptomatic but can cause death, severe disease, or chronic carrier status	Carrier state is indicated by persistence of the viral surface antigen (HbsAg). Affected infants may develop chronic hepatitis. Combined hyperimmune globulin and hepatitis B vaccine is advised for infants of HbsAg positive mothers. Routine testing in pregnancy is advised for mothers at risk

Influenza	It is at present uncertain whether maternal influenza causes congenital abnormalities. Various surveys give conflicting results		Diagnosis is often difficult, many non-specific viral illnesses are incorrectly diagnosed as 'flu. A very high fever or drugs taken to counteract symptoms may possibly cause congenital abnormalities
Measles	May affect the fetus but less often than rubella. Manson et al. (1960) reported eight congenital abnormalities in 103 infants of women infected during pregnancy, 3 times as many as in controls	Rarely maternal measles late in pregnancy may give rise to a neonatal infection, usually a rash and respiratory symptoms	Some cases of rubella are likely to be mis-diagnosed as measles
Mumps	Studies do not show that maternal mumps causes congenital abnormalities: notably Manson et al. (1960) reported on 501 cases of mumps in pregnancy	Mumps in late pregnancy may rarely lead to neonatal infection	

Table 20.2 Cont'd

Virus	Effect on fetus	Clinical features in the newborn	Comments
Infectious mononucleosis (Epstein–Barr virus)	No evidence that this causes congenital defects. Congenital infection is rare		Many infections in young adults are likely to be asymptomatic
Erythema infectiosum (parvovirus)	Virus infects the fetus in one-third of cases. Miscarriage (usually in second trimester) or still-birth with hydrops fetalis, in 9% of cases (PHLS Working Party 1990). No increase in congenital defects.	Babies that are born appear normal	Fetal hydrops may be detected by ultrasound. Causes rubella-like rash and arthralgia. May be asymptomatic. In child causes bright red flushed cheeks ('slapped face' disease) followed by maculo-papular rash on trunk and limbs. Lace-like appearance on fading

3. Symptoms may develop soon after birth in an apparently healthy baby, e.g. in congenital herpes, many cases of congenital syphilis, or gonorrhoea.

The more prevalent the infection, the higher the proportion of women who became immune during childhood, but also the higher the risk to a non-immune susceptible woman of becoming infected during pregnancy.

Maternal infection with the organisms mentioned below is often asymptomatic, so the only sure way to diagnose them is by screening all pregnant women for antibodies. A recent infection is signalled by the presence of IgM, previous infection by the presence of IgG, and re-infection or reactivation of a latent infection by a rising titre of IgG (p. 81).

Maternal IgG crosses the placenta but IgM does not. The presence of IgG antibodies in fetal or umbilical cord blood may therefore reflect maternal immunity, but IgM antibodies in cord blood are diagnostic of recent fetal infection. Their value in prenatal diagnosis is limited by the fact that the normal fetus starts to produce antibodies only at 20–24 weeks' gestation, and fetal infection can further delay the maturation of the immune system.

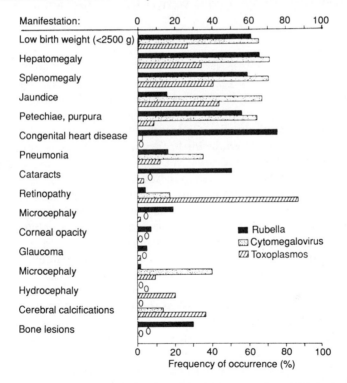

Fig. 20.1 Relative frequency of different clinical manifestations of infection in new-borns with congenital rubella, cytomegalovirus, and toxoplasma infections. (Reproduced from Behrman and Vaughan 1983, with permission.)

Hence prenatal diagnosis of infection based on a fetal immune reaction can usually be done only after 22 weeks of pregnancy (Desmonts *et al.* 1985).

The main maternal infections that can cause congenital abnormality are often known as the 'TORCH complex', standing for *T*oxoplasmosis, *O*ther (= syphilis), *R*ubella, *C*ytomegalovirus, and *H*erpes simplex (Behrman and Vaughan 1983). There is some evidence that varicella or measles may rarely be associated with congenital malformations. Other organisms, such as *Listeria* and the viruses listed in Table 20.2 can cause miscarriage or congenital infection. The evidence about influenza is conflicting. It is rare for immunodeficiency virus (HIV) to damage the fetus: the effects of congenital infection are delayed, and its course is still rather uncertain.

Figure 20.1 summarizes the relative frequency of manifestations in newborns who have been infected with rubella virus, cytomegalovirus, or toxoplasma during fetal life. Many congenital defects are common to all three infections, despite the fact that two of the organisms are viruses while *Toxoplasma* is an intracellular protozoan parasite. Apparently unaffected infants may later develop hearing or sight loss and/or various degrees of mental retardation.

Rubella (German measles)

Rubella is almost completely preventable; but when maternal infection does occur in early pregnancy it is very likely to cause fetal damage (Table 20.3). Cataracts, malformations of the heart, and deafness are the best known complications of fetal rubella. Cataracts may be uni- or bilateral, and although they usually present at birth, they may not become obvious until some weeks later. Pigmented areas may be seen in the retina and this is of diagnostic importance. The commonest cardiac abnormalities are persistent ductus arteriosus, pulmonary artery stenosis, and pulmonary valvular stenosis. Perceptive or sensorineural deafness is the commonest problem, and may occur alone. Hearing loss ranges from a severe disability making normal schooling impossible to a mild loss detectable only by audiometry. Central nervous system symptoms such as abnormal irritability with or without occasional fits and mental retardation can also occur. There are many other manifestations of congenital rubella, some of which may not appear until after birth, including all of those in Figure 20.1.

Congenital rubella affects fetuses at a stage when the immune system is not fully developed (Chapter 7). Thus the virus persists for several months and can be isolated from the throat, nasopharynx, and urine in affected infants, including those without obvious signs of disease. Babies with congenital rubella excrete the virus at birth and are potentially infectious, though to a gradually lessening degree, for about six months or so. IgM in the neonate is evidence of congenital infection, as is the persistance of antibodies beyond

Table 20.3 Risk of congenital rubella defects following confirmed maternal rubella at successive stages of pregnancy (Miller 1990)

Gestation (weeks)	Number followed up	Frequency of infection		Risk of infant with rubella defect (%)
		Number	%	
2–10	20	18	90	90
11–12	12	6	50	34
13–16	36	12	33	17
17–18	15	1	7	3
>19	58	0	–	–

about 6 months, by which time maternal antibodies have usually disappeared from the infant's plasma.

Infants with obvious disease and seemingly healthy babies at risk for congenital rubella need careful follow-up for several years, so that, for example, subtle defects of hearing can be diagnosed as early as possible. It is important to reassure parents early on that this problem will not recur in a future pregnancy.

Before immunization against rubella was available, the incidence of newborns with severe congenital rubella syndrome ranged from very low to as high as 6 per 1000 depending on whether or not there was a rubella epidemic. The triple MMR (measles, mumps, and rubella) vaccine now given in the second year of life produces an antibody response to all three viruses in over 95 per cent of people vaccinated, and rubella antibody remains detectable for over fifteen years. Pregnant women are routinely tested for antibodies to the virus in most Western European countries. Those who are not immune are (or should be) offered immunization soon after delivery. There is no evidence that rubella vaccine inadvertently given during the first trimester of pregnancy is teratogenetic, so though such an event will naturally cause anxiety, there are no scientific grounds for considering termination of pregnancy (DoH 1990).

The report of Smithells *et al.* (1990) on congenital rubella in the UK confirms the effectiveness of the immunization programme. The average annual number of confirmed and expected cases fell from 69 in 1970–74 to 23 in 1985–87, and there was a ten-fold decrease in the number of rubella-associated terminations of pregnancy. Forty-one per cent of affected children had multiple defects and 39 per cent a single one, usually deafness. Only half the mothers had a history of a rubella-like illness, confirming that the fetus is equally likely to be affected whether or not there are maternal symptoms. Seven mothers of affected children had previously been immunized, but it was not possible to say whether there had been re-infection, or failure of immunization. These figures are supported by the Eurocat study, which suggests that the incidence of congenital rubella declined from 3.5 to 0.41 per 100 000 births between 1980 and 1985 in most of Western Europe (de la Mata 1989).

Advice is often sought in early pregnancy by a woman who has a rash or (more commonly) has been in contact with someone, usually a child, who may have rubella. Rubella infection is difficult if not impossible to diagnose from clinical findings alone, as other common virus infections (e.g. parvovirus) produce a similar picture. Even if antenatal testing in a previous pregnancy has suggested that they are immune, such women should be serologically investigated. In the end most will be reassured as, regardless of immunization, more than 80 per cent of most adult populations are already immune.

When rubella is suspected in a pregnant woman, two serum samples should be taken, the first within 2–3 days of the onset of the rash, and the second 8–9 days later, to show whether antibody has developed. Recent infection is indicated by the presence of IgM rubella antibody, which persists for only about 4 weeks. Occasional false positive IgM assays do occur, so the test should be repeated before a confident diagnosis of rubella infection is made. The M-antibody capture radio-immune assay (MACRIA) is a more up to date test of recent infection. If these tests confirm the diagnosis, the risk of congenital rubella should be explained so that the woman or couple can decide whether the pregnancy should be terminated.

Most women who have been infected with rubella virus in early pregnancy opt for termination, but it would be desirable to be able to discriminate between affected and unaffected fetuses. Definitive prenatal diagnosis is possible by measuring IgM antibodies in fetal blood taken after 21 weeks' gestation, but most couples prefer an early abortion to a late prenatal diagnosis. Rubella virus has also been detected by DNA studies of chorionic villus samples obtained in the first trimester (Terry *et al*. 1986), but this approach to early diagnosis is still experimental. If the request is made, it is advisable to contact an expert in fetal medicine.

The value of giving immunoglobulin to pregnant women who have been exposed to rubella is uncertain: Butler and co-workers (1965) found that 20 of 70 women given immunoglobulin at varying stages before and after they developed a rubella rash had an infant with congenital rubella. However, for women who would not countenance a termination of pregnancy, 1500 mg of normal immunoglobulin may be given after taking a serum sample. If the test is negative, another injection of 1500 mg (or a smaller amount of hyperimmune globulin) may be given, and another blood sample collected three or four weeks later (Hanshaw *et al*. 1985).

Cytomegalovirus (CMV)

CMV is one of the herpes group of viruses. As with herpes, infection is very common. The proportion of the population that are immune increases with age, varies very considerably from country to country, and tends to be higher in people in lower socio-economic groups. The virus is transmitted by close contact, and in adults infection may be related to sexual activity. Virus has been isolated from the cervix as well as from saliva and urine. CMV infection can also be acquired from blood transfusions and transplanted organs.

Overt CMV infection resembles glandular fever, with lymphadenopathy, hepatosplenomegaly and fever, atypical lymphocytosis, and abnormal liver function tests, but the heterophile antibody (Paul Bunnell) test is negative and the sore throat usually less severe. Overt infection is particularly common in immunosuppressed people. However, most infections in adults are asymptomatic, and the virus may persist in many tissues in a latent form. A CMV infection can be primary, secondary (the presence of antibody does not necessarily prevent re-infection), or due to reactivation of latent virus.

Table 20.1 shows that up to 1 per cent of non-immune pregnant women acquire a CMV infection during pregnancy, but the risk of congenital infection is not as clear-cut as with rubella. In the prospective study of Stagno and co-workers in the United States (1982), serological studies showed that 17 of 1203 non-immune women (1.5 per cent) developed a primary infection during pregnancy, but only one presented with symptoms. Eight transmitted the virus to the fetus, as shown by the presence of IgM in cord blood and/or excretion of virus in the newborn's urine. Re-infections among the 2330 mothers who were already immune could not be diagnosed as they caused no specific symptoms and antibody was already present. However, the fact that 20 of these mothers (1 per cent) passed the virus on to their infants suggests that maternal re-infection is common. These 28 affected infants were added to five that had been diagnosed retrospectively in a separate series. Only five of the 33 babies had clinically detectable disease. As the mothers of all five had had primary infections, the authors concluded that congenital disease is commoner and more severe following primary maternal infection.

Cytomegalovirus infection appears to have become the commonest viral cause of mental retardation (Best and Banalvala 1990). Severely affected infants have the general signs of congenital infection (p. 298). Additional characteristics are abnormalities of the eye such as chorioretinitis and squint, areas of peri-ventricular cerebral calcification visible on X-ray or ultrasound examination, and deafness. Congenital malformations of the heart, kidneys, musculoskeletal and gastrointestinal systems can occur. A neonate with obvious CNS abnormalities tends to have a bad prognosis. Congenitally infected infants with few or no symptoms may develop psychological and neurological problems later, but may also develop apparently completely normally.

There is no effective treatment for CMV infection of mother or infant, though various antiviral drugs and hyperimmune globulin have been tried with uncertain results. Possibly the only effective approach will be prevention by vaccinating non-immune young women prior to pregnancy, but the efficacy and safety of the vaccine will first have to be demonstrated.

When CMV infection is suspected in a pregnant woman, a definitive diagnosis should be made either by culturing the virus from the urine, or by demonstrating a rising antibody titre to CMV. When the result is positive, the parents should be informed of the implications. Counselling can be quite complicated. Few couples choose to terminate the pregnancy unless the fetus can be shown to be severely affected. Prenatal diagnosis is possible by combined ultrasound examination (for intracranial calcification) and fetal blood sampling to measure IgM, but only after 22 weeks of pregnancy (Daffos *et al.* 1985).

Toxoplasmosis

Toxoplasma gondii is a protozoan intracellular parasite found in many animals, but sexual reproduction and spore production occurs only in the intestine of the cat. It is thought to be transmitted to humans (and other animals) through contact with cat faeces, or by eating under-cooked meat. In adults, overt infection may mimic glandular fever with particularly marked lymphadenopathy, but most infections in pregnancy are asymptomatic. Once acquired, the organism encapsulates in various tissues, and usually remains latent. If reactivated it can damage the eye and lead to blindness. Reactivation is most common with congenital infection but can also occur in immunosuppressed people, as in AIDS or after tissue or organ transplantation. Treatment with spiramycin kills the organism and prevents reactivation or fetal infection. The drug does not, however, cross the placenta, and so cannot cure fetal infection once it has occurred.

The picture of congenital toxoplasmosis is very variable. At one extreme is fetal or neonatal death, or the neonate develops the typical signs of a generalized infection, usually with choroidoretinitis. Other infants appear normal at birth but may later develop chorioretinitis, hydrocephalus, and intracerebral calcification. A mildly affected child may have normal intelligence but poor sight because of retinal damage. It appears that on average one-quarter of infected infants have symptoms leading to diagnosis in the newborn period, which should lead to successful treatment with pyrimethamine, sulphonamides, and folic acid. Spiramycin has also been recommended. However in asymptomatic (and therefore untreated) affected infants, the infection may progress, leading to later damage of the eye or brain. Infected infants need to be carefully followed up over a long period.

Table 20.1 summarizes data on the infection rate, which is highest in warmer countries, and on the risk of congenital infection. The most informative studies on the natural history, treatment, and prevention of toxoplasmosis come from France, where about 1 per cent of pregnant women become infected (Behrman and Vaughan 1983) and screening is now universal. By contrast, in the UK, about 0.5 per cent of women become infected during pregnancy, and screening is not routine. Table 20.4 shows that in France prior to universal screening, about 36 per cent of infected mothers transmitted the infection to the fetus, with serious consequences in 15 per cent of cases (still-birth, neonatal death, severe or mild symptoms). Infection in the first or second trimester is most serious (Desmonts and Couvreur 1974). Koppe and co-workers (1986) followed up 11 infected but untreated infants diagnosed in the 1960s, including seven who had been asymptomatic for the first five years. Infection had been reactivated in most during later childhood, so that by 20 years of age, nine of the 11 had chorioretinitis, and five had severely impaired vision in one or both eyes.

Once screening was started in France, almost all infected mothers were

Table 20.4 Congenital toxoplasmosis in France: outcome of infection before and after screening and introduction of maternal treatment

		Outcome				
	Number of infected mothers observed	Still-birth or neonatal death	Congenitally infected child			Uninfected child
			Severe	Mild	'Normal'	
Before maternal screening and treatment	171	6 (3.5%)	9 (5.2%)	11 (6.4%)	35 (20%)	110 (64%)
			Total = 26 (15%)			
Since routine screening and maternal treatment	468	13 (2.7%)	Total = 13 (2.7%)		3 (0.6%)	439 (94%)
			Total = 26 (5.5%)			
With prenatal diagnosis						98%

(Based on Desmonts *et al.* 1974, 1985)

diagnosed and treated with spiramycin daily until term. The proportion of infants with severe effects fell to about 5 per cent, asymptomatic infections almost disappeared, and the proportion of unaffected infants rose from 64 to 94 per cent (Desmonts *et al.* 1985). But screening means that all infected women become aware of the risk to their fetus, and despite treatment and the greatly improved outcome, some request termination of pregnancy. Prenatal diagnosis is difficult, but can be achieved by a combination of fetal blood sampling and ultrasound scanning in the mid-trimester. Selective abortion reduced the number both of affected infants born, and of abortions of unaffected fetuses, practically to zero. In spite of these results there is still uncertainty about how effective spiramycin is in reducing the rate of fetal infection (Anonymous 1990*a*).

The role of the primary care worker

This differs in the three conditions described above. In Rubella it is to ensure immunization prior to pregnancy and to cope with the large numbers of women in early pregnancy who present with anxieties about Rubella contact.

It is important to have a high index of suspicion for CMV and toxoplasma if a woman has a fever during pregnancy, especially if it is accompanied by a glandular-fever type illness. The risk of toxoplasmosis may be reduced if pregnant women cook meat carefully and wash their hands after handling raw meat. Those with pet cats should also cook the cat's meat well (tins seem to be safe) and wear gloves when cleaning up cat litter. The effectiveness of such advice has not been demonstrated. Another possibility might be to immunize pet cats against toxoplasma. There is a case for offering serological screening for toxoplasmosis to women before pregnancy, and advising those who are not immune to be particularly careful. About 20 per cent of pregnant women in London have evidence of past toxoplasma infection, compared with about three-quarters of Parisians.

Syphilis

Syphilis is a sexually transmitted infection with the bacterium *Treponema pallidum*. In females, primary syphilis is often asymptomatic. Natural immunity does not eliminate the organism, so a mother once infected can transmit infection in every subsequent pregnancy. Mothers with primary or secondary syphilis are most infective for the fetus, and syphilis is transmitted more often to the fetus towards the end of pregnancy. There is a considerable increase in the chance of a still-birth or neonatal death. Twenty per cent of the infants that survive—though usually apparently normal at birth,— develop congenital infection. Treatment of the mother with penicillin before the fetus is damaged eliminates the risk of congenital syphilis, though there may still be immunological evidence of intra-uterine infection. Infants with early congenital syphilis may have some of the symptoms shown in Figure 20.1 (failure to thrive, hepatosplenomegaly, anaemia, and haemorrhages into the skin and other organs). In addition there are lesions of the skin and mucous membranes which may be infectious. Bone is often involved, with painful osteochondritis and periostitis. Infected babies are treated with penicillin. Untreated infants may develop all the severe consequences of active syphilis in childhood. All infants born to mothers with treated or untreated syphilis need to be carefully monitored at regular intervals until the serological tests for treponema antibodies become negative, which may take some months.

Screening for anti-treponema antibodies is routine in antenatal care and congenital syphilis is now extremely rare. It is easy to forget its importance, and to take for granted the screening that has abolished it.

Herpes simplex

Herpes simplex is highly contagious and is one of the commonest of human infections. More than 60 per cent of most populations have been infected before 5–6 years of age. There are two main types of Herpes simplex virus (HSV). HSV1 is found in many epithelia, HSV2 is responsible for genital herpes infection. Most primary infections are asymptomatic, but in infants can cause painful vesicular infection of the mouth and lips. There is a prompt and effective immune response, but immunity to HSV1 does not confer immunity to HSV2, nor vice versa. However, the virus is not eradicated but settles intracellularly in a latent form, and like other herpes viruses (e.g. CMV) may be reactivated. HSV1 favours ganglion cells of the spinal nerves including the facial nerve, but may settle elsewhere. HSV2 remains latent in epithelial cells of the genital tract. The commonest types of reactivation are localized recurrent cold sores (HSV-1) and genital lesions (HSV-2).

Infection of the fetus is very rare but can occur, causing abortion or congenital abnormalities, but the main danger is neonatal infection. If the mother acquires a primary (usually genital) infection in the month or so before delivery, her baby will not be protected by maternal IgG, and is highly likely to be infected with HSV2 during delivery. HSV-1 neonatal illness may also occur if the baby catches it for example from the mother (with a primary infection), or from a nurse with a cold sore. Neonatal herpes can take various forms, generalized disease being very severe and often fatal (Fig. 20.2) (Hanshaw *et al.* 1985). Clinical diagnosis is easiest if there is a vesicular rash from which the virus is usually cultivated. The virus may be isolated from other body fluids. Antibody tests may also be useful. The presence of IgM indicates fetal viral infection. Acyclovir and vidarabine have been used with some success to treat the infected neonate.

Caesarean section is recommended if the mother develops a herpetic genital infection at term. However, what should be done for pregnant

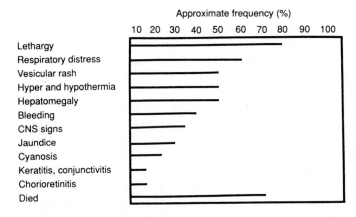

Fig. 20.2 Relative frequency of clinical manifestations of herpes simplex virus infection in the newborn. (Reproduced from Hanshaw *et al.* 1985, with permission.)

women with a past history of genital herpes? The suggestion that they should be monitored in the last weeks of pregnancy, and a Caesarean section recommended if the virus is cultivated from the genital tract is not feasible because infection is too common. Other suggestions include giving such women acyclovir in the final weeks of a pregnancy, or giving the newborn baby a prophylactic course of acyclovir.

Human immunodeficiency virus (HIV) and AIDS

Women may become HIV positive through sexual intercourse with an infected partner, through intravenous drug use, or from contaminated blood or blood products. There is no evidence that asymptomatic infection affects the course or outcome of the pregnancy. The virus is transmitted more frequently to the fetus when the disease is more advanced in the mother, but by no means all babies with HIV positive mothers get AIDS. When a mother is HIV positive, maternal antibody is present in the circulation of the newborn and can persist for up to 16 months. This complicates the diagnosis of infection in the baby. In a prospective European Collaborative Study (1991), only about 13 per cent of 372 infants followed up for more than 18 months had evidence of congenital infection, a rather more optimistic figure than the approximately 30 per cent reported in earlier studies (e.g. Blanche *et al.* 1989).

Infected infants did not differ from the remainder at birth with respect to weight, length, head circumference, or frequency of malformations. By 6 months about half the infected infants had developed clinical or immunological evidence of infection, about 30 per cent had developed AIDS, and almost 20 per cent had died of a condition related to the infection in the first

12 months. The condition progressed more slowly in the second year, but by 16 months over 80 per cent of the infected infants had developed some infection-related signs and symptoms or immunological changes. A small group remained clinically healthy after 2 years of age. It is still uncertain what the outcome would be if all infected infants were diagnosed at birth and treated from the outset.

The clinical picture of HIV infection in infants is similar to that in older age groups. It includes lymphadenopathy and hepatosplenomegaly, parotitis, persistent oral thrush (the most consistent early signs), eczema failure to thrive, pneumonias, diarrhoea, and recurrent fevers. Developmental delay may also occur.

It is not clear what advice to give to HIV positive women detected early in pregnancy, or that they will follow it once given. In Edinburgh, where most infections had been acquired through intravenous drug use by the woman or her partner, informed HIV positive women had no more terminations of pregnancy than women with a similar life-style who were not positive (Johnstone *et al*. 1990). However, the women's decisions may have been influenced by the fact that at the time few overt problems had yet arisen among infected infants in the area.

Anonymous testing for HIV infection in pregnant women attending antenatal clinics, using blood left over from other tests, has been established in some areas in an attempt to monitor the spread of the disease (Anonymous 1990*b*). This raises numerous ethical issues of which perhaps the most important is the inability to trace a positive sample back to the woman it originally came from. Each diagnosis of neonatal HIV infection is important because of the possible benefit of antiviral treatment. Present opinion is that provided patient consent is obtained for the original test, it is reasonable to use any remaining blood for HIV testing. However, it is desirable for women in antenatal clinics to be informed by posters or leaflets that testing is taking place and to have the opportunity to 'opt out' if they want. Counselling should also be available for women who may wish to have an individual test.

As maternal antibody crosses the placenta, neonatal HIV positivity is an indirect indicator of infection in the mother. In a recently published pilot study from South-East England using Guthrie blood spots, the prevalence of anti-HIV antibody in newborn serum ranged from 0.49 per 1000 in inner London to 0.04 per 1000 outside London (Peckham *et al*. 1990). Unfortunately the incidence of HIV positive pregnant women in increasing and has reached 1:250 in some inner London Districts.

Listeriosis

This is a very common infection of many animals including shell fish, domestic and game fowl, sheep, cattle, and flies. It is caused by a widespread intracellular parasite, *Listeria monocytogenes*, a small Gram positive bacillus.

Pregnant women seem to be particularly susceptible to Listeriosis. It usually causes only a mild influenza-like illness; meningitis is very rare. World-wide about a third of severely affected patients die (Jones 1990). If a mother is infected early in pregnancy, transplacental spread to the fetus can cause abortion or still-birth. If infection occurs late in pregnancy, the baby may be acutely ill and die soon after birth with pneumonia, septicaemia, and widespread abscesses. Even with antibiotic treatment, mortality is around 50 per cent. The baby may also aspirate the bacillus at the time of delivery. In this case the illness starts after the first week of life, when the infant develops signs of septicaemia, and particularly meningitis. The mortality rate is about 25 per cent (Jones 1990). The antibiotics of choice are ampicillin or amoxycillin, perhaps with an aminoglycoside.

If listeriosis is suspected, various body fluids such as blood, urine, CSF, and swabs from the mother's vagina and cervix, should be sent for culture. The possibility of listeriosis should be specifically mentioned on the laboratory request form, as *L. monocytogenes* can be confused with other types of bacteria. Serological diagnosis (by measuring antibodies in the blood) does not appear to be very reliable.

Pregnant women are advised to avoid food that may be contaminated with *Listeria*, in order to prevent infection. These include under-cooked poultry or cooked chilled meals, pâté and soft cheese.

Information on other relevant infections in pregnancy is given in Table 20.2.

Alcohol and smoking

About a third of the adult population smoke and more drink alcohol, at least socially. It is well known that cigarettes and alcohol have a deleterious effect on the fetus, and mothers-to-be are advised to reduce both to a minimum before embarking on a pregnancy rather than waiting until pregnancy is confirmed. It is not yet possible definitely to identify a safe limit of drinking and smoking during pregnancy. Women who consume small amounts of alcohol and smoke the occasional cigarette usually have perfectly healthy babies. The common combination of excess alcohol with smoking is probably more potentially damaging to the fetus.

Alcohol

The fetal alcohol syndrome (FAS), summarized in Table 20.5 was first described in 1973. There is good evidence that 30 ml or more of ethanol daily (the amount in two and a half glasses of wine or two pints of beer) affects the fetus to some degree (Fig. 20.3) (Vitez *et al.* 1984). The classical fetal

Table 20.5   Signs and symptoms of fetal alcohol syndrome (FAS)

Mental retardation

FAS is an important cause of identifiable mental handicap ranging from mild to very severe. Other CNS symptoms include tremulousness, hyperactivity, and language and speech disorders

Reduction in height, weight, and head circumference

Below the third centile for severely affected, below the tenth centile for less severely affected

Facial anomalies

Include general under-development of the mid-facial area; close-set eyes: thin upper lips; upturned nose with flat bridge; epicanthic folds; low set ears

Various cardiovascular, renal, and orthopaedic abnormalities

There can be minor abnormalities affecting other system; it is difficult to prove that they are related to maternal alcohol consumption

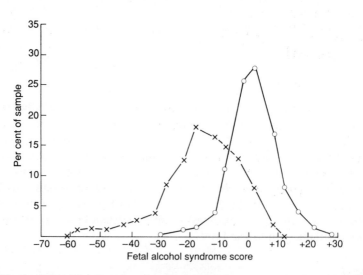

Fig. 20.3 Summary of the results of a controlled study of the children of 301 women treated for alcoholism (30 ml of ethanol per day or more) in Budapest in 1977–9. × = Children of alcoholic mothers; ○ = Controls. All children were carefully examined and scored for the characteristics listed in Table 20.5. Features such as IQ were given particular importance (e.g. IQ < 50 = −10). The aggregate score is shown. Control group of 464 children matched for age and sex. (Data from Vitez *et al.* 1984.)

alcohol syndrome represents the most severe damage. It affects on average about 1.5 per 1000 children, many more in some deprived populations, and may be one of the commoner causes of (mild to moderate) mental retardation and neurological abnormality in the western world (Jones *et al.* 1974, Burd and Martsolf 1989). Breast feeding by an alcoholic mother continues the damage (Little *et al.* 1989), which is also likely to be compounded by environmental deprivation. Parents have a higher chance of dying of alcohol-related causes, and there is an increased chance that affected children will be taken into institutional care.

Burd and Martsolf (1989) report that in one study, none of five women who stopped drinking in the first trimester of pregnancy had a child with fetal alcohol syndrome, but two of seven women who stopped drinking in mid-pregnancy and five out of nine who drank throughout pregnancy had affected children. The authors remark that 'there are few syndromes that have as much potential for prevention as does FAS' and recommend comprehensive team management focusing on identifying women at risk, intervention prior to, during, and after pregnancy, and treatment of affected children. They comment however, that 'one's optimism must be sharply tempered by the long history of society's problems with alcohol'.

In primary care it is not common for an infant to present with full-blown fetal alcohol syndrome. It is more usual to see a child who is small, somewhat slow mentally and physically and comes from a family where excess drinking is suspected. In these cases it is difficult to prove a direct relationship between maternal alcohol consumption and the child's stunted developmental progress, and to separate congenital from environmental factors. An added difficulty in assessing the place of alcohol in infant morbidity is that people tend to under-report the amount they drink. The effect of alcohol is likely to be continuous, from the picture described above to the symptoms in Table 20.5. It will be necessary to point out to parents of an affected child that a high recurrence rate of FAS is to be expected.

Smoking in pregnancy

Several studies in the 1970s indicated that smoking in pregnancy was a risky business, the risk increasing with the number of cigarettes smoked.

Smoking increases the chance of miscarriage, bleeding in pregnancy, and premature birth, with an increase in the perinatal death rate (Meyer and Tonascia 1977). On average, smokers' babies weigh 200 g less than those born to mothers who did not smoke in pregnancy. Passive smoking can also cause fetal growth retardation. Perhaps 200 g is not of much importance for a healthy full-term infant, but it can significantly decrease a premature baby's chance of survival. Although tobacco smoke contains various compounds that can cause fetal hypoxia, there is no evidence that maternal smoking significantly increases the risk of congenital malformations. However, one study suggests that it may increase the child's chance of developing cancer, especially acute lymphoblastic leukaemia (Stjernfeldt *et al.* 1986). Women who smoke in pregnancy often have partners who smoke, and usually continue smoking themselves after the baby is born. Thus either a possible mutagenic effect of paternal smoking on sperm or passive smoking may increase the risk of childhood leukaemia.

There is some evidence from a cohort study of children born in 1958, that the educational attainment of young adults is somewhat less if their mothers smoked during pregnancy (Fogelman and Manor 1988). It has been suggested that letting pregnant women see an ultrasound scan of the fetus can help them give up smoking (Waldenstrom *et al.* 1988).

Maternal metabolic disease: diabetes

Some women develop abnormal glucose tolerance during pregnancy (diabetes of pregnancy) which usually improves after delivery. Their infants do not have a detectable increase in congenital abnormalities.

The majority of diabetic pregnant women will have type I (insulin-dependent) diabetes. Some older mothers may have type II (non-insulin-dependent or maturity onset) diabetes, and this is not associated with increased risk of fetal abnormality. By contrast, in type I diabetics there is an increased risk of pregnancy ending in a still-birth or a sick and often overweight infant. There is also a three-fold increase (to about 6 per cent) in the incidence of congenital malformations in the offspring of diabetic mothers, with a high proportion of fatal and multiple malformations, the risk being highest for mothers with diabetic vascular complications. The pattern of abnormality is non-specific, except for a few rare malformations such as sacral agenesis. Olofsson and co-workers (1984) found that 16 out of 237 infants of diabetic mothers had congenital malformations, mainly of the heart and great vessels. Only a third of these survived infancy.

Meticulous control of the blood sugar throughout pregnancy minimizes maternal and fetal risks in the neonatal period, and reduces the incidence of congenital malformations. In one study of 142 pregnancies in women with type I diabetes, the congenital malformation rate ranged from 6 out of 17 pregnancies in women whose Hb A_1 was 10 per cent or more, to 3 out of 63 pregnancies where the initial Hb A_1 was below 8 per cent (Ylinen *et al.* 1984). The malformations associated with maternal insulin-dependent diabetes arise very early in pregnancy, so women intending to conceive will need to ensure that they enter pregnancy with their blood sugar very well controlled. Diabetic women are usually referred for expert mid-trimester fetal anomaly scanning.

As the morbidity and mortality in the fetus and newborn are related to the standard of clinical care of the mother, the family doctor should refer pregnant diabetic patients to an expert centre, if one is accessible, and emphasize to their younger female diabetics that good long-term control of blood glucose is related not only to their own health, but also to the health of their future children.

The effect of maternal phenylketonuria on the fetus has been discussed on p. 219.

Drugs in pregnancy

Family doctors and other health workers are often consulted by women who are anxious about the effects of medicines taken before they were aware that they were pregnant. Fortunately, relatively few drugs have been involved as causes of congenital malformations: for a list see Table 20.6. Among the most important are anti-epileptic drugs, which cannot be discontinued during pregnancy (Brodie 1989). The risk to the fetus involves the interaction of maternal genetic predisposition (ability to detoxify the anticonvulsant) with environmental exposure (Holtzman 1989). Mothers on these drugs should be offered expert fetal anomaly scanning in the second trimester of pregnancy.

A comprehensive, regularly up-dated table of the effects of drugs taken in different stages of pregnancy and lactation is included in the British National Formulary, which offers the sensible advice that 'Drugs should be prescribed in pregnancy only if the expected benefit to the mother is thought to be greater than the risk to the fetus, and all drugs should be avoided if possible during the first trimester. Drugs which have been extensively used in pregnancy and appear to be usually safe should be prescribed in preference to new or untried drugs; and the smallest effective dose should be used.' 'Drugs and pregnancy' by Hawkins (1987) is a useful small reference book for those seeking further information.

Vitamin A

An excessive intake (more than 10 000 IU per day) of vitamin A or related compounds immediately before or during early pregnancy increases the risk of congenital abnormality, typically causing facial dysmorphology, eye, ear, and palatal defects. Liver normally contains a relatively large amount of vitamin A in addition to many other valuable nutrients, and modern feeding methods for farm animals increase the concentration of the vitamin. Nelson (1990) suggests that, for the time being, pregnant women should not eat more than 50 g of liver a week or 100 g of liver sausage or paté.

The vitamin A analogue isoretinoin used in moderate to severe acne is also a teratogen, and should never be given to a woman who may become pregnant whilst taking it.

Table 20.6 Teratogenic effects of commonly used agents

Agent	Effect
Anti-convulsants	Incidence of congenital malformations in children born to epileptic mothers is about 6%. This appears to be largely due to teratogenic effects of anti-convulsants. Combining drugs increases the risk. Incidence of cleft lip and palate and cardiac anomalies rises to 18/1000 live births
Sodium valproate	Increases risk of neural tube defect to about 1/1000 pregnancies
Lithium carbonate	Increase especially in cardiovascular abnormality
Warfarin	Various congenital malformations including abnormalities of the CNS, and the nose, and bony epiphyses
Corticosteroids	No clear evidence of teratogenesis. Possible very small risk of cleft palate and intra-uterine growth retardation
Live vaccines (rubella, polio, measles, mumps)	Risk of congenital abnormalities is only theoretical but these vaccines are best avoided
Narcotics (e.g. heroin)	Incidence of congenital malformations in drug addicts' babies is at the upper limit of normal, at 2.7–3.2% (Hawkins, 1987)
Trimethoprim and pyrimethamine	Possible teratogenesis risk from interferance with folate metabolism. Folic acid supplements suggested
Oral contraceptives	Very small risk (1/500–1/2000) of fetal abnormalities affecting mainly males. May not apply to current preparations

This list does not include drugs as penicillin, erythromycin, commonly used antihistamines, paracetamol, magnesium trisilicate mixture, common laxatives, aspirin, tricyclic antidepressants, and salbutamol (British National Formulary 1991).
  This last three of these should be avoided or used with caution during the third trimester of pregnancy.

Radioactivity

Radioactivity is an intrinsic component of the natural world, and of living matter. The internal heat of the earth is generated from the decay of radioactive elements, mainly uranium, thorium and potassium, into more stable forms. Many elements (nuclides) exist as a set of isotopes containing the same number of positively charged protons but different numbers of neutrons, and so having different molecular weights. For instance, potassium, one of the most abundant ions in living cells, exists as ^{39}K, ^{40}K, and ^{41}K. Some isotopes are radioactive, i.e. they are unstable and break down to other elements, while emitting particles of energy.

Radioactive particles may be α- β- or γ-particles, which have different energies and travel at different speeds. When radioactive particles pass through living tissue, some are absorbed (i.e. trapped by the nucleus of an atom), the proportion absorbed depending on the speed and energy of the particles. Absorption results in loss of electrons and 'ionization' of the atom, which in turn can lead to chemical reactions. These can cause two types of genetic effect, either point mutations in DNA, or, with higher doses, chromosomal aneuploidy. Such mutations can lead to cancer as well as to damage to an embryo.

The background level of natural radiation differs with area, and Figure 20.4 summarizes the proportion due to different factors in modern life. The millisievert (mSV), one thousandth of a sievert, is the most commonly used measure of the energy deposited in the various tissues of the body by ionizing radiation. The higher the dose of radiation, the more mutations will be observed. Also the period of time over which a given dose is received is important. The longer this is, the lower the rate of mutation. Environmental sources of radiation contribute on average only 2.5 mSV per year.

By far the most important source of radiation for most people is radon gas from underlying rock entering houses, the lung being the most exposed organ. Though better insulation and reduced ventilation makes for a warmer house in winter, it can increase the amount of radon in the family's environment. Radon emits α-rays which are more damaging to tissues than β- or γ-radiation. Clarke and Southwood (1989) suggest that a life-long exposure to radon at an average dose of 20 mSV a year leads to a 5 per cent risk of dying of lung cancer. The risk can be reduced by installing specific ventilation systems to increase positive pressure in the home. It has been estimated that natural radiation, excluding radon, is responsible for about 2000 out of 160 000 cancer deaths a year in the UK (1.3 per cent) (Dunster 1990).

Though there is a great deal of information from animals, most of our knowledge of the effects of large doses of radiation on man is based on studies of the effects of the atomic bombs dropped in Japan. The deleterious effects of radiation are similar to other mutagenic or teratogenic agents. Exposure to a large dose shortly after conception can lead to death of the

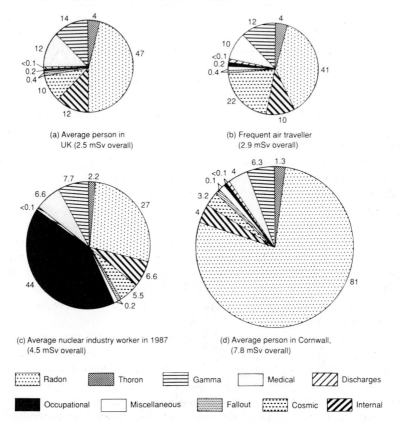

(a) Average person in
UK (2.5 mSv overall)

(b) Frequent air traveller
(2.9 mSv overall)

(c) Average nuclear industry worker in 1987
(4.5 mSv overall)

(d) Average person in Cornwall,
(7.8 mSv overall)

Radon   Thoron   Gamma   Medical   Discharges

Occupational   Miscellaneous   Fallout   Cosmic   Internal

Fig. 20.4 Pie charts showing the total annual exposure to environmental radiation of individuals in different parts of the United Kingdom, and the proportions due to various sources. Workers in the nuclear industry have twice the exposure of the average person in the UK. However, in certain areas the dose from natural sources of radiation is higher than that of nuclear industry workers. The largest contribution is from radon from the ground, and gamma rays from building materials. Frequent air travellers have an increased dose from cosmic radiation. (Reproduced from Clarke and Southwood 1989, with permission.)

embryo, intrauterine growth retardation, or the birth of a mentally handicapped child. Abnormalities of the CNS such as microcephaly with abnormalities of the brain or eye can occur.

The most frequent anxiety of mothers-to-be will be whether inadvertent diagnostic radiology could have injured their fetus. The mutagenic effect of X-rays in pregnancy was initially signalled in the 1960s by an increase in cases of acute leukaemia, but diagnostic doses were then relatively high. Though it is advisable to restrict non-essential X-rays to the first few days following the onset of a period (the '10 day rule'), the present risk of damage appears to be slight. Mole (1979) in a useful review concluded that if the embryo is exposed to 50 mSV of radiation there would be 0–2 extra cases of congenital handicaps, usually mental retardation, and five extra cases of childhood cancer per 1000 live births. About 2 mSV are absorbed from a single chest X-ray; the upper limit of diagnostic radiation is usually 10 mSV. This means that the overall risk of serious damage is probably 0–1 case per 1000 women X-rayed in early pregnancy. This opinion has been confirmed by subsequent studies.

Treatments such as cancer chemotherapy or organ transplantation involve exposure to large doses of radiation or cytotoxic drugs, and could cause mutations. Harper (1988) recommends that a couple should wait some months after the man has had such treatment to allow exposed sperms to be shed, before embarking on a pregnancy. Even then, CVS or amniocentesis is advised to detect any chromosomal abnormality. It is now becoming commonplace for men to deposit sperm at a sperm bank prior to treatment, so that it can be used for artificial insemination by husband (AIH) at a later date. Studies of the children of parents who had successful cancer treatment in childhood have not shown any increase in abnormalities, though it is always possible that some problems could emerge in the future (Mulvihill *et al.* 1987).

In the UK there has been considerable unease in recent years about the larger than expected number of children living near nuclear installations who develop leukaemia and non-Hodgkin lymphoma. A recent study of the population around the nuclear processing plant at Sellafield in Cumbria (Gardner *et al.* 1990) suggested a positive association between the amount of radiation received by fathers working at the plant, and the incidence of leukaemia in their children. Though the numbers in the study were small, it appeared that children of fathers who had received a total dose of more than 100 mSV before conception were to six to eight more times more likely to get leukaemia than a group of controls (though it must be remembered that the risk is still low). Subsequent studies have cast doubt on the association. If the relationship is confirmed, it seems likely to be exerted by a mutagenic effect of ionizing radiation on spermatozoa. The same studies showed that children of older mothers were more at risk of developing leukaemia, and that the relative risk following maternal abdominal X-rays in pregnancy was about 1.5. × normal.

Key references

Hanshaw, J.B., Dudgeon, J.A., and Marshall, W.C. (1985). Viral diseases of the fetus and newborn. *Major problems in clinical pediatrics*, Vol. 17, (2nd edn). Saunders, New York.

Macgregor, R. (1990). Thames Television's *The Treatment. Preconception and pregnancy planning*. Available from the Community Education Officer, Thames

Television p.l.c., 149 Tottenham Court Rd, London W1P 9LL.

Behrman, R. and Vaughan, V.C. III (eds). (1983). *Nelson textbook of pediatrics*, (12th edn) WB Saunders Company. Philadelphia etc.

Hawkins, D.F. (ed.). (1987). *Drugs and pregnancy*. (2nd edn). Churchill Livingstone, Edinburgh.

21. Pregnancy (II): sporadic and genetic risks

Table 21.1 lists risks to the fetus that are largely independent of environment, and the screening methods used to detect them.

Miscarriage

We use the term miscarriage rather than abortion to describe spontaneous loss of a pregnancy, because it is unambiguous.

Causes

Miscarriage usually occurs because a pregnancy is non-viable. This may be because no embryo has developed in the gestation sac, or an embryo has started to develop but fails to progress (usually seen as an empty sac on ultrasound examination) or an embryo that appears to be developing normally dies. Miscarriage of a live embryo or fetus is relatively uncommon.

Miscarriage prevents the birth of most infants with chromosomal abnormalities, especially the more severe ones. At least 50 per cent of early

Table 21.1 Risks that can be detected by screening during pregnancy

Risk	Primary screening method	% detectable
Miscarriage	Ultrasound	*ca.* 80
Chromosomal abnormality	Maternal age	15–50
	Maternal serum screening	*ca.* 60
Congenital malformation	Fetal anomaly scan	*ca.* 75
	Maternal serum AFP	70 of NTDs
Inherited disease	Family history;	5–most[1]
	Carrier screening for	
	haemoglobin disorders	100
	Tay–Sachs disease	100
	cystic fibrosis	*ca.*85
Rhesus haemolytic disease	Maternal Rh testing	100

[1] The proportion of carriers of an inherited disease that can be identified through the family history varies from a few to most, depending on the mode of inheritance, mutation rate, etc. (see Chapter 8).

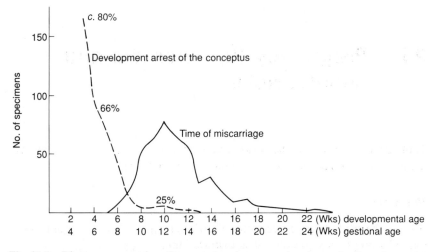

Fig. 21.1 Diagram comparing the gestational age at the time of miscarriage with the development age of the embryo obtained from the abortion material. Failure of development or embryonic death often precedes spontaneous abortion by several weeks. Numbers show the percentage of chromosomally abnormal specimens. (Reproduced from Boue and Boue 1978, with permission.)

miscarriages, and more in mothers over 35, are chromosomally abnormal (Alberman and Creasey 1977), the commonest abnormality being 45 X, resulting from loss of the paternal X chromosome. The later in pregnancy a miscarriage occurs, the lower the chance that it is due to a chromosomal abnormality. Some miscarriages are associated with congenital malformations, some may be due to metabolic abnormalities of the fetus. Some may be due to new mutations, or to recessively-inherited disorders, like α^0-thalassaemia (p. 201).

Usually a number of weeks pass between failure of the pregnancy and miscarriage, because placental tissue is nourished by maternal blood and can survive for weeks, even if the embryo is absent or dead (Fig. 21.1). Therefore, though most miscarriages occur at around 12 weeks' gestation, most non-viable pregnancies can be diagnosed prospectively by ultrasound examination at 7–9 weeks' gestation. When no embryo is visible on ultrasound this could simply mean there is a mistake in the dates and the pregnancy is not as advanced as expected, so the woman is usually seen again one week later to make a definitive diagnosis. Alternatively, an embryo may be visible but without a heart beat: this finding is unambiguous.

Frequency

It is difficult to give a figure for the proportion of pregnancies that miscarry, because the incidence of miscarriage varies greatly with factors such as maternal age, social class, maternal smoking, and other genetic and obstetric factors. In addition, most studies underestimate the number of miscarriages because they are based on women attending hospitals. Many conceptions fail before the fertilized ovum is implanted, and others abort shortly afterwards leading to a slightly prolonged menstrual period which may not even be delayed, so the woman does not realize that conception occurred. According to Lachelin (1985), 10–30 per cent of conceptions are lost (unrecognized) before 4 weeks' gestation, 10–20 per cent of recognized pregnancies miscarry between 4 and 13 weeks' gestation, and 2–3 per cent miscarry between 13 and 28 weeks' gestation (Fig. 21.2).

The incidence of miscarriage rises steeply with maternal age (Fig. 21.3), partly associated with the increased rate of fetal chromosomal abnormality (p. 63). The general risk for recognized pregnancies is about 10 per cent for women aged 20–35, about 20 per cent in women of 35–40, and at least 30 per cent for women over 40. These facts should be more widely known, especially to older women, as they can influence peoples' decisions about when to plan a pregnancy, and parents' expectations during early pregnancy.

Chromosomal rearrangements quite often lead to miscarriage, but are relatively rare (p. 65). Other factors associated with an increased risk of miscarriage are: heavy smoking and perhaps drinking; abnormalities of

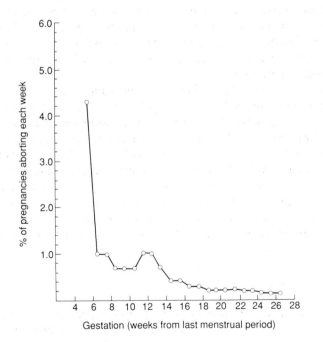

Fig. 21.2 Incidence of spontaneous abortion of recognized pregnancies in relation to gestational age. (Based on Stein *et al.* 1980.)

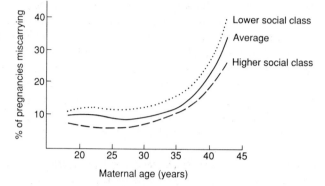

Fig. 21.3 Per cent of recognized pregnancies miscarrying in relation to maternal age and social class. (Based on Stein *et al.* 1980).

the uterus such as malformations or fibroids; poorly controlled diabetes; pelvic inflammatory disease; some forms of auto-immune disease; pregnancy with an intrauterine device *in situ*; the maternal infections described in Chapter 21; urinary tract infections; and severe generalized infections (Simpson 1989). A preventable cause of second trimester abortion is an incompetent cervical os, usually due to excessive cervical dilation during a previous termination of pregnancy. An interesting study by Regan *et al.* (1990) has shown that a high pre-pregnancy level of luteinizing hormone (LH) is associated with decreased fertility as well as with an increased number of miscarriages.

In theory, it might seem likely that immunological factors could be involved in miscarriage, but this has not been demonstrated. The mother's immune system is adapted to tolerate the developing embryo in spite of the fact that approximately half the fetal antigens are 'foreign' (inherited from the father). Maternal/fetal tolerance is probably complex: trophoblast cells do not express fetal HLA antigens and there is also some depression of the mother's immune system. Progesterone has an important role in maintaining pregnancy, but it is still uncertain whether progesterone deficiency is significantly associated with recurrent miscarriages.

Drugs are more often associated with fetal abnormality than with abortion.

Recurrent miscarriage

There is a probable recurrence risk of 25–30 per cent after one miscarriage, somewhat higher if there has been no previous live-born baby. Women who have had two miscarriages seem to be at a slightly increased risk of a third one. However, the chance of a successful outcome seem reasonably high even after recurrent abortion. In one study of 24 women who had had three or more consecutive spontaneous abortions, 80 per cent of subsequent pregnancies ended in a live-born infant (Vlaanderen and Treffers 1987), and in the study of Regan *et al.* referred to above, 32 out of 49 women with one or two previous miscarriages had successful pregnancies without hormonal treatment.

Nevertheless, referral for further investigations such as chromosome studies, blood grouping and antibodies, immunological studies, investigation for corpus luteum deficiency and luteinizing hormone levels, and a hysterosalpingogram is usually indicated after two or three successive miscarriages. A physical examination and reassessment of the personal and family history may reveal a possible genetic cause, or past history of a relevant illness such as pelvic inflammatory disease. It can be helpful to reassure the couple that there is no evidence that working hard, sexual intercourse, having an argument, or lifting heavy weights etc. cause vaginal bleeding in pregnancy.

Management of a miscarriage

Miscarriage is alarming and painful, and is usually treated as an emergency. The immediate danger is of excessive blood loss. The long-term risk is incomplete expulsion of the 'products of conception', which can lead to prolonged bleeding, infection, and pelvic disease. Rhesus-negative women who miscarry or have an induced abortion should have an injection of anti-D globulin to prevent iso-immunization (p. 338).

However, we are remarkably ignorant about the correct management of miscarriage. There are divergent views about the need for hospital admission. All the obstetricians contacted in a study by Ann Oakley and co-workers (1984) thought that every women who miscarries should have a dilatation and curettage (D and C) to ensure complete removal of the products of conception, but most family doctors thought that a D and C was required only for positive indications. This apparent disagreement may make little difference in practice. Ninety-four per cent of the study sample contacted a doctor before the miscarriage was complete, 78 per cent were admitted to hospital, and 72 per cent had a D and C. If these figures apply generally, which they probably do, there are probably over 50 000 hospital admissions for miscarriage in the UK every year. Miscarriage is big business in a health service, and its management deserves proper scientific evaluation.

Role of the primary care team

'Threatened' miscarriage (a painless loss of fresh blood) is quite common in the first trimester, but only sometimes progresses to miscarriage. Some doctors suggest complete bed-rest until bleeding stops or the loss turns brown. Others believe that there is no way to influence whether a miscarriage occurs or not, and suggest carrying on as normal. Increased understanding of the causes of miscarriage tends to support the latter approach.

Support and reassurance that everything possible is being done is important. An early speculum examination is necessary to diagnose other causes of slight vaginal bleeding such as a cervical erosion, and to visualize a dilating os, which would indicate an almost inevitable abortion.

In view of what has been said above, the first step for a woman who is threatening to miscarry should, when possible, be an ultrasound examination. The woman should be encouraged to try bed-rest only if ultrasound shows a viable fetus of the appropriate gestational age. There is no evidence that hormones such as human chorionic gonadotrophin or progesterone reduce the chance of threatened miscarriage progressing to loss of the fetus (Lachelin 1985). When bleeding stops, an ultrasound examination is useful to confirm that the pregnancy is still viable.

If the pregnancy is non-viable on ultrasound examination, the woman may be offered admission for a D and C to forestall an often prolonged, painful, and messy miscarriage at home (though some women may prefer to wait for miscarriage to start spontaneously). Another advantage of a 'cold' admission is that clinic staff have more time to provide counselling than staff on an emergency ward.

Bleeding often continues for 1–3 weeks after miscarriage or D and C. It may be like a normal period, gradually lessening in amount with the colour of the blood changing from red to brown. If heavy bleeding or clots occur or recur, then a second D and C to remove retained products is likely to be necessary. A period usually occurs after 4–6 weeks, but ovulation can occur immediately after miscarriage so some form of contraception is advised until the couple wish to try to conceive again. It is usually advised that intercourse is avoided until the bleeding has stopped because of a theoretical risk of infection, and to wait 2–3 months before embarking on another pregnancy.

The later in pregnancy a miscarriage occurs, the more likely is it to be associated with milk coming into the breasts. This can be very upsetting as well as producing tender, tense, lumpy breasts, but usually settles after a few days as long as the milk is not expressed. Drugs such as bromocriptine are seldom indicated, by contrast with the situation after induced mid-trimester abortion (p. 153).

Psychological aspects

The emotional effects of a miscarriage are influenced by many factors – how sensitively the miscarriage was handled, the stage of pregnancy at which it occurred, how difficult it had been to conceive, the relationship between the couple, and their previous awareness of the risk. Not every pregnancy is wanted, and miscarriage can sometimes come as a relief.

Health workers are sometimes unwilling to tell a woman that a fetal chromosomal abnormality probably caused her miscarriage, for fear of generating anxiety. However, if people are not given the true explanation they tend to invent one. In the study of Oakley *et al.* (1984), only 20 per cent of women thought the miscarriage was due to fetal abnormality, more than 30 per cent had seized on an explanation with a needless element of self-blame, and only 23 per cent said they did not know. The true explanation, with its implication that probably the fetus could never have become the wanted healthy child, is much more helpful.

Even after an early miscarriage, many couples go through a period of mourning and need to be reassured that it is natural to feel flat and depressed for a time. Some find it helpful to have an opportunity to express their emotions and be given 'permission' not to resume normal activities immediately as if nothing had happened. For couples who have lost a fetus at a late stage of pregnancy, the bereavement process may be helped if they can see and hold the fetus, keep a photograph, or even arrange for a burial. The wish of some women to become pregnant as soon as possible to replace the lost fetus should not be discouraged, especially with older women. A simple explanation of what has happened may help older children in the family understand why their parents are so upset.

We recommend two paperbacks: *Miscarriage* by Ann Oakley, Ann McPherson, and Helen Roberts, and *Miscarriage* by Gillian Lachelin, for women (and their partners) who would like fuller information.

Prevention of congenital malformations

Routine fetal anomaly scanning

This is discussed on p. 147.

Maternal serum alpha-fetoprotein (AFP) estimation

The level of AFP in maternal serum can be used for screening for some fetal malformations that disrupt the continuity of the fetal skin. The results of maternal serum AFP assay are usually expressed as 'multiples of the median' (MoMs) for each centre, rather than in absolute figures because normal values vary with method, ethnic group, and gestational age. A maternal serum AFP of more than about 2.5 MoMs (Fig. 21.4) should lead to further action, as this group includes all pregnancies where the fetus has anencephaly and most where it has spina bifida. As it also includes multiple pregnancies, non-viable pregnancies, growth-retarded fetuses, and pregnancies with under-estimated gestational age, a routine basic ultrasound scan (p. 144) soon became a necessary part of the investigation. In about 50 per

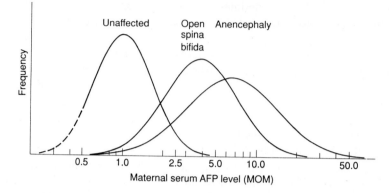

Fig. 21.4   Maternal serum alphafetoprotein (AFP) ranges at 16–18 weeks of gestation, in singleton pregnancies. Note the logarithmic scale. The extensive overlap between normal and abnormal ranges means that a certain proportion of abnormal pregnancies will be missed at most practical cut-off levels. Most of these will have open spina bifida, as AFP levels are higher in anencephaly. (Reproduced from Cuckle and Wald 1984, with permission.)

cent of cases no cause for the raised maternal serum AFP is found, but there is an increased perinatal mortality in this group (Ward *et al.* 1981). Maternal serum AFP screening is usually organized through the antenatal clinic, but can be done in collaboration with general practitioners.

Initially, the next step after the maternal serum AFP was found to be raised was amniocentesis, to assay amniotic fluid AFP and enzymes specific for neural tube defects. However, expert ultrasound examination is increasingly reliable for detecting malformations, so amniocentesis is now uncommon. Twenty to thirty per cent of cases with closed spina bifida, especially smaller lesions, are still missed because the maternal serum AFP level is normal and they are difficult to see on a routine fetal anomaly scan.

As yet, routine fetal anomaly scanning is complementary to, rather than a substitute for, maternal serum AFP screening. In a follow-up study of all neural tube defects diagnosed in the North-Western region of the UK in 1986–7, 63 per cent of affected pregnancies had been terminated, but at least 11 affected cases had been missed because ultrasound screening was not accompanied by maternal serum AFP screening (Bernard *et al.* 1988).

Maternal serum AFP screening is routinely available to pregnant women in most of the UK and in Hungary, the two European countries with the highest incidence of neural tube defects. In Hungary, the incidence of neural tube defects has since fallen by 62 per cent, from 2.8 to 1.06 in 1000 (Czeizel 1988b). In the UK, the birth incidence of anencephaly decreased by 94 per cent and that of spina bifida by 68 per cent between 1972 and 1985. Rather less than half of this fall may have been due to a decrease in background incidence, but over half was definitely due to screening and selective abortion (Cuckle *et al.* 1989).

Prevention of chromosomal abnormalities

The offer of fetal karyotyping to older mothers is the largest and longest established of prenatal diagnosis services. All fetuses with Down syndrome could be detected by offering karyotyping to all pregnant women, but as there are obstetric risks and as facilities are limited, fetal karyotyping is usually offered only to women with a more than 1 in 100 to 1 in 200 (1–0.5 per cent) risk of having an affected child. Until recently, the only two groups of pregnant women who could be shown to have this level of risk were mothers over 35–37 years old (see Figure 5.3), and mothers who had already had an affected child (or fetus). Both these groups of women should be encouraged to report a pregnancy at a very early stage.

The service, started in the late 1960s, has gradually become more acceptable both to women of child-bearing age and to the medical profession, but is still far from equally available to all at risk. Though from 60–80 per cent of women at risk request karyotyping when they are fully informed (Kuliev *et al.* 1985; Knott *et al.* 1986), in most of Europe in 1987 fewer than half the women in accepted risk groups actually had fetal karyotyping (Modell *et al.* 1991). The shortfall was mainly due to inadequate counselling by obstetricians and primary care workers. However, the number of tests is rising rapidly, reflecting a continuing shift in social attitudes.

Prenatal diagnosis has had a limited impact on the total birth incidence of infants with chromosomal disorders so far, partly because even when it is offered, only a limited proportion of Down syndrome babies are born to older women. Maternal age distribution is a major determinant of the total birth incidence of chromosomal abnormalities, and also determines the proportion of affected births that can be avoided when prenatal diagnosis is offered to older women. Figure 23.5 shows that uptake of testing by *all* pregnant women over 35 could prevent the birth of from 15–55 per cent of Down syndrome fetuses in different European countries, the figure being lowest in Eastern Europe, where parents are predominantly young (Modell *et al.* 1991).

The main problem in screening for Down syndrome revolves around the inefficiency of the primary screen for identifying women at increased risk (p. 165). Over 95 per cent of the 'at risk' older women have unaffected infants, and around two-thirds of affected infants are born in the group excluded. In addition, the second step, — amniocentesis or chorionic villus sampling for fetal karyotyping — is both invasive and expensive.

Improved and less invasive methods of screening pregnant women for Down syndrome in the fetus seem to be on the way. It was first reported in 1984 that in most pregnancies where the fetus ultimately turns out to have

Down syndrome, the maternal serum alpha-fetoprotein (AFP) level is in the lower half of the normal range (Merkatz 1984). Since maternal serum AFP screening is routine in much of the UK, this rapidly made it possible to improve our ability to identify mothers at risk by integrating maternal age and maternal serum AFP level (Cuckle *et al.* 1984).

More recently it has been noted that when the fetus has Down syndrome, unconjugated oestriol in maternal serum is also relatively low, and human chorionic gonadotrophin (hCG) is relatively raised, the hCG being the most reliable indicator (Wald *et al.* 1988). If two or all of these assays are routinely done at 16–17 weeks of pregnancy and integrated with maternal age, it is possible to give a statistical risk for every pregnant woman that her fetus actually has Down syndrome. This risk can vary from 1 in 6 to 1 in 10 000: about 5–6 per cent of pregnant women of all ages have a risk of 1 in 200 or more (Table 21.2). It seems possible that about 60 per cent of affected fetuses could be identified if amniocentesis were offered to women identified in this way. The approach has the additional benefit that some older women can be shown to be at low risk, and may be able to avoid CVS

Table 21.2 Proportion of fetuses with Down syndrome that could be detected using maternal age and different combinations of serum factors

Detection rate (%) of fetuses with Down's Syndrome	% of mothers who would have to accept amniocentesis to achieve the indicated detection rates when maternal age is integrated with:	
	α fetoprotein concentration	α fetoprotein, unconjugated oestriol, and human chorionic gonadotrophin concentrations
80	44	16
75	37	12
70	30	8.6
65	25	6.4
60	20	4.7
55	16	3.4
50	12	2.5
45	9.8	1.7
40	7.3	1.2
35	5.3	0.8
30	3.6	0.5
25	2.2	0.3
20	1.3	0.2

The least effective screen is unconjugated oestriol.
Calculations for the UK population (Based on Wald *et al.* 1988)

or amniocentesis. Measurement of maternal urea resistant neutrophil alkaline phosphatase activity may eventually become a useful additional test for a fetus with Down syndrome (Cuckle *et al.* 1990).

This type of screening is far from simple. It depends on precise and rather specialized tests, and the results have to be interpreted by computer to arrive at an individual risk for each pregnant woman. However, it may be well worth the additional organization and cost: identifying the 5 per cent of women at highest risk in this way could double the effectiveness of screening in Western Europe and triple it in Eastern Europe (p. 361). However, even with the best screening methods available, 40 per cent of women bearing a Down syndrome fetus will still be missed, because all the test results will be in the normal range.

Fetal karyotyping is done primarily to exclude Down syndrome, but inevitably reveals any other chromosomal abnormality that is present in the fetus. When for example a sex chromosome aneuploidy is detected, it can be difficult for the parents to decide on the outcome of the pregnancy (p. 274).

At present there is an apparent conflict between the two main developments in screening for fetal chromosomal anomalies. On the one hand, the introduction of CVS means that early testing can be offered to women at risk, so they should be identified, informed, and referred for prenatal diagnosis (if they wish) as early in pregnancy as possible. On the other hand, maternal serum testing is usually done at 16 weeks or later, and is more efficient, but allows only mid-trimester prenatal diagnosis. Though initial studies suggest that maternal screening may also be useful in early pregnancy, permitting early risk assessment for everyone (Brambati *et al.* 1986), it will be a long time before this will be fully worked out.

Role of the primary care worker

This area illustrates the wide range of knowledge of different aspects of prenatal screening needed by a general practitioner or health visitor. They will frequently be consulted by pregnant women who, in their late 30s, are undecided whether to undergo invasive diagnostic procedures, or unaware of their increased risk of producing a child with a chromosomal aneuploidy, or have been found to have a low serum AFP level. As biochemical methods for assessing the risk of carrying a Down syndrome fetus are refined, the primary care worker will also need to advise younger women with test results in the mid-trimester that put them in a risk group.

When a woman in a group at risk for one of the inherited diseases listed in Table 21.1 presents to her family doctor with a confirmed pregnancy, if her carrier status is not known, blood should be taken for testing and sent to the laboratory at the same time as a referral letter is written for the antenatal clinic. If she is a carrier, this will give the best chance of getting her partner tested and completing any other necessary investigations before the end of the first trimester.

Haemolytic disease of the newborn due to blood group incompatibility

In addition to A, B, and O blood group substances, red cells carry numerous other antigens such as Rhesus, MN, Secretor, Duffy and Kell on their surfaces. However, only the Rhesus antigen is a common cause of important incompatibility between mother and fetus.

In about 15 per cent of pregnancies, mother and fetus are *ABO incompatible* (i.e. the mother is group O and the fetus group A, B, or AB), but the only result is mild neonatal jaundice, rarely needing treatment, in about 2 per cent of infants. The following facts probably explain its mildness relative to Rhesus blood group incompatibility (described below).

1. The mother's naturally occurring anti-A and anti-B antibodies are a mixture of IgM, IgA, and IgG. Only the IgG fraction can cross the placenta.
2. A and B antigens occur on most types of cell, and fetuses have less A or B substance on their red cells than adults. Other tissues mop up most of the maternal anti-A or -B that enters the fetal circulation.
3. ABO incompatibility does not usually get worse with subsequent pregnancies, because the mother's natural antibodies destroy incompatible fetal cells that leak across the placenta, before they can stimulate further (IgG) antibody production.

Haemolytic disease due to Rhesus incompatibility

This is more serious, and is almost completely preventable. In the past it led to a fatal outcome or severe residual brain damage in about 1.5–3.5 per 1000 births (Bowman and Pollock 1965), and was a recurrent problem for women who had become immunized.

The Rhesus locus includes three closely linked genes (C, D, and E) of which D is the most important. They direct the synthesis of a surface protein that occurs only on red cells. Many people carry the Rhesus negative d allele, which represents absence of the D protein. Everyone is either DD (homozygous Rhesus positive), Dd (heterozygous Rhesus positive), or dd (Rhesus negative). About 15 per cent of Caucasian populations, about 1 per cent of Africans, and even fewer Chinese or Japanese are Rhesus negative (dd).

The presence of D substance on the red cells of the fetus of a Rhesus negative (dd) mother can lead to serious complications because:

1. D is highly antigenic, i.e. a Rhesus negative person is easily stimulated to produce anti-Rhesus antibodies. Most immunization occurs when a small number of red cells leak from a Rhesus positive fetus to a Rhesus negative mother at delivery. Cells may also leak during pregnancy, or

during diagnostic interventions such as amniocentesis or chorionic villus sampling, or at miscarriage or medical abortion. People may also be sensitized by transfusion of Rhesus positive blood.

2. The maternal antibody made after the first exposure to fetal Rhesus positive red cells is IgM, which cannot cross the placenta. However, a second exposure stimulates the production of IgG, which crosses the placenta.

3. Because Rhesus antigens occur only on red cells, no antibody is absorbed by other tissues.

4. Anti-Rhesus antibodies are lytic antibodies: once attached to antigen they react with complement and destroy the Rhesus positive cell.

Sensitization usually occurs near the end of pregnancy, so the first pregnancy is usually unaffected, but future pregnancies are at risk. The following factors are known to influence the chance that a Rhesus negative mother will be immunized by a Rhesus positive fetus.

1. The genotype of her partner. The risk of sensitization is less if he is dD than if he is DD. Once a woman is sensitized, all future pregnancies are at risk when the father DD, and 50 per cent are at risk when he is dD.

2. The ABO blood groups of mother and fetus. In the absence of prevention, the risk of maternal immunization following delivery is about 16 per cent when fetus and mother are ABO compatible, but only about 1.5 per cent when they are incompatible, because fetal red cells entering the maternal circulation are rapidly destroyed by natural maternal anti-A or anti-B (Woodrow 1970).

3. The mother's transfusion history. A previous transfusion of Rhesus positive blood guarantees immunization.

4. The mother's obstetric/gynaecological history—the more pregnancies, the greater the risk.

The observation that ABO incompatibility protects against Rhesus haemolytic disease led to the idea of protecting Rhesus negative mothers by injecting anti-D globulin after delivery, in order to mop up any Rhesus positive cells that have entered the mother's circulation, before they can stimulate antibody production. Isoimmunized women are the main source of human anti-D globulin.

Clinical picture   IgG antibody production by the mother causes increased destruction of fetal red cells with increased bilirubin production and sometimes anaemia. The fetus is not jaundiced (excess bilirubin is metabolized by the mother) but the amniotic fluid contains dark bile pigments. These can be assayed in amniotic fluid samples, and the results used to assess the need for early delivery or intra-uterine transfusion of the fetus. This was the first clinical application of amniocentesis (Bowman and Pollack 1965).

The main threat to the fetus *in utero* is anaemia leading to compensatory hypertrophy of liver, spleen, and bone marrow (which all produce red cells at this stage). Severe anaemia causes heart failure and oedema, leading to still-birth or neonatal death with classical hydrops fetalis. Less severely affected infants are born with varying degrees of haemolytic anaemia. After birth, jaundice leading to kernicterus is the main danger.

In the newborn, severe anaemia is indicated by a haemoglobin level of less than 12 g/dl (normal newborn level = 16–20 g/dl) and large numbers of nucleated red cells (erythroblasts) in the blood film. The baby usually has an enlarged liver and spleen. There is a strongly positive direct Coombs test. This means the baby's red cells are agglutinated by serum containing anti-human globulin, which demonstrates the presence of antibody on their surface. The bilirubin level is normal at birth but usually rises rapidly, and most affected infants are visibly jaundiced within 24 hours of birth — earlier than is usual for physiological jaundice.

The main danger of neonatal jaundice is brain damage. In a mature brain, myelinated neurones are protected from a high bilirubin level, but in the newborn, myelination is incomplete and neonatal jaundice can lead to kernicterus (yellow nuclei). The name comes from the fact that at post mortem the brain-stem nuclei are seen to be stained yellow. The earliest signs of kernicterus may be non-specific. The infant appears unwell and becomes progressively more floppy and dehydrated. If the baby survives, the end result may be mental handicap with uncontrolled involuntary movements, fits, dysarthria, and perhaps hearing loss.

In severe cases, intra-uterine transfusion with Rhesus negative red cells may be necessary, often on several occasions. After birth, phototherapy can control the jaundice only in less severely affected infants: exchange transfusion is often necessary. The objectives are

(1) to replace the baby's Rhesus positive red cells with ABO compatible Rhesus *negative* red cells (unaffected by the antibody);
(2) to wash out bilirubin and
(3) to wash out anti-D antibody.

Prevention of Rhesus haemolytic disease   All pregnant women are routinely screened for their ABO and Rhesus status. Rhesus negative women are tested for anti-Rhesus antibodies at intervals during pregnancy, and their partner is screened. It is routine to give Rhesus negative women an injection of anti-D globulin within 72 hours of delivery, miscarriage, or induced abortion, and after interventions such as amniocentesis or chorionic villus sampling. Nevertheless, the 1–2 per cent of Rhesus negative women who become immunized by leakage of a few fetal red cells across the placenta during pregnancy will not be protected by injection of anti-D after delivery.

Therefore in some centres, pregnant Rhesus negative women are given anti-D at 28 and 34 weeks' gestation.

Women should be informed of their Rhesus status so that they can avoid sensitization by checking that they are given anti-D globulin at the appropriate times (it is sometimes forgotten), and help to avoid incompatible blood transfusion.

Results of prevention programmes   In the UK since 1950, perinatal mortality from Rhesus haemolytic disease has fallen by over 95 per cent from about 560 per year to about 23 per year, i.e. from 0.75 per 1000 to less than 0.03 per 1000 (Clarke and Whitfield 1984). There must have been a parallel fall in the number of surviving, brain-damaged children. If similar figures held for the rest of Europe (13.6 million total annual births), this would represent a gain of between 10 000 and 20 000 healthy children per year in the whole region.

Key references

Lachelin, G. L. L., (1985). *Miscarriage. The facts.* Oxford University Press.
Oakley, A., MacPherson, A., and Roberts, H. (1984). *Miscarriage.* Fontana Paperbacks, Glasgow.
Wald, N. J. (ed.). (1984). *Antenatal and neonatal screening.* Oxford University Press.

22. Neonatal screening

Neonatal screening means organized examination of all newborns in order to diagnose specific disorders so that they can be treated. It is one of the best established forms of screening, and covers a wide range of activities.

Still-birth and neonatal death

Infants who are still-born or die in the neonatal period should be examined by a paediatric pathologist. The examination includes X-rays of the skeleton, examination of blood and urine for evidence of metabolic disease, and, when relevant cell culture for karyotyping and storage of a DNA sample. Psychological aspects have been discussed on p. 289. The parents should be told the results of post mortem examinations, and when relevant they should be offered an appointment with a clinical geneticist. They may benefit from being put in touch with a support association (see Appendix).

Clinical examination of the newborn

The aims are to see how well the infant is adapting to extra-uterine life, to check for obvious disorders and, when appropriate to reassure the parents that their baby is normal. Early postnatal discharge from hospital increases the responsibility of primary care workers for early diagnosis of congenital disorders. These may be missed unless the baby is examined carefully a few days after coming home. It is too late to leave this examination until the infant is 6 weeks old, by then she will have passed the most vulnerable period. Many of the early problems will be identified by the midwife who visits the home daily in the immediate postnatal period.

Careful observation of the baby may be very informative. Most infants with Down syndrome or other severe chromosomal disorders are identified soon after birth because most have a typical appearance (p. 269). Abnormal pallor or jaundice needs to be noted. Blue hands and feet are seen in any cold baby, but a blue tongue is always a sign of a congenital heart defect or severe lung disease. A full check for malformations is needed. Even minor ones such as an umbilical hernia need to be discussed with the parents.

Observation of the lower spine may reveal a sinus, which usually ends blindly, or a tuft of hair which may indicate a spina bifida occulta. This is usually of no significance, and should be simply noted, but rarely it is associated with a defect of the underlying spinal cord and nerve roots. There should be symmetrical movements of the limbs, well demonstrated by the Moro reflex. If the baby is picked up face downwards, its tone can be assessed. There should be some flexion of all limbs. An infant's foot is very mobile and it does not matter if it points somewhat outwards or inwards. A club foot cannot be put through the full range of movements, and in the commonest form the foot is so twisted that the sole faces inward. These infants need referral for early correction.

The head and neck should be palpated for abnormal lumps, and the head circumference should be measured. Babies with a head circumference above the 97th percentile or below the 3rd need referral or early reassessment, unless the head size is in keeping with that child's weight and length, and the head size of the parents. A head circumference crossing the centile lines during a period of observation is particularly significant (p. 258).

The genitals need to be carefully examined. In 1–3 in 1000 male babies, the urethral meatus opens on the underside of the penis (hypospadias). This is one of the commonest congenital malformations. If it is present, other renal tract abnormalities need to be excluded (p. 266). In boys a poor urinary stream and a palpable bladder may indicate posterior urethral valves. In girls an enlarged clitoris may occasionally point to the adrenogenital syndrome (p. 226).

The neonatal period is a good time for checking that both testes are present, as the scrotal cremaster muscle is relaxed at this stage. One or both testes are undescended at birth in 5.5 per cent of full-term baby boys and in up to 30 per cent of low birth weight boys. The percentage falls to 1.6 per cent of boys by 3 months (John Radcliffe Hospital Cryptorchidism Group 1986). The recurrence rate of undescended testes and hypospadias in siblings is about 10 per cent.

A congenital cataract will be excluded if the infant has a normal red reflex (not obscured by an opacity) when a light is shone on the pupil from an ophthalmoscope through a 1+ to 3+ lens from a distance of 16–30 cm. Roving nystagmus is an important sign of visual pathology.

The normal heart rate in newborns is 120–140 beats per minute. Up to 50 per cent of infants have a systolic murmur. The vast majority of such murmurs are benign and many later disappear. The louder a murmur, the more likely it to be pathological, especially when a diastolic murmur is also present.

The respiratory rate of a resting newborn is about 40 beats per minute. A constant rate of over 50 beats per minute when the baby is not crying or feeding is abnormal, and is likely to be accompanied by other signs of respiratory difficulties, such as sub-costal recession, use of the sternomastoids, and indrawing the intercostal spaces with each inspiration. A sustained increase in respiratory rate is a most useful sign of heart failure.

It may be difficult to distinguish a child in heart failure from one with severe lower respiratory tract infection. In the former there are likely to be other associated signs of heart disease, such as a cardiac murmur, and a liver edge felt more than the normal 2 cm below the costal margin.

If there has been delay in passing meconium, a rectal examination will be necessary to exclude Hirschprung's disease (p. 49). The rectum is usually empty on examination. Hirschprung's disease is commoner in boys, and male siblings have about a 10 per cent, and female siblings a 5 per cent, risk of developing the condition.

Abdominal masses are uncommon. A normal spleen can occasionally be tipped below the left costal margin.

A general discussion with the mother about the progress of her child is as important as the physical examination. Has breast-feeding become satisfactorily established? Are there any problems with taking milk from a bottle? What is the child's sleeping pattern?

Finally the doctor needs to assess whether the baby falls into a high risk group that needs extra attention; for example was the baby premature or small for dates, or has he failed to regain his birth weight during the early part of the second week? Does persistent jaundice make it essential to carry out further tests? Is the family in a socially high risk group?

Any newborn who becomes more than moderately jaundiced should have a bilirubin estimation. Levels above 300 μmol/litre appear to be dangerous, and these infants should be referred for diagnosis and appropriate treatment. Observation of neonatal jaundice is a form of screening for conditions shown in Figure 8.7.

The six week check

If the infant has not been thoroughly examined since discharge from hospital, a full check is necessary at six weeks, when the examination described above should be carried out. In addition, a six week old baby will usually have started to smile and respond to the mother's overtures, or at least to gaze intently at her face, and within a week or two will begin to vocalize. A baby of this age can hold the head in the same plane as the rest of the body for a few seconds when suspended face downwards. The elbows and knees will still be somewhat flexed. The six week check is a good opportunity to confirm that the infant is growing properly (average weight gain being 200 g per week after the first ten days), that the mother has had her postnatal examination, and to discuss contraception and the baby's immunization schedule.

Deafness

The incidence of congenital sensorineural deafness is about 1 in 1000 live births. About half of these are non-genetic. The genetics of congenital deafness are complicated. In about one-third of cases it is part of a particular syndrome and about three-quarters of the remainder are recessively inherited. The situation can be complicated when deaf people marry each other. The early detection of severe deafness is valuable because the infant can be fitted with a hearing aid which will help the acquisition of language. At present, screening by various sensitive physiological tests is reserved for newborns considered to be at increased risk of hearing loss, those with a positive family history or abnormal appearance, or who have spent time in a neonatal special care unit (Haggard 1990; Reardon and Pembrey 1990). In addition, a baby who is not startled by loud noises may be deaf.

Congenital dislocation of the hip (CDH)

This preventable condition occurs with very different frequency in different populations. As with neural tube defects it seems to involve both genetic and environmental factors. Genetic predisposition is due to differences in the development of the hip. Various environmental factors may favour displacement of the hip of the fetus or newborn from the acetabulum. For example, CDH occurs more often in infants with indications of intra-uterine compression such as postural deformities of the feet or a history of oligohydramnios and fetal growth retardation. Other recognized risk factors include breech delivery, Caesarian section and family history. CDH is more common in premature babies (because the hip is immature) and in girls than boys (as maternal hormones may relax the ligaments around the hip), and in populations where infants are swaddled with extended legs (Wilkinson 1985). Some dislocatable hips can become dislocated when weight-bearing begins (Berman and Klenerman 1986).

Dislocatable hips can be diagnosed clinically in newborns by the Ortolani and Barlow manoeuvres. They are most easily picked up in the first 2–3 days of life. This is a rather uncomfortable procedure, which should be performed gently lest it damage a normal hip. The newborn baby's pelvis is held still with one hand. The other hand grasps the thigh, with the middle fingers placed over the greater trochanter and the thumb pressing on the inner aspect of the thigh. The knee and hips are flexed and the hip abducted to about 45°. Stability is tested in this position by attempting to move the femur in and dislocate it out of the acetabulum by applying forwards and backwards pressure with the finger and thumb. If the head is felt to move, with or without an audible clunk, the hip is dislocatable and the child should be referred. Treatment at this stage is by splinting the infant with the legs apart to hold the head of the femur in the acetabulum until further development has occurred, and is simple and successful.

About 2 per cent of newborns are found to have clinically unstable hips. Since the incidence of congenital dislocation of the hip in the UK is about 2 per 1000, it is thought that about 10 per cent of dislocatable hips become permanently dislocated if untreated, so for every case prevented, nine children who would not have developed any problem will be treated.

Limited abduction is the most important sign in infants. With the infant supine and hips flexed to 90°, the thigh normally abducts to 75°. If there is limited abduction, the child must be referred. From about 6 weeks of age onwards, diagnosis is by the 'classical' signs of limited abduction together with any of the following: a shortened knee-to-pelvis distance, a flat buttock, extra skin creases, persistent external rotation, or a wide perineum.

If it is not diagnosed early, dislocation of the hip usually presents in the second year, after the child begins to walk, with a limp or waddling gait. Weight-bearing increases the dislocation of a dislocated hip, and may dislocate a dislocatable hip with a poorly formed acetabulum. Dislocation diagnosed at this stage requires operative treatment. Bilateral dislocation of the hip is more difficult to spot than unilateral. If a mother is worried about her toddler's 'funny walk', her anxieties should not be dismissed out of hand.

Results of neonatal screening

Because it is relatively common, causes severe problems if untreated, can be diagnosed, and if diagnosed early can be simply and successfully corrected, CDH is one of the most important abnormalities to detect in the newborn period. Accordingly, doctors with responsibility for newborn babies were encouraged to screen them for congenital dislocation of the hip in the first 3 months of life. However, subsequent studies showed that the incidence of congenital dislocation of the hip presenting later in infancy may actually have increased after universal screening for CDH was started (Fig. 22.1) (Catford *et al.* 1982). In addition, cases of epiphysitis of the femoral head began to show up among children who had been splinted. It seems possible that clumsy screening might actually dislocate some hips, and that inexpertly tight splinting might damage some femoral heads—another demonstration of the fact that in the absence of quality control (which is particularly difficult for clinical procedures), screening can do more harm than good.

Three research studies suggest that the problem may be to do with the clinical nature of the screening test. For example, the evidence on which the decision to screen was based was obtained at expert centres, but a universal clinical screening procedure is practised with different levels of expertise at different levels of the health system. There may also be differences of opinion about what is significant. For example, one group of experts concluded that 'ligamentous clicks *without* movement of the head of the femur in or out of the acetabulum may be elicited in 5–10 per cent of hips and should be disregarded' (special report 1986): but a study of referrals in a non-teaching hospital (Catford *et al.* 1982) found that less expert examiners may have difficulty distinguishing movement and a click, and need to refer babies in

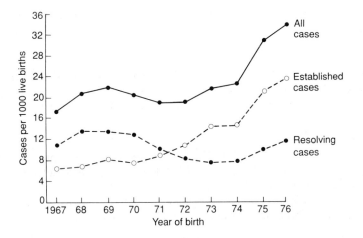

Fig. 22.1 The incidence of congenital dislocation of the hip in five year averages in the Southampton district of England (Catford *et al.* 1982). The trend to increasing incidence is representative of the UK in general. The number of resolving cases (i.e. cases diagnosed and treated at birth) has remained fairly constant, but the number of established cases (i.e. those detected in the first year of life or even later, where dislocation has already occurred) has been rising. (Reproduced from Catford *et al.* 1982, with permission.)

which they detect either. One proposed solution was to recommend careful examination of the hip at five different points before 2 years of age. An alternative may be ultrasound or sound transmission screening of the hip in all newborns (Berman and Klenerman 1986, Stone *et al.* 1990). These look promising, but their value is yet to be proved. Hip X-ray is of little use until 3 or 4 months of age when the femoral head becomes calcified.

Biochemical screening of newborns

An infant with an inherited metabolic disorder often presents with non-specific symptoms of acute illness, and these conditions are often difficult to diagnose. Table 22.1 highlights features that may arouse suspicion that a baby has an inherited metabolic disorder: many are identical with the symptoms of a severe neonatal infection (Association of Clinical Biochemists Newsheet 1983).

Biochemical screening is a supplement to physical examination, to allow early diagnosis of some metabolic diseases. It was first introduced for phenylketonuria (PKU) in 1966 after it had been shown that severe mental retardation could be prevented if a low phenylanine diet is started in the first weeks of life (p. 219). Blood samples are usually taken by heel-prick by midwives between 5 and 10 days after birth, when the baby's metabolism has

Table 22.1   Presentation of an infant who may have an inherited metabolic disorder

- Sick baby who was well at birth
- Alteration in level of consciousness, fits, or increasing hypotonia; occasional babies are irritable with increased tone
- Failure to thrive, perhaps with persistent jaundice; vomiting may occur
- Hypoglycaemia, particularly in association with an enlarged liver. (It can occur in small-for-dates babies and babies of diabetic mothers, and is unusual in previously healthy full-term babies.)
- Metabolic disorder should be suspected if no cardiac or respiratory cause is found for persistent breathlessness
- Suspicion should be raised if a sibling has died unexpectedly in the neonatal period, or a sibling has an unexplained neurological problem

stabilized sufficiently for testing to give reliable results. Several drops of blood are collected in four rings marked on a special filter paper (the Guthrie card). It is essential to fill the ring completely with blood. The cards are posted to the central newborn screening laboratory for the area.

Phenylketonuria

For diagnosis of PKU (p. 217), a standard disc is punched out from the dried blood spot and incubated on a culture medium rich in all requirements except phenylalanine and containing spores of *Bacillus subtilus* (which requires phenylalanine for growth). The radius of the ring of bacterial growth around the disc after incubation of the culture is directly proportional to the pheny-lalanine concentration in the baby's blood − good growth indicating a high phenylalanine level. The first test is positive in about 1 in 3000 babies. In these cases a repeat sample is collected. The second result is positive in one-third of those re-tested. Two positive tests mean that the odds are 99:1 in favour of phenylketonuria or benign phenylalaninaemia (p. 218). At present in Europe, nearly 500 infants with phenylketonuria are diagnosed and treated annually.

Once the system was set up for PKU, it became simple and cheap to screen for other conditions. Biochemical screening is routine (in various places) for the disorders listed in Table 22.2.

Congenital hypothyroidism

The European incidence of congenital hypothyroidism (CH) is about 1 in 4000. Over twice as many females as males are affected. Most cases are the result of a sporadic congenital malformation of the thyroid, but in a few the disease is a recessively inherited disorder of thyroxine synthesis. Prolonged neonatal jaundice may be an early sign, but only a small minority of cases are suspected clinically in the neonatal period. By six months an untreated baby will have very slow mental and physical development with a dry skin, large tongue, low temperature, and history of poor feeding, lethargy, and constipation — the typical picture of cretinism. The picture subsequently is of very severe mental retardation (classical idiocy) associated with a difficult personality. Joseph Conrad's story *The Idiots* describes the devastating impact of the untreated (inherited) disease on the family.

However, thanks to biochemical screening, the classical picture of cretinism is now seldom seen. While being as serious as PKU, CH is almost three times as common, can be detected by assaying thyroid stimulating hormone level (Table 22.2), and can be easily and effectively treated with thyroxine. Since the late 1970s, filter paper blood spots have been used to screen for CH in most (but not all) countries where PKU screening is already established.

Production of thyroid hormone (T4) is controlled by thyroid stimulating hormone (TSH) secreted by the pituitary gland. Lack of T4 causes an increased level of serum TSH because of failure of feedback inhibition. TSH can be reliably measured by radio-immunoassay using dried Guthrie blood spots. The first measurement is positive in about 1 in 170 infants. About 4 per cent of these have a positive repeat test, and most are subsequently found to have CH (Layde 1984): more than 97 per cent of infants with CH are detected. Grant and Smith (1988) reported that in the UK only four out of 493 cases were missed by neonatal screening from 1982 to 1984, and retesting the original blood spot showed a normal TSH in two of the four. The incidence of other major congenital anomalies in children with CH was 7 per cent.

Treatment is by oral thyroxine (lifelong) coupled with regular monitoring of the child's T4 and TSH. In Europe, over 1250 affected children are diagnosed and treated annually. A collaborative study of outcome (Illig *et al.* 1987) has shown essentially normal development, providing treatment is started within the first seven weeks of life. Thereafter, the later treatment is begun, the lower the IQ.

Other reasons for biochemical screening

Indications for neonatal screening for the above disorders are clear-cut according to the accepted criteria for screening (p. 169). The system has also been used for screening babies for HIV infection in the mother (p. 314). Might it be possible to take advantage of the existing infrastructure to extend neonatal screening for other genetic conditions?

There is now some experience of neonatal screening for

Table 22.2 Neonatal screening tests

Disorder	Abbreviation	Birth incidence	Neonatal assay/test	First screening test			Comments
				Positives /1000	Detection[*] rate (%)	False positive[*] rate (%)	
Phenylketonuria	PKU	1/10 000	Phenylalanine	0.3	98	60	Established value. Permits largely successful treatment
Congenital hypothyroidism	CH	1/4000	TSH	6.0	98	96	Established value. Permits successful treatment
Sickle cell disease	SCD	0–1/50[1]	Hb electrophoresis	Varies with population	Can be[2] 100%	Can be[2] <10%	Established value. Avoids many infant deaths
Cystic fibrosis	CF	1/2000	Immunoreactive trypsin	50	>70	99	Value uncertain. May improve prognosis. Permits genetic counselling

Duchenne muscular dystrophy	DMD	1/6000[3]	Creatine phosphokinase	1.8	?50–90	96	Value mainly for genetic counselling for parents
Congenital adrenal hyperplasia	CAH	1/10 000	17-OH progesterone	0.2–2	98?	95	Value in males clear. Little value in females with normal external genitalia
Congenital dislocation of the hip	CDH	1/400	Ortolani and Barlow manoeuvres	20	Varies with examiner	90	Requires expertise and repeated examination.

[1] In certain ethnic groups (p. 195).
[2] Depends on method used.
[3] 1/3000 boys.
• see Table 13.2 for terminology.

(1) conditions where the evidence of benefit is less clear (cystic fibrosis);

(2) conditions where carrier as well as affected infants are detected (e.g. sickle cell disease);

(3) less predictably severe conditions (e.g. α1-antitrypsin deficiency) (p. 228);

(4) conditions where the parents could benefit from advice about reproductive risks in subsequent pregnancies (this may also include groups 1 and 2).

This experience suggests that there may be quite narrow limits to the indications for neonatal screening.

Screening when the benefit is unclear: cystic fibrosis

It has been proposed that neonatal identification of infants with cystic fibrosis (CF) may improve their prognosis, because preventive management can be started before the lungs are damaged (Bowling *et al.* 1988). Parents can also be offered prenatal diagnosis in subsequent pregnancies. Pilot studies of neonatal screening by assay of immunoreactive trypsin in dried blood spots showed undesirably high false positive and false negative rates, so universal extension of screening has not been recommended (Roberts *et al.* 1988).

It has now been shown that dried blood spots can be used for DNA diagnosis (Williams *et al.* 1988), and this would increase the accuracy of neonatal screening. However, DNA methods introduce the complication (or advantage) that they detect heterozygotes as well as homozygotes. The implications have already be encountered with sickle cell disease.

Screening that also detects carriers: sickle cell disease (SCD)

The abnormal haemoglobins (S, C, E, D, and others) can be detected relatively simply and cheaply in newborns by electrophoresis of cord blood, or by examining haemoglobin eluted from dried Guthrie blood spots. Neonatal screening for SCD reduces its childhood mortality (p. 201) and also allows parents to be counselled about recurrence risk in subsequent pregnancies, so neonatal screening is recommended for populations with a significant incidence of SCD (Neonatal screening for sickle cell disease 1989). β-thalassaemia major may sometime be detected, but thalassaemia trait cannot be reliably diagnosed in newborns using these methods. In several parts of the UK all newborns regardless of ethnic origin are now screened for SCD, and in some other areas babies are tested if the mother has been found to carry an abnormal haemoglobin on antenatal screening.

However, problems arise because 10–40 heterozygotes are detected for every affected infant found, but there is no provision within the paediatric service for counselling on such a scale. (In screening for cystic fibrosis, about

100 heterozygotes (1 in 20) would be identified per homozygote (1 in 2000) detected.) Since it is better not to inform families at all than to give out such results without counselling, in some places a decision has been made to identify only the infants with sickle cell disease, and not to inform the families of carriers. However, both the ethics and the cost-effectiveness of such decisions are doubtful.

Table 8.4 shows the chance, for several common recessively inherited conditions, that *both* the parents of a carrier child are carriers, and the chance that their next child will be a homozygote. The relatively high risk in some instances shows that it is important to provide reproductive counselling to parents of carrier infants. Counselling such families also leads to detection of other carriers, and raises the level of awareness in the community, while failure to counsel them greatly reduces the cost–benefit ratio of the service. However, adequate counselling can be provided only if the primary care team is involved.

Neonatal screening seems a very unsatisfactory way to detect couples at risk for infants with recessively inherited disorder in order to offer them reproductive counselling. By the time they are identified, 25 per cent of at-risk couples will already (avoidably) have had an affected child, and 25 per cent will have had a normal child and so escaped detection, but are still at risk in the next pregnancy. Only screening before or during pregnancy allows risk to be avoided in every pregnancy.

Screening for less predictably severe problems: α-1 antitrypsin deficiency

Neonatal screening for less predictably severe conditions raises other new issues, illustrated by the Swedish experience of screening for α-1 antitrypsin deficiency (p. 228), which can be precisely diagnosed in the newborn by both protein and DNA methods. The objective of the Swedish study was to detect homozygotes (1) in order to set up a long-term follow-up study to establish the natural history of the condition, and (2) to advise the parents of the child's vulnerability to environmental pollution, with the particular aim of discouraging smoking in the family.

Following the convention of PKU screening, parents were not given prior information, nor asked if they wished screening to be carried out. At follow-up, most parents of the homozygous children identified were upset and angry, presumably because anxiety had been raised without being allayed; and they smoked no less than before. The experience underlines the importance of consulting parents before screening is carried out. It was decided to discontinue neonatal screening, but to study the value of screening in high schools (McNeil *et al.* 1988; Gustavson 1989). Recent evidence that breast-feeding might reduce the incidence of severe liver disease in affected infants could, if confirmed, make it desirable to reconsider neonatal screening.

This work brings out several important points of principle. Families with an infant with PKU are always glad the diagnosis has been made early and are very co-operative. The different reaction in the case of α-1 antitrypsin deficiency shows that, from the families' point of view, the information received is quite different in the two conditions. In the case of PKU, the doctor informs the parents that their child has a potentially devastating disorder that can fortunately be satisfactorily treated. But in the case of α-1 antitrypsin deficiency, the information sprung on the parents is vague, and so in some ways more alarming. Their apparently normal child had a hidden vulnerability that they cannot be certain of controlling, and might be exacerbated by their smoking habit. The only practical help available was the advice to give up smoking to avoid its secondary effects.

Screening so that the parents can be offered genetic counselling: Duchenne muscular dystrophy (DMD)

This severe disorder (p. 247) can be detected simply and reliably in the newborn (Dellamonica *et al.* 1978). Early diagnosis does not affect outcome, but if all affected infants were identified early, it is estimated that 15–20 per cent of cases could be avoided through family studies and genetic counselling (Gardner-Medwin 1983). There has, however, been little enthusiasm for neonatal screening with the primary objective of reproductive counselling, partly because it entails informing the parents of their son's fatal illness long before he has any symptoms. An alternative approach based on testing infants who could not walk by 18 months, whose parents would already be worried, was relatively ineffective (Smith *et al.* 1989). Neonatal screening for DMD is practised in one region of France, and evaluation of the results is awaited.

G6PD deficiency

Neonatal screening for G6PD deficiency (p. 249) using a simple dye discoloration test is practised in parts of Southern Europe and Singapore where the condition is a common cause of neonatal jaundice with the risk of kernicterus (WHO 1989). Affected (hemizygous) males and many female carriers can be detected. G6PD deficiency now affects a significant proportion of the population of most North-West European countries (Table 15.2) but is not systematically screened for, ostensibly because infants with jaundice due to G6PD deficiency, and children and adults with haemolytic crises due to broad beans or drugs, are already diagnosed and treated within the health system. However, screening at least males for G6PD deficiency could easily be added to existing programmes for screening for haemoglobin disorders, since the ethnic groups involved are the same. This would allow hemizygotes to avoid some potentially disastrous haemolytic crises.

Key references

Wald, N. J. (ed.) (1984). *Antenatal and neonatal screening*, Oxford University Press.
Modell, M. and Boyd, R. (1989) *Paediatric problems in general practice*, (2nd edn). Oxford University Press.

Part 5

Long-term implications

23. Bad medicine for the human race?

It is commonly believed that modern medicine prevents natural selection by protecting the unfit, and so must be causing a slow but inexorable increase in the population frequency of bad genes (Muller 1950), which will eventually lead to a genetic 'Gotterdammerung' (Fig. 23.1). It is also feared that if people are allowed to exercise choice about the genotype of their children (through prenatal diagnosis or other methods), they will increase genetic and social uniformity by choosing only 'perfect' children. However, examination of the facts gives a much more optimistic view of future of the human race, and shows no tendency to genetic uniformity. To discuss these concepts, we need to review the ideas of human fitness and of a 'bad' gene, consider some of the factors presently influencing the human gene pool, and look at how parents actually use genetic information in practice.

Genetic fitness?

The term 'fitness' as used by population geneticists refers to an individual's relative ability to hand on their genes to subsequent generations. To be fit, people must occupy a place in the world that allows them to survive and bring up offspring (an ecological 'niche'). We are inclined to visualize human

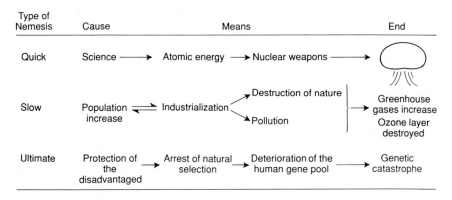

Fig. 23.1 The 'Götterdämmerung' outlook. The downfall of the human race is assured: even if global catastrophe due to nuclear weapons or the greenhouse effect can be overcome, an ultimate genetic catastrophe is unavoidable.

fitness in terms of the natural (= wild) world, but this is quite inappropriate for the world we now live in.

Human population density depends on the level of technical development (McEvedy and Jones 1978). Throughout most of history the main factor limiting population numbers has been food supply, and the main control mechanism has been infant and maternal mortality. War is a relatively ineffective method of population control when other conditions favouring population growth are present. Until recently, for instance, it killed predominantly the male sex. As the number of children that can be born depends on the number of females, the numerical effects of even the worst wars have been overcome within a very few generations in growing populations

Hunter-gatherers require many square miles per person for survival. The introduction of agriculture permitted population densities hundreds of times higher, and modernization permits previously unimaginable densities. Furthermore, in 'developed' societies, population number is controlled for the first time by birth control rather than high early mortality. The present vast numbers of the human race mean that we are irreversibly committed to a highly technological, urbanized, and man-made society. Such societies offer many new niches in which deaf, blind, physically disabled, and dwarfed people, for example, who would be hopelessly handicapped in more primitive conditions, can earn a living, reproduce, bring up children successfully, and make a valuable contribution to the community in which they live. For example, in the past achondroplasia (p. 237) led to low 'fitness', partly because of social isolation, and partly because pregnancy was often lethal for females. Most cases were due to new mutations. However, in large urban societies achondroplasics can work normally and associate with each other: most now marry, and many reproduce (Murdoch *et al.* 1970). The fact that a female achondroplasic requires a Caesarian section cannot even be considered unusual in societies where from 5–12 per cent of 'normal' pregnancies end in Caesarean delivery (*Having a baby in Europe* 1987). The proportion of familial cases is therefore increasing. The same principle applies to many other genetic conditions.

The effect of social change on the human gene pool

The objective in treating people with a chronic disability is to achieve a quality and length of life that is as near normal as possible, an aim that includes sexual relationships, and reproduction for those able to undertake the responsibilities of parenthood. However, it is often feared that if people who would previously have died or been infertile because of a major genetic disorder now survive and reproduce, genetic disease will become increasingly common. On p. 4 it was shown that the *prevalence* of people with chronic congenital and genetic problems inevitably rises as treatment improves.

Here, though, we are concerned with the *birth incidence*, the objective measure of frequency of people with a genetic disadvantage.

There is little real evidence that treating affected people so that they survive and have children when they can, will prove disadvantageous for the human race. Relatively few congenital malformations are simply inherited, and as relatively few are handed on to offspring, there is as yet little evidence that correcting them has any effect on the human gene pool. People with the common chromosomal aneuploidies rarely hand them on to offspring. Treatment often fails to restore reproductive fitness in inherited diseases (Hayes *et al.* 1985) and patients who can reproduce often choose prenatal diagnosis in order to have healthy children. However, we still know too little about the reproduction of people with genetic disorders relative to population norms, to be able reach any definite conclusions.

The high infant mortality that prevails in less developed conditions is often seen as a rigorous form of natural selection in which the unfit are weeded out. But survival in conditions where 50 per cent of children die before the age of 12 depends more on social and environmental factors than on innate biological fitness, and most of the survivors end up less fit than they would be in better conditions. There is no real evidence that allowing such circumstances to continue is good for the human gene pool. On the other hand, improving the environment leads to substantial improvements in phenotype as genotype is given the chance to express itself more fully. In developed

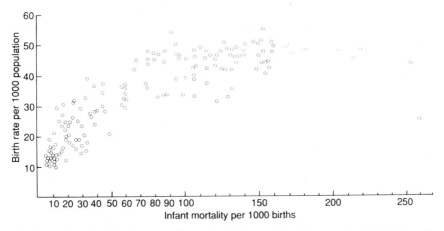

Fig. 23.2 Relationship between the crude birth rate (per thousand population) and the infant mortality (per thousand births) in about 1980. Each point represents one country (all are shown). When infant mortality is high, birth rate remains high in order to compensate. The birth rate falls when infant mortality falls, whether or not there is a national family planning policy. However, a family planning programme greatly accelerates the fall. Points in the bottom left-hand corner represent the developed countries with low infant mortality and a low birth rate, including the European countries, North America and Japan. (Based on data in the UN Demographic Yearbook 1984.)

countries, recent generations are on average taller than their parents (Tanner 1989), and there has also been a steady increase in IQ (Lynn *et al.* 1987), both presumably the result of improved nutrition and protection from infections.

In addition, there is evidence that improving public health is positively good for the human gene pool. One of its principal benefits is reduced infant mortality and the availability of family planning, which lead to a greatly decreased birth rate (Fig. 23.2) — a rational response to increased confidence in one's children's survival. This in turn greatly alters parental age distribution, as couples stop reproducing once they have adequate surviving children (Fig. 23.3). Since the end of the second world war, the birth rate has fallen

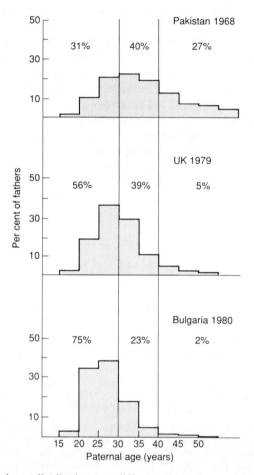

Fig. 23.3 Paternal age distribution can differ markedly between countries. Pakistan is typical of many contemporary developing countries, with a majority of fathers over 30 years old. In the UK, just under 50% of fathers are over 30 and in most Eastern European countries the majority are below 30.

throughout Europe, details varying with country and time (Modell and Kuliev 1989). The peak age for reproduction in Western Europe is now 25–30 years, and in Eastern Europe 20–25 years, and far fewer couples than previously have children in their late 30s and 40s. The proportion of mothers over 35 years old fell from over 20 per cent in the 1950s to about 6 per cent in the early 1970s (but has since rebounded to 10 per cent or more in many Western European countries). There has been a similar fall in the proportion of older fathers.

Such changes tend to reduce the incidence of abnormalities related to parental age. Firstly, the birth incidence of infants with chromosomal aneuploidies has fallen by about 50 per cent. It is estimated that the incidence of

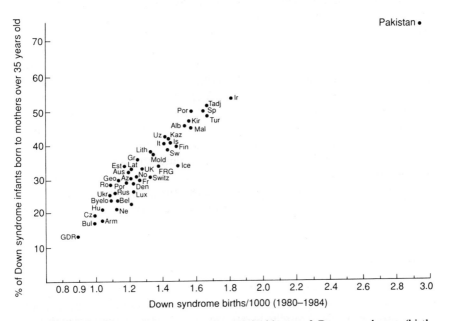

Fig. 23.4 Relationship between the estimated incidence of Down syndrome (births per 1000) in different European countries (Pakistan is shown for comparison with a developing country), and the per cent of infants with Down syndrome born to mothers over 35 years old. The latter figure determines the efficiency of prenatal screening for Down syndrome using maternal age as a primary indicator. (Alb = Albania; Arm = Armenia; Aus = Austria; Az = Azerbaidjan; Bel = Belgium; Bul = Bulgaria; Byelo = Byelorussia; Cz = Czechoslovakia; Den = Denmark; Est = Estonia; Fin = Finland; Fr = France; FRG = Federal Republic of Germany; GDR = German Democratic Republic; Geo = Georgia; Gr = Greece; Hu = Hungary; Ice = Iceland; Ir = Ireland; Is = Israel; It = Italy; Kaz = Kazakhstan; Kir = Kirgizia; Lat = Latvia; Lith = Lithuania; Lux = Luxembourg; Mal = Malta; Mold = Moldavia; Ne = Netherlands; No = Norway; Po = Poland; Por = Portugal; Ro = Romania; Rus = Russia; Sp = Spain; Sw = Sweden; Switz = Switzerland; Tadj = Tadjikistan; Tur = Turkmenia; UK = United Kingdom; Ukr = Ukraine; Uz = Uzbekistau). (Reproduced from Modell and Kuliev 1989, with permission).

Down syndrome in Europe in the past with no family planning may have been as high as 2.5 in a 1000 (as in some contemporary developing countries). Today the figure ranges from 0.9–1.8 per 1000 in different European countries (Fig. 23.4). There must have been a corresponding fall in the incidence of spontaneous abortion, which is similarly related to maternal age.

Mutation rate is similarly related to paternal age (Fig. 5.1). A recent calculation suggests that in Europe in the past 50 years, the mutation rate could have fallen by as much as half, as a result of changed paternal age distribution (Modell and Kuliev 1990). In the long term, this could have quite a marked effect on the incidence of inherited disease.

Most severe inherited disorders are rare because they reduce reproductive fitness, and so are passed on to fewer of the next generation. The frequency of genes for such rare conditions is maintained by new mutation. At any given time, therefore, their collective birth incidence represents a balance between net mutation rate (tending to increase it) and the effects of natural selection (tending to reduce it). Thus, in very general terms, the total frequency of rare deleterious genes reflects the recent human mutation rate. If the mutation rate decreases because there are fewer older fathers, the incidence of rare inherited diseases should also start to decrease.

However, increased exposure to environmental mutagens could counteract such effects: for example, smoking is still increasing in parts of Europe (WHO 1987), and tobacco smoke may cause mutations in germ as well as somatic cells (International commission for protection against environmental mutagens and carcinogens, 1979). Clearly the human mutation rate is a dynamic entity that varies from place to place and time to time as a result of changes in human behaviour.

What is a bad gene?

As yet, we know far too little about most genes to be able to label any but a few as completely undesirable or desirable. For example, intelligence, one of the most desirable of human assets, is to some extent associated with some mildly disadvantageous or unusual characteristics, such as short sightedness (Fig. 23.5) (Karlsson 1975), left-handedness (Persson Benbow 1987) and gout (Sofaer and Emery 1981). It is likely that as we learn more about them, we will find that most genes interact with many others in rather unexpected ways, with both harmful and beneficial results, the balance between the two varying with the environment.

Some genes have relatively little disadvantage when average life expectancy is short but become increasingly 'bad' as life expectancy increases and allows disorders of later life such as Alzheimer's disease, Huntington's disease and others to manifest themselves. This is causing increasing concern, and at the same time alerts families to risk. Genes for such highly disadvantageous late onset conditions are not rapidly removed by natural

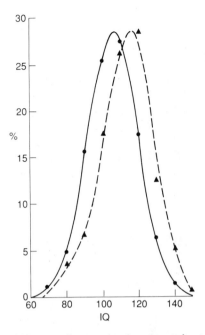

Fig. 23.5 Distribution of IQ scores for people who are not short-sighted (solid line) and short-sighted people (broken line) on the Lorge–Thorndike group test at age 17–18 years. Forty-four per cent of subjects with an IQ of 135 and over had started wearing glasses before 10 years of age. (Reproduced from Karlssen 1975, with permission.)

selection, which operates on disorders that cause disability before or during the reproductive years. However, some at-risk couples are now beginning to use carrier and prenatal diagnosis to select against genes that are harmful in later life.

In fact, most really common (recessively inherited) disorders are positively advantageous to heterozygotes (carriers) (Table 23.1). Since their gene frequency represents a balance between their advantage to carriers and their disadvantage to homozygotes (p. 68), it will be influenced by changes in either of these factors.

A heterozygote advantage usually means that under natural conditions there is greater mortality amongst 'normals' than among carriers, so public health initiatives (such as sanitation, immunization, and malaria eradication) that reduce the mortality of all infants reduce heterozygote advantage. This can lead to quite rapid changes in gene frequency. For example, in Cyprus normal infants were far more vulnerable to malaria than infants with β-thalassaemia trait, and this led to a 17 per cent incidence of thalassaemia trait in the adult population (Angastiniotis and Hadjiminas, 1981). In 1947 malaria was eradicated from Cyprus, and this differential mortality was eliminated. Though the care of patients with thalassaemia major has

Table 23.1 Possible selective advantage associated with traits for some common inherited disorders

Condition		Proposed advantage to heterozygote	Reference
Recessive	Sickle cell	Protection against malaria	Livingstone (1967)
	Thalassaemia	Protection against malaria	
	Tay-Sachs disease	Protection against tuberculosis	Rotter and Diamond (1987)
	Cystic fibrosis	Protection against gastroenteritis	Quinton (1982)
	Haemochromatosis	Protection against iron deficiency	
X-linked	G6PD deficiency	Protection against malaria (mainly female carriers)	Luzzatto and Mehta (1989)
Dominant	Huntington's disease	Increased final family size	Walker et al. (1983)

improved enormously, few are yet able to reproduce, so thalassaemia genes are still leaving the population. In theory, the frequency of β-thalassaemia trait should consequently fall from 17 to 15.5 per cent in a single generation, and the change has been observed in practice (M. A. Angastiniotis, personal communication). Even if future treatment for thalassaemia major becomes so good that patients reproduce normally, there would be no *increase* in thalassaemia gene frequency. It would simply stabilize because neither normal nor thalassaemia genes would be leaving the population disproportionately. A similar argument probably applies for most conditions in Table 23.1.

Genetic traits undesirable at one stage of life can be beneficial at another. Carriers of thalassaemia trait or G6PD deficiency seem to have a decreased risk of coronary heart disease (Crowley *et al.* 1987, Long and Wilson 1967). This probably had little relevance in the past, because relatively few people in populations where these genes are common lived long enough or had a rich enough diet to die from coronary heart disease. However, in modern societies one of these genes could be a highly desirable asset.

Effects of genetic counselling

On its own, genetic counselling can reduce the frequency of severe dominant and X-linked disorders, if people who understand their risk then decide against having (further) children (Harper *et al.* 1981). Prenatal diagnosis considerably reinforces such effects because many people who might previously have remained childless or 'taken the chance' use prenatal diagnosis to have only healthy children (Modell 1988).

It has been proposed that prenatal diagnosis for recessively inherited conditions will, paradoxically, increase the frequency of deleterious recessive genes, because it allows homozygotes (who would not have reproduced) to be replaced with healthy individuals, two-thirds of whom are heterozygous and will pass on their genes (Harris 1974). This calculation is wrong because it assumes that couples undergoing prenatal diagnosis attain the same final family size as the rest of the population, but most couples at high genetic risk aim for a small healthy family, and then use contraception very effectively. For instance, it seems that a significant proportion of the decrease in thalassaemic births in Cyprus may be due to at risk couples having fewer children than others (Angastiniotis *et al.* 1986).

In the future, if reliable diagnosis prior to implantation becomes available (p. 161), couples might select normal rather than carrier fertilized ova for implantation. If so, the frequency of severely deleterious genes would gradually fall.

Population migrations

The large scale population movements now common in many parts of the world can also alter the pattern of genetic disease. For example, ethnic groups with a high frequency of genes for haemoglobin disorders now constitute 1–9 per cent of North-West European populations and up to twice as many births (WHO 1988). The groups involved mostly have a relatively low incidence of cystic fibrosis, and population mixing will reduce the frequency of both types of gene. In the UK complete mixing might finally reduce the birth incidence of children with cystic fibrosis by 16 per cent and of children with haemoglobin disorders by 90 per cent. Some migrant groups also have a convention of consanguineous marriage (Chapter 9). One ultimate effect could be a slight reduction in the frequency of deleterious genes in the whole population.

These considerations all suggest that a decreased incidence of genetic disease is an important side-effect of improving primary care. But could the tendency go too far?

Effects on the human gene pool of allowing parents control over the genotype of their children

Will the availability of prenatal diagnosis and reproductive choice militate against the current increasing acceptance of handicapped or unusual people, and generate pressures to accept only 'perfect' babies? A 'slippery slope' can be visualized, leading from selective abortion for severe handicaps, through abortion for milder disorders, to abortion of fetuses with only minimal disadvantages. However, let us see what actually happens with prenatal diagnosis at present.

Both uptake of prenatal diagnosis and the decision to terminate an affected pregnancy depend on the couple's perception of numerous factors discussed in Chapter 14, on counselling. People do not rush into prenatal diagnosis, nor do they rush to terminate a pregnancy when an abnormality is diagnosed. It is true that most couples choose termination of pregnancy when the fetus is severely affected, but if the situation is less clear, many choose to continue the pregnancy. When a congenital malformation is diagnosed by ultrasound scanning, for example, the parents usually request more information rather than opting for immediate abortion. In most cases their primary concern is the well-being of the fetus and of their family, and many continue an affected pregnancy if they judge that their family circumstances and prevailing social and economic conditions will allow the affected child to lead a fulfilling life. For example, there is at present relatively little demand in Western Europe for selective abortion of fetuses with phenylketonuria (p. 217). In practice this means that people are starting to decide for themselves on the appropriate range of fit phenotypes for present society.

In conclusion, there is no evidence that primary health care and medical genetics are bad for the human gene pool. On the contrary, there is positive evidence that replacing natural selection by improved care for both normal and handicapped people, actually reduces the incidence of genetic disease. As it gradually becomes possible to identify people at genetic risk and allow them to make their own decisions by placing information in their hands, human evolution is naturally and unavoidably becoming a human responsibility.

Key references

Bodmer, W.F. and Cavalli-Sforza, L.L. (1976). *Genetics, evolution, and man*. W.H. Freeman and Co., San Francisco.

Modell, B. and Kuliev, A.K. (1988). The impact of public health on human genetics. *Clinical Genetics*, **36**; 286–98.

Vogel, F. and Motulsky, A.G. (1986). *Human genetics: problems and approaches* (2nd edn). Springer-Verlag, Berlin.

Appendix

Support groups for patients and families with genetic and congenital disorders, or related problems

We are grateful to Christine Lavery of Contact a Family, and Ann Hunt of the Genetic Interest Group, for providing the following information. The list is not complete. If you cannot find the appropriate contact for a patient, you can obtain advice from:

Contact a Family
16 Strutton Ground
London
SW1P 2HP
Tel: 071–222–2695

The Genetic Interest Group (GIG) is the umbrella organization for support groups for genetic disorders. It can be contacted through the above telephone number.

Abortion for fetal abnormality

SATFA (Support After Termination
for Abnormality)
29/30 Soho Square
London
W1V 6JB

CARE (The Scottish Association for
Care and Support after Termination
for Fetal Abnormality)
32 Linkwood Press
Lawthorn
Irvine
Ayrshire

Agenesis of corpus callosum

CORPAL
4 Harcourt Road
Dawney Road
Maidenhead
Berks.
SL6 0DU

Alzeimer's disease

Alzheimer's Disease Society
158/160 Balham High Road
London
SW12 9BN

Angelman syndrome

Angelman Syndrome Support Group
15 Place Crescent
Waterlooville
Portsmouth
Hants.
PO7 5UR

Ataxia telangiectasia

Ataxia Telangiectasia
33 Tuffnells Way
Harpenden
Herts.
AL5 3HA

Talangiectasia Self-help Group
39 Sunny Croft
Downley
High Wycombe
Bucks.
HP13 5UP

Autism

National Autistic Society
276 Willesden Lane
London
NW2 5RB

Cancers in childhood

BACUP (British United Cancer
Parents)
121/3 Charterhouse Street
London
EC1M 6AA

Charcot–Marie–Tooth disease

CMT International UK
(Charcot–Marie–Tooth International
UK)
34 Burleigh Close
Strood
Kent
ME2 3TQ

Chromosome disorders

Down's Syndrome Association
12–13 Clapham Common
Southside
London
SW4 7AA

Rare Unspecified Chromosome
Disorder Support Group
160 Locket Road
Harrow Weald
Middx.
HA3 7NZ

Chronic granulomatous disease

CGD Society
Seafield
Shootersway Lane
Berkhampstead
Herts.
HP4 3NP

Cockayne syndrome

Cockayne Syndrome Support Group
18 Edenway
Brickhill
Bedford
Beds.
MK41 7EP

Cystic fibrosis

Cystic Fibrosis Research Trust
5 Blyth Road
Bromley
Kent
BR1 3RS

Deafness

Hearing Research Trust
330/332 Grays Inn Road
London
WC1X 8EE

Down syndrome

see chromosome disorders

Dystonia

The Dystonia Society
Omnibus Workspace
39–41 North Road
London
N7 9DP

Ectodermal dysplasias

Ectodermal Dysplasias
37 Cambridge Close
Haverhill
Suffolk
CB9 9HP

Ehlers Danlos syndrome

Ehlers Danlos Support Group
2 High Garth
Richmond
North Yorkshire
DL10 4DG

Epidermolysis bullosa

DEBRA (Dystrophic Epidermolysis
Bullosa Research Association)
1 Kings Road
Crowthorne
Berks
RG11 7BG

Fanconi's anaemia

FAB-UK (Fanconi Anaemia
Breakthrough UK)
4 Pateley Road
Woodthorpe
Nottingham
Notts.

Fragile X

Fragile X Association
11 Radlet Avenue
London
SE26 4BZ

Friedreich's ataxia

Friedreich's Ataxia Group
Copse Edge
Thursley Road
Elstead Godalming
Surrey
GU8 6DJ

Galactosaemia

Galactosaemia Parent Support Group
18 Nuthurst
Sutton Coldfield
West Midlands
B75 7EZ

Gaucher's disease

Gaucher's Disease Group
31 Delaney Close
Tilehurst
Reading
Berks.
RG3 4UY

Glaucoma

International Glaucoma Association
King's College Hospital
Denmark Hill
London
SE5 9RS

Growth

Restricted Growth Association
103 St Thomas Avenue
Hayling Island
Hants.
PO11 OEU

Haemophilia

Haemophilia Society
123 Westminster Bridge Road
London
SE1 7HR

Heart

Heart Care
112 Irish Street
Downpatrick
County Down
BT30 6BT

Huntington's disease

Huntington's Disease Association
108 Battersea High Street
London
SW11 3HP

Hydrocephalus

ASBAH (Association for Spina
Bifida and Hydrocephalus)
22 Upper Woburn Place
London
WC1H OEP

Hypogammaglobulinaemia

HGG Society
(Hypogammaglobulinaemia Society)
74 Beverley Road
Whyteleafe
Surrey CR3 DP

Lowes syndrome

Lowes Syndrome Association
29 Gleneagles Drive
Penwortham Preston
Lancs.
PR1 0JT

Marfan

Marfan Association
70 Greenways
Courtmoor
Fleet
Hants.
GU13 9XD

Mental handicap

MENCAP
123 Golden Lane
London
EC1Y ORT

Undefined Mental Handicap Contact Group
7 Springwell Close
Dean's Park
Billingham
Cleveland

Metabolic diseases

RTMDC (Research Trust for
Metabolic Diseases in Children)
53 Beam Street
Nantwich
Cheshire
CW5 5NF

Mucopolysaccharidosis

Mucopolysaccharidosis Society
7 Chessfield Park
Little Chalfont
Bucks.
HP6 6RU

Muscular dystrophy

Muscular Dystrophy Association
35 Macaulay Road
Clapham
London
SW4 OQP

Naevus

Naevus Support Group
58 Necton Woad
Wheathampstead
St Albans
Herts.
AL4 8AU

Neurofibromatosis

Neurofibromatosis Association (LINK)
BO9 Surrey House
34 Eden Street
Kingston
Surrey
KT1 1ER

Osteopetrosis

Osteopetrosis Contact Group
10 Cumberland Avenue
Fixby
Huddersfield

Phenylketonuria

The National Society for
Phenylketonuria (UK) Ltd
26 Towngate Grove
Mirfield
West Yorks.

Retinitis pigmentosa

BRPS (British Retinitis Pigmentosa
Society)
Greens Norton Court
Greens Norton
Towcester
Northants

Rett syndrome

UK Rett Syndrome Association
Hartspool
Golden Valley
Castlemorton
Malvern
Worcs.
WR13 6AA

Rubella

SENSE (The national deaf-blind and
Rubella association)
311 Grays Inn Road
London
WC1X 8PT

Rubinstein Taybi syndrome

Rubinstein Taybi Syndrome Support Group
46 Windsor Road
Great Harwood
Blackburn
Lancs.
BB6 7RR

Schizophrenia

National Schizophrenia Fellowship
78 Victoria Road
Surbiton
Surrey
KT6 4NS

Sickle cell disease

OSCAR (Organization for sickle cell
anaemia research)
103 Cotswold Way
Tilehurst
Reading
Berks.
RG3 6SR

Sickle Cell Society
Green Lodge
Barretts Green Road
London
NW10

National Sickle Cell Programme
PO Box 322
London
SE25 4BW

SCARF (Sickle Cell Anaemia Research
Foundation)
PO Box 2055
London
W12 7JE

Sotos

Sotos Group
Kilndown House
Kilndown
Cranbrook
Kent
TN17 2SG

Spina bifida and hydrocephalus

ASBAH (Association for Spina Bifida
and Hydrocephalus)
22 Upper Woburn Place
London
WC1H OEP

Spinal muscular atrophy (SMA)

Jennifer Trust for SMA
11 Ash Tree Close
Wellesbourne
Warwicks.
CV35 9SA

Tay–Sachs disease

British Tay–Sachs Foundation
44a New Cavendish Street
London
W1M 7LG

Tay–Sachs and Allied Disease Association
17 Sydney Road
Barkingside
Ilford
Essex
IG6 2ED

Termination of pregnancy

see Abortion for fetal abnormality

Thalassaemia

UK Thalassaemia Society
107 Nightingale Lane
London
N8 7QT

Treacher Collins

Treacher Collins Family Support
Group
114 Vincenty Road
Thorpe Hamlet
Norwich
Norfolk
NR1 4HH

Tuberous sclerosis

Tuberous Sclerosis Association of
Great Britain
Little Barnsley Farm
Catshill
Bromsgrove
Worcs.
B61 0NQ

Von Hippel Lindau

Von Hippel Lindau Contact Group
The Old Farmhouse
Dawson Fold
Lyth
Kendal
Cumbria
LA3 8DE

Werdnig–Hoffman disease

see Spinal muscular atrophy

Glossary

Alleles   Alternative possible genes found at the same locus on homologous chromosomes. They usually code for alternative versions of the same protein.

Aneuploid   Not having a normal number of chromosomes or an exact multiple of this number. One or more chromosomes are present in too many, or too few copies.

Antibody   An immunoglobulin which is produced by B lymphocytes after contact with a specific antigen. The antibody then binds that antigen.

Antigen   Any substance that can stimulate antibody production, or another type of immune response.

Autosome   A chromosome that is not one of the sex chromosomes.

B lymphocytes   White blood cells that produce antibodies after contact with a specific antigen.

Barr body   The inactivated X chromosome, which can be seen as sex chromatin in somatic cells of females.

Base pair   A pair of complementary bases in double-stranded DNA. Adenine (A) pairs with thymine (T) and guanine (G) with cytosine (C).

Carrier   A symptomless heterozygote for a mutant allele.

Cell-mediated immunity   Immunity mediated by thymus-derived lymphocytes (T cells) which are activated when they come into contact with a host cell.

Centromere   The point on the chromosome at which two chromatids are joined together when a chromosome has duplicated. Spindle fibres attach to the centromere.

Chimera   An individual with cells of two different genotypes.

Chromatids   The two strands produced when a chromosome replicates during cell division.

Chromosome   (= coloured body). A thread-like structure in the nucleus of the cell, which contains a tightly packed portion of an individual's DNA. It is acidic and therefore picks up basic dyes.

Clone   A group of cells descended from a single cell and genetically identical

Codon   A sequence of three DNA or RNA bases which code for one amino acid

Complementary DNA (cDNA)   DNA built up on an mRNA template using the enzyme reverse transcriptase.

Concordance (in twins)   The presence of the same trait in both members of a pair of twins.

Congenital   Present at birth

Congenital anomaly   Any structural, functional or biochemical abnormality present at birth, whether detected at that time or not.

Consanguineous marriage   A marriage in which the partners are related.

Crossing over   Exchange of segments of homologous chromosomes during meiosis.

Deoxyribonucleic acid (DNA)   Molecule made of two paired strands composed of subunits made of an organic base and a pentose sugar, linked by phosphate bonds. It is the acidic component of chromosomes and carries the genetic code.

Discordance (in twins) Presence of a trait in only one member of a pair of twins.

Diploid Having two complete sets of chromosomes. In humans, diploid cells (the somatic cells) have 46 chromosomes.

DNA polymerase The enzyme responsible for replication of DNA.

Dominant trait A trait which is clinically expressed when one copy of a mutant gene is present.

Embryonic disc The cells that will form the actual embryo itself.

Empirical risk Estimate of recurrence risk made by analysing observations in a large number of cases, but without understanding the underlying causes.

Euploid Having one or more complete chromosome sets. A single normal human set consists of 23 chromosomes. Normal human gametes are haploid (1×23 chromosomes), other cells in normal humans are diploid (2×23 chromosomes).

Exon A part of the DNA sequence of a gene which codes for the final protein product. All human genes contain several exons separated by non-coding 'intron' sequences.

Expression (of a genetic disorder) The clinical severity of a genetic disorder.

Fitness (biological) Relative ability to pass one's genes on to future generations. Measured by the number of viable live births relative to the population norm.

Founder effect In small populations, a pathological gene carried by only one of the founding members can be passed on to a disproportionately high number of descendants, and a high frequency of a particular condition may arise.

Frame-shift A DNA mutation that results in deletion or insertion of one or two base pairs and so alters the triplet coding from that point on. A frame-shift completely alters the sense of a message. For example, the previous sentence typed with one shift to the right on the keyboard reads as follows: S gtsmr = dhigy vp,[;ryr;y s;yrtd yjr drmdr pg s ,rddshr.

Gametes Ova or sperm.

Gene Portion of a DNA molecule with the sequence that codes for a specific protein.

Genetic Determined by the hereditary material.

Genetic code In coding sections of DNA, each set of three nucleotides (triplet) codes for one amino acid. The specific sequence of each triplet determines which of the 20 amino acids will be selected for incorporation into the protein.

Genetic engineering The alteration of DNA sequences artificially to produce a different version of a gene or part of a gene.

Genotype A person's genetic constitution, i.e. the pattern in their DNA.

Germ cells Ova and sperm and their precursors.

Guthrie card An absorbent card which is a vehicle for blood spots for neonatal screening (e.g. for phenylketonuria).

Haploid Having one complete chromosome set. In humans, one set consists of 23 chromosomes. Gametes are haploid.

Haplotype A number of identifiable DNA sites close together on a single chromosome and therefore usually inherited together.

Hemizygous Having 1 instead of 2 copies of a gene. Males have only one X chromosome, and so only 1 copy of all genes on the X chromosome. They are hemizygous for all genes on the X chromosomes.

Heritability   The proportion of a characteristic that is the result of genetic rather than environmental factors.

Heterochromatin   The parts of chromosomes that stain with basic dyes when other parts do not.

Heterozygote   An individual who carries two different versions of the same gene on the two relevant homologous chromosome.

Histones   Proteins, scarcely altered during evolution, which are associated with DNA in chromosomes.

HLA glycoproteins (MHC glycoproteins).   A set of polymorphic proteins found on the surface of most cells, which play an important part in immunity. They are called the HLA (= human leucocyte antigen) complex, because they were first identified in white cells.

Homologous chromosomes   Corresponding chromosomes which pair during meiosis. One is derived from each parent.

Homozygote   An individual who carries identical versions of the gene under consideration on the two relevant homologous chromosomes.

Human genome   The characteristic pattern, i.e. the common content and order, of the DNA of the human species. The term refers to the haploid DNA complement, as only one set of DNA is necessary to define the sequence characteristic for the species.

Hybridization (in DNA technology)   Specific binding of a probe to complementary DNA (or RNA) to re-form a double strand. This is the basis for identifying the presence of specific known sequences for which a probe exists.

Intron   Non-coding DNA sequence contained within a gene. Introns are initially transcribed into RNA in the nucleus, but are removed before messenger RNA enters the cytoplasm.

Karyotype   The chromosomal make-up of an individual. A photomicrograph of a person's chromosomes.

Kilobase   One thousand base pairs.

Linkage   Two genes that are close together on a chromosome, and therefore are seldom separated during crossing over in meiosis, tend to be inherited together and are said to be 'linked'. An example is the association of specific HLA genes with diseases such as diabetes and ankylosing spondylitis.

Lyonization   The phenomenon of random inactivation of one X chromosome in every cell of females.

Meiosis   The type of cell division that occurs only in germ cells and halves the number of chromosomes, to produce haploid cells from diploid ones.

Mitochondria   Cytoplasmic organelles responsible for aerobic respiration. They contain their own DNA.

Mitosis   The usual type of cell division which results in two daughter cells each with a diploid chromosome complement.

Monosomy   Absence of one member of a pair of chromosomes. For example, Turner's syndrome = 45 X.

Mosaic   A person or tissue whose cells are partly of one genetic make-up and partly of another.

Multifactorial conditions   Disorders which result from the additive effect of a number of factors, some genetic, others environmental or unknown. This very large and heterogeneous group includes most congenital malformations and most major chronic diseases of later life.

Mutagen A substance which can cause an alteration (mutation) in the sequence of the DNA.

Mutation A spontaneous change in the structure or arrangement of DNA, or in the number or structure of the chromosomes.

Non-disjunction The failure of two homologous chromosomes to separate during meiosis, so that both enter the same daughter cell.

Nucleolus A nuclear structure where ribosomal RNA is made. It is formed by aggregation of the short arms of chromosomes 13, 14, 15, 21, and 22 that carry the genes for ribosomal RNA.

Nucleotide The subunit of DNA and RNA. It consists of a nitrogenous base, a pentose sugar, and a phosphate group.

Oligonucleotide probe Short DNA probe, usually about 19 base pairs long.

Oncogene A modified growth-controlling gene that may cause cancer.

Organelle A specialized structure inside a cell.

Penetrance (of a gene) The proportion of people carrying a (dominant) gene who are clinically affected.

Phenotype The physical, physiological, and biochemical characteristics of an individual. The phenotype is the result of interaction of genotype with environment.

Plasmid A small self-replicating circle of DNA within a bacterium or yeast, into which DNA probes can be inserted and so multiplied.

Polymerase chain reaction (PCR) A simple method of amplifying a short length of DNA a million times or more.

Polymorphism An alternative harmless form of a gene, which is common.

Polyploidy Presence of more than two whole sets of chromosomes. For example, triploid ($3n$) = three sets (69 in humans); tetraploid ($4n$) = four sets (92 in humans).

Probe A length of labelled DNA with a specific sequence which is complementary to the DNA sequence being sought, and will pair with it when it is single-stranded.

Prospective diagnosis Detection of a couple at risk of having a child with a genetic disorder, before they start their family.

Propositus or proband The first family member in whom a genetic condition is suspected.

Proto-oncogene Normal gene which codes for a protein involved in control of cell growth. When modified, it can become an oncogene and lead to cancer.

Recessive trait A trait which is clinically evident only in individuals homozygous for the mutant gene.

Restriction fragment length polymorphisms (RFLPs) Common mutations in DNA, which alter the length of the DNA fragment cut by a specific restriction enzyme.

Restriction enzyme A bacterial enzyme which can cut DNA at a specific base sequence, and therefore produces specific lengths of DNA.

Retrovirus RNA virus which is transcribed to DNA on entering a cell and can be integrated into the host's DNA.

Retrospective diagnosis Detection of a couple at risk of having a child with a genetic disorder, after the diagnosis of the first affected child.

Reverse transcriptase An enzyme which can catalyse the synthesis of DNA from an RNA template.

Ribonucleic acid (RNA) A nucleic acid which contains the sugar ribose. It is the characteristic nucleic acid of the cytoplasm and the nucleolus. It is the main interpreter of the genetic code of the DNA to protein. Genetic information from DNA is transcribed into messenger RNA (mRNA), which binds the ribosomes in the cytoplasm, and transfer RNA (tRNA) brings individual amino acids to the ribosomes to be incorporated into protein, according to the code of the bound mRNA.

Ribosome   Large cytoplasmic particle consisting of RNA and proteins, which contains the machinery for protein synthesis.

Sentinel phenotype   A clinical disorder of significant frequency and low fitness, which occurs sporadically as a consequence of a single highly penetrant mutation, and leads to a dominant or X-linked trait that is uniformly expressed and accurately diagnosable with minimal effort, at or near birth.

Sex chromatin   Inactivated X chromosome = Barr body.

Somatic   All cells of the body except ova and sperm.

Sporadic   Occurring at random.

Stem cells   Cells that can multiply throughout the life of an organism to produce more stem cells, or terminally differentiated cells.

T lymphocytes   White blood cells of the immune system responsible for cell-mediated immunity.

Teratogen   A substance which increases the rate of congenital malformations.

Transcription   Copying the sequence of the coding strand of DNA to complementary single-stranded messenger RNA.

Translocation   Transfer of part of a chromosome to another non-homologous chromosome at meiosis.

Transposable element   A mobile segment of DNA which can on rare occasions move to another chromosome in the same cell.

Trisomy   Presence of an additional copy of one of a pair of homologous chromosomes.

Zona pellucida   The thick, non-adhesive gelatinous capsule that surrounds the ovum.

Zygote   The fertilized ovum.

Further reading

We recommend the following books to readers who wish to pursue various aspects further.

Alberts, B., Bray, D., Lewis, J., Raff, M., Roberts, K., and Watson, J. D. (1989). *The molecular biology of the cell*. (2nd edn). Garland Publishing Inc., New York. (Absolutely essential for anyone interested in the scientific aspects.)

Scriver, C. R., Beaudet, A. L., Sly, W. S., and Valle, D. (eds) (1989). *The metabolic basis of inherited disease*. (6th edn). McGraw Hill Inc., New York. (Limited to conditions whose molecular basis is reasonably well understood. Contains useful clinical as well as scientific sections — but the book is very expensive!)

Weatherall, D. J. (1991). *The new genetics and clinical practice*. (3rd edn). Oxford University Press, for the Nuffield Provincial Hospitals Trust, Oxford.

Harper, P. S. (1988). *Practical genetic counselling*. (3rd edn). Wright, Bristol. (Very clearly written.)

Emery, A. E. H. and Rimoin, D. L. (2nd edn) (1990). Volumes 1 and 2. *Principles and practice of medical genetics*. Churchill Livingstone, Edinburgh.

Wald, N. J. (ed.) (1984). *Antenatal and neonatal screening*. Oxford University Press.

Ferguson-Smith, M. A. (ed.) (1983). *Early prenatal diagnosis*. British Medical Bulletin, vol. 39.

Rodeck, C. H. and Nicolaides, K. H. (ed.) (1983). *Prenatal diagnosis*. John Wiley and Sons, Chichester.

Brambati, B., Simoni, G., and Fabro, S. (eds) (1986). *Chorionic villus biopsy: fetal diagnosis in the first trimester of pregnancy*. Inc., New York.

Modell, M. and Boyd, R. (1989). *Paediatric problems in general practice*. (2nd edn). Oxford University Press. (Clearly written practical guide for family doctors.)

Gelehrte, T. D. and Collins, F. S. (1990). *Principles of medical genetics*. Williams and Wilkins, Baltimore. Attractive, simple illustrations.

References

Alberman, E.D. and Creasy, M.R. (1977). Frequency of chromosomal abnormalities in miscarriages and perinatal deaths. *Journal of Medical Genetics*, **14**, 313–15.

Alberts, B., Bray, D., Lewis, J., Raff, M., Roberts, K., and Watson, J.D. (1989). *The molecular biology of the cell*, (2nd edn). Garland Publishing Inc., New York.

Allen, L.D., Crawford, D.C., Chita, S.K., and Tynan, M. (1986). Prenatal screening for congenital heart disease. *British Medical Journal*, **292**, 1717–19.

Angastiniotis, M.A. and Hadjiminas, M.G. (1981). Prevention of thalassaemia in Cyprus. *Lancet*, **i**, 369–70.

Angastiniotis, M.A., Kyriakidou, S., and Hadjiminas, M. (1986). How thalassaemia was controlled in Cyprus. *World Health Forum*, **7**, 291–7.

Anionwu, E.N., Patel, N., Kanji, G., Renges, H., and Brosovic, M. (1987). Counselling for prenatal diagnosis of sickle-cell disease and β-thalassaemia major. A four year experience. *Journal of Medical Genetics*, **25**, 769.

Anonymous (1981). Children born as a result of incest. *British Medical Journal*, **282**, 250.

Anonymous (1989). Cerebral palsy, intrapartum care, and a shot in the foot. *The Lancet*, **ii**, 1251–2.

Anonymous (1990a). Antenatal screening for toxoplasmosis in the UK. *The Lancet*, **336**, 346–7.

Anonymous (1990b). Anonymous HIV testing. *The Lancet*, **i**, 575–6.

Association of Clinical Biochemists Newsheet (1983). Proposals for a co-ordinated scheme for the post-natal diagnosis of inherited metabolic diseases in infancy. Volume 237, pp. 6–14.

Asthana, J.C., Sinha, S., Haslam, J.S., and Kingston, H.M. (1990). Survey of adolescents with severe intellectual handicap. *Archives of Disease in Childhood*, **65**, 1133–6.

Baird, P.A. and Sadovnick, A.D. (1988). Life expectancy in Down syndrome adults. *The Lancet*, **ii**, 1354–6.

Baird, P.A., Anderson, T.W., Newcombe, H.B., and Lowry, R.B. (1988). Genetic disorders in children and young adults: a population study. *American Journal of Human Genetics*, **42**, 677–93.

Bakker, E., Hofker, M.H., N., Goor, Maudel, J.L., Wrogemann, K., and Davies, K.E. *et al.* (1986). Prenatal diagnosis and carrier detection of Duchenne muscular dystrophy with closely linked RFLPs. *The Lancet*, **ii**, 655–8.

Barker, D.J.P., Winter, P.D., Osmond, C., Margetts, B., and Simmonds, S.J. (1989). Weight in infancy and death from ischaemic heart disease. *The Lancet*, **ii**, 577–80.

Barker, D.J.P., Bull, A.R., Osmond, C., and Simmonds, S.J. (1990). Fetal and placental size and risk of hypertension in adult life. *British Medical Journal*, **301**, 259–62.

Behrman, R. and Vaughan, V. C. III (ed.) (1983). *Nelson Textbook of Pediatrics*, (12th edn). W. B. Saunders Company, Philadelphia.

Berman, L. and Klenerman, L. (1986). Ultrasound screening for hip abnormalities: preliminary findings in 1001 neonates. *British Medical Journal*, **293**, 719–22.

Bernard, S. H., Walsworth-Bell, J. P., Super, M., Read, A. P., and Harris, R. (1988). A survey of neural tube defect pregnancies in the north-western region of England. *Surgery in Infancy and Childhood*, **43** (Supplement 1), 15–16.

Best, J. M. and Banalvala, J. E. (1990). Congenital virus infections. *British Medical Journal*, **300**, 1151–2.

Bianchi, D. W., Flint, A. F., Pizzimenti, M. F., Knoll, J. H. M., and Latt, S. A. (1990). Isolation of fetal DNA from nucleated erythrocytes in maternal blood. *Proceedings of the National Academy of Sciences*, **87**, 3279–83.

Bicknell, J. (1989). Consent and people with mental handicap. *British Medical Journal*, **299**, 1176–7.

Bittles, A. H. (1980). Inbreeding in human populations. *Biochemical Reviews*, **50**, 108–17.

Bittles, A. H., Mason, W. M., Greene, J., and Rao, A. R. (1991). Reproductive behaviour and health in consanguineous marriages. *Science*, in press.

Black, J. and Laws, S. (1986). *Living with sickle cell disease. An enquiry into the need for health and social service provision for sickle cell sufferers in Newham.* Unpublished document of the East London Branch of the Sickle Cell Society, c/o Durning Hall, Earlham Grove, Forest Gate, London E7, UK.

Blanche, S., Rouzioux, C., Moscato, M.-L. G., Veber, F., Mayaux, M.-J., and Jacomet, C. (1989). A prospective study of infants born to women seropositive for human immunodeficiency virus type 1. *The New England Journal of Medicine*, **320**, 1643–8.

Bodmer, W. F. and Cavalli-Sforza, L. L. (1976). *Genetics, evolution, and man*. W. H. Freeman and Co., San Francisco.

Bothwell, T. H., Charlton, R. W., and Motulsky, A. G. (1989). Hemochromatosis. In *The metabolic basis of inherited disease*. (6th edn) (ed. C. R. Scriver, A. L. Beaudet, W. S. Sly, and D. Valle), pp. 1433–62. McGraw Hill Inc., New York.

Boue, A., Gallano, P., Serre, J. L., Feingold, J., and Boue, J. (1983). Genome and chromosome mutations: balance between appearance and elimination. In *Issues and reviews in teratology*. (ed. H. Kalter). Vol. 1. Plenum Press, New York.

Boue, A., Muller, F., Nezelhof, C., Oury, J. F., Duchatel, F., and Dumez, Y. (1986). Prenatal diagnosis in 200 pregnancies with a 1-in-4 risk of cystic fibrosis. *Human Genetics*, **74**, 288–97.

Bovicelli, L., Orsini, L. P., Rizzo, N., Montacuti, V., and Bachetta, M. (1982). Reproduction in Down Syndrome. *Obstetrics and Gynaecology*, **59** (Supplement), 13s–17s.

Bowling, F., Cleghorn, G., Chester, A., Curran, J., Griffin, B., Prado, J., Francis, P., and Shepherd, R. (1988). Neonatal screening for cystic fibrosis. *Archives of Disease in Childhood*, **63**, 196–8.

Bowman, J. M. and Pollack, J. M. (1965). Amniotic fluid spectrophotometry and early delivery in the management of erythroblastosis fetalis. *Pediatrics*, **35**, 815–32.

Boyd, J. D. and Hamilton, W. J. (1970). *The human placenta*, (2nd edn). Heffer, Cambridge.

Brambati, B., Simoni, G., Bonacchi, I., and Piceni, L. (1986*a*). Fetal chromosome aneuploidies and maternal serum alpha-fetoprotein levels in the first trimester. *The Lancet*, **ii**, 165–6.

Brambati, B., Simoni, G., and Fabro, S. (ed.). (1986*b*). *Chorionic villus biopsy: fetal diagnosis in the first trimester of pregnancy*. Marcel Dekker Inc., New York.

British National Formulary No 21 (1991). British Medical Association and the Pharmaceutical Society of Great Britain, London.

British Society for Haematology (1988). Guidelines for haemoglobinopathy screening. *Clinical and Laboratory Haematology*, **10**, 87–94.

Brock, D. J. H., Mennie, M., Curtis, A., Millan, F. A., Barron, L., Raeburn, J. A., Dinwoodie, D., Holloway, S., Crosbie, A., Wright, A., and Pullen, I. (1989). Predictive testing for Huntington's disease with linked DNA markers. *The Lancet*, **ii**, 463.

Brodie, M. J. (1989). Epilepsy, anticonvulsants and pregnancy. *Prescriber's Journal*, **29**, 251–8.

Buckley, S. and Sacks, B. (1987). *The adolescent with Down's syndrome. Life for the teenager and the family*. Portsmouth Down's Syndrome Trust, Psychology Department, Portsmouth Polytechnic, King Charles St., Portsmouth PO1 2ER, UK.

Burd, L. and Martsolf, J. T. (1989). Fetal alcohol syndrome: diagnosis and syndromal variability. *Physiology and Behavior*, **46**, 39–43.

Burton, L. (1975). *The family life of sick children: a study of families coping with chronic childhood disease*. Routledge and Kegan Paul, London.

Butler, N. R., Dudgeon, J. A., Hayes, K., Peckham, C. S., and Wybar, K. (1965). Persistence of rubella antibody with and without embryopathy. A follow-up study of children exposed to maternal rubella. *British Medical Journal*, **2**, 1027–9.

Campbell, S. and Smith, P. (1984). Routine screening for congenital abnormalities by ultrasound. In *Prenatal diagnosis* (ed. C. H. Rodeck and K. H. Nicolaides). John Wiley and Sons, Chichester.

Canadian Collaborative CVS-Amniocentesis Clinical Trial Group. (1989). Multicentre randomised clinical trial of chorion villus sampling and amniocentesis. First report. *The Lancet*, **i**, 1–6.

Carothers, A. D., McAllion S. J., and Paterson, C. R. (1986). Risk of dominant mutation in older fathers: evidence from osteogenesis imperfecta. *Journal of Medical Genetics*, **23**, 227–30.

Catford, J. C., Bennet, G. C., and Wilkinson, J. A. (1982). Congenital hip dislocation: an increasing and still uncontrolled disability? *British Medical Journal*, **285**, 1527–30.

Chang, H. J., Clark, R. D., and Bachman, H. (1989). Prenatally diagnosed 45, X/46, XY and normal phenotype. *The Lancet*, **i**, 961–2.

Chiu, H. Y., Flynn, D. M., Hoffbrand, A. V., and Politis, D. (1986). Infection with *Yersinia enterocolitica* in patients with iron overload. *British Medical Journal*, **292**, 97.

Clark, C., and Whitfield, A. G. W. (1984). Deaths from rhesus haemolytic disease in England and Wales during 1980 and 1981 and a comparison with earlier years. *Journal of Obstetrics and Gynaecology*, **4**, 218–22.

Clarke, R. H. and Southwood, T. R. E. (1989). Risks from ionising radiation. *Nature*, **339**, 197–8.

Cohen-Overbeek, T. E., Hop, W. C., den Ouden, M., Pijpers, L., Jahoda, M. G. J., and Wladimiroff, J. W. (1990). Spontaneous abortion rate and advanced maternal age: consequences for prenatal diagnosis. *The Lancet*, **336**, 27–9.

Cox, D. W. (1989). α-1 Antitrypsin deficiency. In *The metabolic basis of inherited disease*, (6th edn.) (ed. C. R. Scriver, A. L. Beaudet, W. S. Sly, and D. Valle). McGraw Hill Inc., New York.

Craufurd, D., Dodge, A., Kerzin-Storrar, L., and Harris, R. (1989). Uptake of presymptomatic predictive testing for Huntington's disease. *The Lancet*, **ii**, 603–5.

Crowley, J. P., Sheth, S., Capone, R. J., and Schilling, R. F. (1987). A paucity of thalassaemia trait in italian men with myocardial infarction. *Acta Haematologica*, **798**, 249–51.

Cuckle, H. S. and Wald, N. J. (1984). Principles of screening. In *Antenatal and neonatal screening*, (ed. N. J. Wald). Oxford University Press.

Cuckle, H. S., Wald, N. J., and Lindenbaum, R. H. (1984). Maternal serum alpha-fetoprotein measurement: a screening test for Down syndrome. *The Lancet*, **i**, 926–9.

Cuckle, H. S., Wald, N. J., and Cuckle, P. M. (1989). Antenatal diagnosis of neural tube defects in England and Wales. *Prenatal Diagnosis*, **9**, 393–400.

Cuckle, H. S., Wald, N. J., Goodburn, S. F., Sneddon, J., Amess, J. A. L., and Dunn, S. C. (1990). Measurement of activity of urea resistant neutrophil alkaline phosphatase as an antenatal screening test for Down's syndrome. *British Medical Journal*, **301**, 1024–6.

Cuthbert, A. W. (1989). Defects in epithelial transport in cystic fibrosis. In *Cystic fibrosis* (ed. P. Goodfellow). Oxford University Press.

Czeizel, A. (1988a). *The right to be born healthy. The ethical problems of human genetics in Hungary*. Akademiai Kiado, Budapest.

Czeizel, A. (1988b). Neural tube defects. *Journal of the American Medical Association*, **259**, 3562.

Czeizel, A. (1989). Ethics of a randomised trial of periconceptual vitamin supplementation. *Journal of the American Medical Association*, **262**, 1634.

Czeizel, A. and Kis-Varga, A. (1987). Mutation surveillance of sentinel anomalies in Hungary, 1980–1984. *Mutation Research*, **186**, 73–9.

Czeizel, A. and Racz, J. (1990). Evaluation of drug intake during pregnancy in the Hungarian case control surveillance of congenital anomalies. *Teratology*, **42**, 505–12.

Czeizel, A. and Sankanarayanan, K. (1984). The load of genetic and partly genetic disorders in man. 1. Congenital anomalies: estimates of detriment in terms of years of life lost and years of impaired life. *Mutation Research*, **128**, 73–103.

Czeizel, A., Dudas, I., Fritz, G., Tecsoi, A., and Bod, M. (1990). Methods and results of the optimal family planning program in Hungary (In press.)

Daffos, F., Capella-Pavlovsky, M., and Forestier, F. A. (1985). Fetal blood sampling during pregnancy with use of a needle guided by ultrasound. A study of 606 consecutive cases. *American Journal of Obstetrics and Gynaecology*, **153**, 655–60.

Dalgaard, O. Z. (1957). Polycystic disease of the kidneys: a follow-up study of 284 patients and their families. *Acta Medica Scandinavica*, **158**, (Supplement), 328.

Dalgaard, O. Z. and Norby, S. (1989). Autosomal dominant polycystic kidney disease in the 1980s. *Clinical Genetics*, **36**, 320–25.

Darr, A. (1990). The social implications of thalassaemia among Muslims of Pakistani origin in England – Family experience and service delivery. Ph.D. thesis, University of London.

Darr, A. and Modell, B. (1988). The frequency of consanguineous marriage among British Pakistanis. *Journal of Medical Genetics*, **25**, 186–90.

David, T. J. (1990). Cystic fibrosis. *Archives of Disease in Childhood*, **65**, 152–157.

de la Mata, I., de Wals, P., Dolk, H., Lechat, M. F., Beckers, R., Borlee, I. *et al.* (1989). Incidence of congenital rubella syndrome in 19 regions of Europe in 1980–1986. *European Journal of Epidemiology* 5: 106–109.

Dellamonica, Ch., Robert, J. M., Cotte, J., Collombel, C., and Dorche, C. (1978). Systematic neonatal screening for Duchenne muscular dystrophy. *The Lancet*, **ii**, 1100.

de Wals P., Weatherall, J. A. C. and Lechat, M. F. (1985). *Registration of Congenital anomalies in Eurocat centres 1979–1983*. An EEC Concerted Action Project. Cabay, Belgium.

Desmonts, G. and Couvreur, J. (1974). Congenital toxoplasmosis. A prospective study of 378 pregnancies. *The New England Journal of Medicine*, **290**, 1110–6.

Desmonts, G., Daffos, F., Forestier, F., Capella-Pavlovsky, M., Thulliez, Ph., and Chartier, M. (1985). Prenatal diagnosis of congenital toxoplasmosis. *The Lancet*, **i**, 500–504.

DHSS (1986). Special Report. Standing Medical Advisory Committee and Standing Nursing and Midwifery Advisory Committee. Screening for the detection of congenital dislocation of the hip. H. M. Government, London.

DoH (DHSS), WO, and SHHD. (1990). *Immunisation against infectious disease*. Prepared by the Joint Committee on Vaccination and Immunisation.

Drug and Therapeutics Bulletin. (1990). Simvastatin to lower cholesterol, **28**, 29–30.

Drummond, M. F. (1980). *Principles of economic appraisal in health care*. Oxford University Press.

Dunster, H. J. (1990). The appreciation of radiation risks. *Journal of the Royal College of Physicians of London*, **24**, 154–6.

Edwards, J. H., Lyon, M. F., and Southern, E. M. (ed.). (1988). *The prevention and avoidance of genetic disease*. The Royal Society, London.

Emery, A. E. H. (1988). *Duchenne muscular dystrophy*. (Revised edition). Oxford University Press.

Epstein, C. J. (1989). Down syndrome. In *The metabolic basis of inherited disease* (6th edn). (ed. C. R. Scriver, A. L. Beaudet, W. S. Sly, and D. Valle). McGraw Hill Inc., New York.

Eriksson, A. W., Forsius, H. R., Nevanlinna, H. R., Workman, P. L., and Norio, R. K. (ed.) (1980). *Population structure and genetic disorders*. Academic Press, London.

EUROCAT working group (1989). Eurocat report 3. *Surveillance of congenital anomalies years 1980–1986*. Department of Epidemiology, Catholic University of Louvain, Brussels, Belgium.

European Collaborative Study. (1991). Children born to women with HIV infection: natural history and risk of transmission. *The Lancet*, **337**, 253–60.

Evans, M. I., Fletcher, J. C., and Rodeck, C. R. (1988). Ethical problems in multiple gestations: selective termination. In *Fetal diagnosis and therapy: science, ethics and the law*. (ed. P. A. Harper). Lippincot, New York.

Ferguson-Smith, M. A. (ed.) (1983*a*). Early prenatal diagnosis. *British Medical Bulletin*, **39**.

Ferguson-Smith, M. A. (1983*b*). Prenatal chromosome analysis and its impact on the birth incidence of chromosome disorders. In: Early prenatal diagnosis. *British Medical Bulletin*, **39**, 355–64.

Ferguson-Smith, M. A. and Ferguson-Smith, M. E. (1990). Relationships between patient, clinician, and scientist in prenatal diagnosis. In *Doctor's decisions: ethical conflicts in medical practice*. (ed. G. R. Dunstan and E. A. Shinebourne). Oxford University Press.

Ferguson-Smith, M. A. and Yates, J. R. W. (1984). Maternal age-specific rates for chromosomal aberrations and factors influencing them: report of a collaborative European study on 52 965 amniocenteses. *Prenatal Diagnosis*, **4**, 5–44.

Flatz, G. (1989). The genetic polymorphism of lactase activity in adult humans. In *The metabolic basis of inherited disease*, (6th edn). (ed. C. R. Scriver, A. L. Beaudet, W. S. Sly, and D. Valle). McGraw Hill Inc., New York.

Fletcher, J. C. (1982). *Coping with genetic disease. A guide for clergy and parents*. Harper and Row, San Francisco.

Fletcher, J. C., Berg, K., and Tranoy, K. E. (1985). Ethical aspects of medical genetics. A proposal for guidelines in genetic counselling, prenatal diagnosis and screening. *Clinical Genetics*, **27**, 199–205.

Fogelman, K. R. and Manor, O. (1988). Smoking in pregnancy and development into early adulthood. *British Medical Journal*, **297**, 1233–6.

Fox, R. (1967). *Kinship and marriage*. Penguin books, London.

Fraccaro, M., Simoni G., and Brambati B. (ed.) (1985). First trimester fetal diagnosis. Springer-Verlag, Berlin.

Fucharoen, S., Ong-sangkun, T., Vaewsorn, V., Suthiwan, I., Kanokpongsak, S., and Modell, B. (1991). Hb Bart's hydrops fetalis. An analysis of 65 cases. Submitted for publication.

Gardner, M. J., Snee, M. P., Hall, A. J., Powell, C. A., Downess, S., and Terrell, J. D. (1990). Results of a case-control study of leukaemia and lymphoma among young people near Sellafield nuclear plant in West Cumbria. *British Medical Journal*, **300**, 423–9.

Gardner-Medwin, D. (1983). Recognising and preventing Duchenne muscular dystrophy. *British Medical Journal*, **287**, 1083–4.

Gill, M., Murday, V., and Slack, J. (1986). An economic appraisal of screening for Down's syndrome in pregnancy using maternal age and serum alpha fetoprotein concentration. *Social Science and Medicine*, **24**, 725–31.

Goldstein, J. L. and Brown, M. S. (1989). Familial hypercholesterolaemia. In *The metabolic basis of inherited disease*, (6th edn) (ed. C. R. Scriver, A. L. Beaudet, W. S. Sly, and D. Valle). McGraw Hill Inc., New York.

Goodchild, M. and Dodge, J. A. (1985). *Cystic fibrosis. Manual of diagnosis and management*, (2nd edn). Baillere Tindall, London.

Goodfellow, P. (ed.) (1989). *Cystic fibrosis*. Oxford University Press.

Grant, D. B. and Smith, I. (1988). Survey of neonatal screening for primary hypothyroidism in England, Wales and Northern Ireland 1982–4. *British Medical Journal*, **296**, 1355–8.

Guidelines for the management of acute urinary tract infection in childhood. Report of a Working Group of the Royal College of Physicians. (1991). *Journal of the Royal College of Physicians of London*, **25**, 36–43.

Gusella, J. F., Wexler, N. S., Coneally, P. M., Naylor, S. L., Anderson, M. A., Tanzi, R. E. *et al.* (1983). A polymorphic DNA marker genetically linked to Huntington's disease. *Nature*, **306**, 234–8.

Gustavson, K.-H. (1989). The prevention and management of autosomal recessive conditions. Main example: α-1 antitrypsin deficiency. *Clinical Genetics*, **36**, 327–32.

Haggard, M. P. (1990). Hearing screening in children — state of the art(s). *Archives of Disease in Childhood*, **65**, 1193–8.

Hall, J. G. (1990). Genomic imprinting. *Archives of Disease in Childhood*, **65**, 1013–16.

Hamerton, J. L., Jacobs, P. A., and Klinger, H. P. (ed.) (1973). Standardisation in human cytogenetics. *National Foundation March of Dimes Original Articles Series*.

Handyside, A. H., Kontogianni, E. H., Hardy, K., and Winston, R. M. L. (1990). Pregnancies from human preimplantation embryos sexed by Y-specific DNA amplification. *Nature*, **344**, 768–70.

Hanshaw, J. B., Dudgeon, J. A., and Marshall, W. C. (1985). *Viral diseases of the fetus and newborn*. Major problems in clinical pediatrics, Vol. 17, (2nd edn). Saunders, New York.

Harper, P. S. (1983). Genetic counselling and prenatal diagnosis. In *Early prenatal diagnosis*, (ed. M. A. Ferguson-Smith.) British Medical Bulletin, Vol. 39.

Harper, P. S. (1988). *Practical genetic counselling*, (2nd edn). Wright, Bristol.

Harper, P. S., Tyler, A., Smith, S., Jones, P., Newcombe, R. G., and McBroom, V. (1981). Decline in the predicted incidence of Huntington's chorea associated with systematic genetic counselling and family support. *The Lancet*, **ii**, 411–13.

Harris, A. L. (1991). Telling changes of base. *Nature* 350, 377–8.

Harris, H. (1974). *Prenatal diagnosis and selective abortion*. The Nuffield Provincial Hospitals Trust, UK.

Harrison, M. R., Golbus, M. S., and Filly, R. A. (1984). *The unborn patient. Prenatal diagnosis and treatment*. Grune and Stratton Inc., New York.

Harrison, M. R., Slotnick, R. N., Crombleholme, T. M., Golbus, M., Tarantal, A. F., and Zanjani, E. D. (1989). *In utero* transplantation of fetal liver haemo-poietic stem cells in monkeys. *The Lancet*, **ii**, 1425–7.

Having a baby in Europe (1985). Public Health in Europe Series no. 26. World Health Organization Regional Office for Europe, Copenhagen.

Hawkins, D. F. (ed.) (1987). *Drugs and pregnancy*, (2nd edn). Churchill Livingstone, Edinburgh.

Hayden, M. R. (1981). *Huntington's chorea*. Springer Verlag, Berlin.

Hayes, A., Costa, T., Scriver, C., and Childs, B. (1985). The effect of Mendelian disease on human health II: response to treatment. *American Journal of Medical Genetics*, **21**, 243–55.

Herschko, C. (1988). Oral iron chelating drugs: coming but not yet ready for clinical use. *British Medical Journal*, **296**, 1081.

Holmes-Siedle, M., Ryynanen, M., and Lindenbaum, R. H. (1987). Parental decisions regarding termination of pregnancy following prenatal detection of sex chromosome abnormality. *Prenatal Diagnosis*, **7**, 239–44.

Holtzman, N. A. (1989). *Proceed with caution. Predicting genetic risks in the recombinant DNA era*. The Johns Hopkins University Press, Baltimore, MD.

Holzgreve, W., Miny, P., Gerlach, B., Westendorp, A., Ahlert, D., and Horst, J. (1990). Benefits of placental biopsies for rapid karyotyping in the second and third trimesters (late chorionic villus sampling) in high-risk pregnancies. *American Journal of Obstetrics and Gynaecology*, **162**, 1188–92.

Hughes, I. A. (1988). Management of congenital adrenal hyperplasia. *Archives of Disease in Childhood*, **63**, 1399–404.

Hunt, G. M. (1990). Open spina bifida: outcome for a complete cohort treated unselectively and followed into adulthood. *Developmental Medicine and Child Neurology*, **32**, 108–18.

Huttenlocher, P. R. (1983). The nervous system. In *Nelson Textbook of Pediatrics*. (12th edn). (ed. R. E. Behrman and V. C. Vaughan). W. B. Saunders Co., Philadelphia, PA.

Illig, R., Largo, R. H., Qin, Q., Torresani, T., Rochiccioli, P., and Larsson, A. (1987). Mental development in congenital hypothyroidism after neonatal screening. *Archives of Disease in Childhood*, **62**, 1050–5.

Ingham, P. W. (1988). The molecular genetics of embryonic pattern formation in *Drosophila*. *Nature*, **335**, 25–34.

International commission for protection against environmental mutagens and carcinogens (ICPEMC) publication no. 3 (1979). Cigarette smoking — does it carry a genetic risk? *Mutation Research*, **65**, 71–81.

John Radcliffe Hospital Cryptorchidism Study Group (1986). Cryptorchidism. An apparent substantial increase since 1960. *British Medical Journal*, **293**, 1401–4.

Johnstone, F. D., Brettle, R. P., MacCallum, L. R., Mok, J., Peutherer, J. F., and Burns, S. (1990). Women's knowledge of their HIV antibody state: its effect on their decision whether to continue the pregnancy. *British Medical Journal*, **300**, 23–4.

Jones, D. (1990). Foodborne listeriosis. *The Lancet*, **336**, 1171–4.

Jones, M. and Ratcliffe, S. G. (1990). IQ test results and educational achievements in unselected children with sex chromosome aneuploidy. Unpublished paper delivered at a meeting of the UK Clinical Genetics Society, London, November 1990.

Jones, K. L., Smith, D. W., Streissguth, A. P., and Myrianthopoulos, N. C. (1974). Outcome in offspring of chronic alcoholic women. *The Lancet*, **i**, 1076–8.

Jones, P. (1990). *Living with haemophilia*, (3rd edn). Castle House, Tumbridge Wells.

Kaback, M. M., Zeigler, R. S., Reynolds, L. W. and Sonneborn, M. (1974). Approaches to the control and prevention of Tay–Sachs disease. *Progress in Medical Genetics*, **10**, 103–34.

Kan, Y. W. and Dozy, A. M. (1978). Antenatal diagnosis of sickle cell anaemia by DNA analysis of amniotic fluid cells. *The Lancet*, **ii**, 910–12.

Karlssen, J. L. (1975). Influence of the myopia gene on brain development. *Clinical Genetics*, **8**, 314–8.

Khlat, M., Halabi, S., Khudr, A., and Der Kaloustian, V. M. (1986). Perception of consanguineous marriages and their genetic effects among a sample of couples from Beirut. *American Journal of Medical Genetics*, **25**, 299–306.

King's Fund Centre. Blood cholesterol measurement in the prevention of coronary heart disease. The 6th King's Fund Forum Consensus Statement.

Kline, J., Shrout, P., Stein, Z., Susser, M., and Warburton, D. (1980). Drinking during pregnancy and spontaneous abortion. *The Lancet*, ii, 176–80.

Knott, P. D. (1988). Patient acceptability of chorionic villus sampling. Unpublished poster at 4th International Conference on Chorionic Villus Sampling and Prenatal Diagnosis.

Knott, P. D., Ward, R. H. T., and Lucas, M. K. (1986). Effect of chorionic villus sampling and early pregnancy counselling on uptake of prenatal diagnosis. *The British Medical Journal*, **293**, 471–4.

Knowlton, R. G., Cohen-Hagenauer, O., Nguen Van Cong, Frezal, J., Brown, V. A., and Barker, D. (1985). A polymorphic DNA marker linked to cystic fibrosis is located on chromosome 7. *Nature*, **318**, 380–2.

Knox, W. E. (1966). Phenylketonuria. In *The metabolic basis of inherited disease*, (2nd edn) (ed. J. B. Stanbury, J. B. Wyngaarden, and D. S. Frederickson). McGraw Hill, New York.

Knox-MacAulay, H. H. M., Weatherall, D. J., Clegg, J. B., and Pembrey, M. E. (1973). Thalassaemia in the British. *British Medical Journal*, **3**, 150.

Koneig, M., Hoffman, E. P., Bertelson, C. J., Monaco, A. P., Feener, C., and Kunkel, L. M. (1987). Complete cloning of the Duchenne muscular dystrophy (DMD) cDNA and preliminary organisation of the DMD gene in normal and affected individuals. *Cell*, **50**, 509–17.

Koppe, J. G., Loewer-Sieger, D. H., and de Roever-Bonnet, H. (1986). Results of 20-year follow-up of congenital toxoplasmosis. *The Lancet*, i, 254–5.

Kuliev, A. (1986). Thalassaemia can be prevented. *World Health Forum*, 7, 286–90.

Kuliev, A. M., Modell, B., and Galjaard, H. (1985). Perspectives in fetal diagnosis of congenital disorders. *Serono symposia review*, No. 8. Ares Serono Symposia, Rome, Italy.

Kulozik, A. E., Wainscoat, J. S., Serjeant, G. R., Kar, B. C., Al-Awamy, B., Essan, G. J. F. *et al.* (1986). Geographical survey of β^s-globin gene haplotypes: evidence for an independent Asian origin of the sickle cell mutation. *American Journal of Human Genetics*, **39**, 239–44.

Lachelin, G. C. L., (1985). *Miscarriage. The facts*. Oxford University Press.

Lamont, M. A. and Dennis, N. R. (1988). Aetiology of mild mental retardation. *Archives of Disease in Childhood*, **63**, 1032–8.

Lansdown, R. (1980). *More than sympathy. The everyday needs of sick and handicapped children and their families*. Tavistock, London.

Larsson, S. A. (1985). Life expectancy of Swedish haemophiliacs. *British Journal of Haematology*, **59**, 593–602.

Larsson, A. (1987). Mental development in congenital hypothyroidism after neonatal screening. *Archives of Disease in Childhood*, **62**, 1050–55.

Laurence, K. M. (1983). Prevention of neural tube defects. In *Prenatal diagnosis*. Proceedings of the Eleventh Study Group of the Royal College of Obstetricians and Gynaecologists. (ed. C. H. Rodeck and K. H. Nicolaides). John Wiley and Sons, Chichester.

Layde, P. M. (1984). Congenital hypothyroidism. In *Antenatal and neonatal screening*. (ed. N. J. Wald). Oxford University Press.

Lenke, R. R. and Levi, H. L. (1980). Maternal phenylketonuria and hyperphenalaninaemia. An international survey of the outcome of untreated and treated pregnancies. *New England Journal of Medicine*, **303**, 1202-8.

Lerner, D. J. and Kannel, W. B. (1986). Patterns of coronary heart disease morbidity and mortality in the sexes: a 26-year follow-up of the Framingham population. *American Heart Journal*, **111**, 383-90.

Little, R. E., Anderson, K. W., Ervin, C. H., Worthington-Roberts, B., and Clarren, S. K. (1989). Maternal alcohol use during breast feeding and infant mental and motor development at one year. *New England Journal of Medicine*, **321**, 425-30.

Livera, L. N., Brookfield, D. S. K., Egginton, J. A., and Hawnaur, J. M. (1989). Antenatal ultrasonography to detect fetal renal abnormalities: a prospective screening programme. *British Medical Journal*, **298**, 1420-2.

Livingstone, F. A. (1987). *Abnormal hemoglobins in human populations.* Aldine, Chicago.

Lo, Y.-M. D., Wainscoat, J. S., Gillmer, M. D. G., Patel, P., Sampietro, M., and Fleming, K. A. (1989). Prenatal sex determination by PCR amplification from maternal peripheral blood. *The Lancet*, **ii**, 1363-5.

Long, W. K. and Wilson, S. W. (1967). Associations between red cell G6PD variants and vascular diseases. *American Journal of Human Genetics*, **19**, 35.

Loukopolous, D. (ed). (1988). *Prenatal diagnosis of thalassaemia and the haemoglobinopathies.* CRC Press Inc., Boca Raton, Florida.

Lucarelli, G., Polchi, P., Galimberti, M., Izzi, T., Delphini, C., Manna M. *et al.* (1985). Marrow transplantation for thalassaemia following busulphan and cyclophosphamide. *The Lancet*, **i**, 1355-7.

Luzzato, L. and Mehta, A. (1989). Glucose-6-phosphate dehydrogenase deficiency. In *The metabolic basis of inherited disease*, (6th edn) (ed. C. R. Scriver, A. L. Beaudet, W. S. Sly and D. Valle). McGraw Hill Inc., New York.

Lynn, R., Hampson, S. L., and Mullineux, J. C. (1987). A long-term increase in the fluid intelligence of English children. *Nature*, **328**, 797.

McEvedy, C. and Jones, R. (1978). *Atlas of world population history.* Penguin Books Ltd, London.

Macgregor, R. (1990). Thames Television's *The Treatment. Preconception and pregnancy planning.* Available from the Community Education Officer, Thames Television p.l.c., 149 Tottenham Court Road, London W1P 9LL, UK.

McKusick, V. A. (1988). *Mendelian inheritance in man.* (8th edn). The Johns Hopkins University Press, Baltimore, Maryland.

McKusick, V. A. (1988). *The morbid anatomy of the human genome. A review of gene mapping in clinical medicine.* Reprinted from *Medicine*, with appendices. The Howard Hughes Medical Institute.

MacLaren, A. (1990). What makes a man a man? *Nature*, **346**, 216-7.

MacLaren, A. (1991). The making of male mice. *Nature*, **351**, 96.

McNeil, T. F., Sveger, T., and Thelin, T. (1988). Psychosocial aspects of screening for somatic risk: the Swedish α-1-antitrypsin experience. *Thorax*, **43**, 505-507.

Manson, M. M., Logan, W. P. D., and Loy, R. M. (1960). Rubella and other virus infections during pregnancy. *Reports on public health and medical subjects*, No. 101. HMSO, London.

Marmot, M. G. (1985). Interpretation of trends of coronary heart disease mortality. *Acta Medica Scandinavica* (Supplement), **701**, 58-65.

Medical Research Council trial of chorion villus sampling (1991). *The Lancet*, **337**, 1491–9.

Merkatz, I. R., Nitowsky, H. M., Macri, J. M., and Johnson, W. E. (1984). An association between low maternal serum alpha-fetoprotein and fetal chromosome abnormalities. *American Journal of Obstetrics and Gynaecology*, **14**, 886–92.

Meuller, U. W., Hawes, C. S., Wright, A. E., Petropoulos, A., deBoni, E., Firgaira, F. A. *et al.* (1990). Isolation of fetal trophoblast cells from peripheral blood of pregnant women. *The Lancet*, **336**, 197–200.

Meyer, M. B. and Tonascia, J. A. (1977). Maternal smoking, pregnancy complications, and perinatal mortality. *American Journal of Obstetrics and Gynaecology*, **128**, 494–502.

Mikkelsen, M. (1988). The incidence of Down's syndrome and progress towards its reduction. In *The prevention and avoidance of genetic disease* (ed. J. H. Edwards, M. F. Lyon, and E. M. Southern). The Royal Society, London.

Miller, E. (1990). Rubella infection in pregnancy. In *Modern antenatal care of the fetus*, (ed. G. Chamberlain). Blackwell Scientific Publications, Oxford.

Modell, B. (1985). Chorionic villus sampling: evaluating safety and efficacy. *The Lancet*, **i**, 737–40.

Modell, B. (1988). Ethical and social aspects of fetal diagnosis for the haemoglobinopathies: a practical view. In *Prenatal diagnosis thalassaemia and the haemoglobinopathies* (ed. D. Loukopolous). CRC Press Inc., Boca Raton, Florida.

Modell, B. (1990). Consanguineous marriage. *British Medical Journal*, **300**, 1662–3.

Modell, B. and Berdoukas, V. (1984). *The clinical approach to thalassaemia*. Grune and Stratton, New York.

Modell, B. and Bulyzhenkov, V. (1988). Distribution and control of some genetic disorders. *World Health Statistics Quarterly*, **41**, 209–18.

Modell, B. and Kuliev, A. M. (1989). Impact of public health on human genetics. *Clinical Genetics*, **36**, 286–98.

Modell, B. and Kuliev, A. K. (1990). Changing paternal age distribution and the human mutation rate in Europe. *Human Genetics*, **86**, 198–202.

Modell, M. and Modell, B. (1990). Genetic screening for ethnic minorities. *British Medical Journal*, **300**, 1702–4.

Modell, B., Ward, R. H. T., and Fairweather, D. V. I. (1980). Effect of introducing antenatal diagnosis on the reproductive behaviour of families at risk for thalassaemia major. *British Medical Journal*, **2**, 737.

Modell, B., Petrou, M., Ward, R. H. T., Fairweather, D. V. I., Rodeck, C., Varnavides, L. A., and White, J. M. (1985). Effect of fetal diagnostic testing on the birth-rate of thalassaemia in Britain. *The Lancet*, **ii**, 1383–6.

Modell, B., Kuliev, A. K., and Wagner, M. (1991). *Community genetics services in Europe*. WHO Regional Office for Europe, WHO Regional Publications. European Series No 38.

Mole, R. H. (1979). Radiation effects on prenatal development and their radiological significance. *British Journal of Radiology*, **52**, 89–101.

Monk, M. and Holding, C. (1990). Amplification of a β-haemoglobin sequence in individual human oocytes and polar bodies. *The Lancet*, **335**, 985–8.

Morris, M. J., Tyler, A., Lazarou, L., Meredith, L., and Harper, P. S. (1989). Problems in genetic prediction for Huntington's disease. *The Lancet*, **ii**, 601–3.

Mouzouras, M., Camba, L., Ioannou, P., Modell, B., Constantinides, P., and Gale, R. (1980). Thalassaemia as a model of recessive genetic disease in the community. *The Lancet*, **ii**, 574–8.

MRC Vitamin Study Research Group (1991). Prevention of neural tube defects: results of the Medical Research Council vitamin study. *The Lancet*, **338**, 131–7.

Multinovic, J., Fialkow, P. J., Phillips, L. A., Agoda, L. Y., Boyaut, J. I., Denney, J. D. (1980). Autosomal dominant polycystic kidney disease. Early diagnosis and data for genetic counselling. *The Lancet*, **i**, 1203–06.

Muller, H. J. (1950). Our load of mutations. *American Journal of Human Genetics*, **2**, 111–76.

Mulvihill, J. J., Myers, M. H., Connelly, R. R., Byrne, J., Austin, D. F. and Bragg, K. (1987). Cancer in offspring of long-term survivors of childhood and adolescent cancer. *The Lancet*, **ii**, 813–7.

Murdoch, J. L., Walker, B. A., Hall, J. G., Abbey, H., Smith, K. K., and McKusick, V. A. (1970). Achondroplasia — a genetic and statistical survey. *Annals of Human Genetics* (London), **33**, 227–44.

Nelson, M. (1990). Vitamin A, liver consumption, and risk of birth defect. *British Medical Journal*, **301**, 1176.

Neonatal screening for sickle cell disease. (1989). *Pediatrics*, **83**, (Supplement).

New, M. I., White, P. C., Pang, S., Dupont, B., and Speiser, P. (1989). The adrenal hyperplasias. In *The metabolic basis of inherited disease*, (6th edn). (ed. C. R. Scriver, A. L. Beaudet, W. S. Sly, and D. Valle). McGraw Hill Inc. New York.

Nicolaides, K. H., Rodeck, C. R., and Gosden, C. M. (1986). Rapid karyotyping in non-lethal fetal malformations. *The Lancet*, **i**, 283–6.

Norio, R. (1966). Heredity in the congenital nephrotic syndrome. *Annals of Clinical Research 12* (Supplement 27).

Norio, R. (1980). Congenital nephrotic syndrome of Finnish type (CNF). In *Population structure and genetic disorders*. (ed. A. W. Eriksson). Academic Press, London.

Oakley, A., McPherson, A., and Roberts, H. (1984). *Miscarriage*. Fontana Paperbacks, Glasgow.

Office of Health Economics (1986). *Cystic fibrosis*. 12 Whitehall, London SW1A 2DY, UK.

Old, J. M., Ward, R. H. T., Petrou, M., Karagozlu, F., Modell, B., and Weatherall, D. J. (1982). First-trimester fetal diagnosis for the haemoglobinopathies: three cases. *The Lancet*, **ii**, 1413–6.

Olofsson, P., Liedholm, H., Sartor, G., Sjoberg, N-O., Svennigsen N. W., and Ursing D. (1984). Diabetes in pregnancy. A 21 year Swedish material. *Acta Obstetrica Gynaecologica Scandinavica*, **122**, (Supplement), 1–62.

Owen, M. J. and Murray, R. M. (1988). Blue genes. *British Medical Journal*, **297**, 871–2.

Pain in sickle cell disease. Proceedings of a symposium held at the Central Middlesex Hospital. (1983). Published for the Sickle Cell Society by the National Extension College, Cambridge.

Palmiter, R. D., Brinster, R. L., Hammer, R. E. *et al.* (1982). Dramatic growth of mice that develop from eggs microinjected with metallothionein-growth hormone fusion genes. *Nature*, **300**, 611–15.

Pang, S., Wallace, M. A., Hofman, L. *et al.* (1988). Worldwide experience of newborn screening for classical congenital adrenal hyperplasia due to 21-hydroxylase deficiency. *Pediatrics*, **81**, 866–74.

Pang, S., Pollack, M.S., Marshall, R.N., and Immken, L. (1990). Prenatal treatment of congenital adrenal hyperplasia due to 21-hydroxylase deficiency. *The New England Journal of Medicine*, **322**, 111–15.

Peckham, C.S., Tedder, R.S., Briggs, M., Ades, A.E., Hjelm, M., Wilcox, A.H., *et al.* (1990). Prevalence of maternal HIV infection based on unlinked anonymous testing of newborn babies. *The Lancet*, **335**, 516–19.

Pekkanen, J., Nissenen, A., Puska, P., Punsar, S., and Karvonen, M.J. (1989). Risk factors and 25 year risk of coronary heart disease in a male population with a high incidence of the disease: the Finnish cohorts of the seven countries study. *British Medical Journal*, **299**, 81–5.

Persson Benbow, C. (1987). Possible biological correlates of precocious mathematical reasoning ability. *Trends in Neurosciences*, **10**, 17–20.

Piga, A., Gabutti, V., Sandri, A., Sacchetti, L., Mandrino, M., and Modell, B. (1991). The changing pattern of infections in homozygous β-thalassaemia. Submitted for publication.

Plachot, M., de Grouchy, J., Junca, A.-M., Mandelbaum, J., Turleau, C., Couillin, P. *et al.* (1987). From oocyte to embryo: a model, deduced from *in vitro* fertilisation, for natural selection against chromosome abnormalities. *Annales de Genetique*, **30**, 22–32.

Public health laboratory sevice working party on fifth disease. (1990). Prospective study of human parvovirus (B19) infection in pregnancy. *British Medical Journal*, **300**, 1166–70.

Pullen, I. (1984). Physical handicap. In *Psychological aspects of genetic counselling* (ed. A.E.H. Emery and I Pullen). Academic Press Inc., London.

Quarrell, O., Meredith, A.L., Tyler, A., Youngman, F.S., Upadhyaya, M., and Harper, P. (1987). Exclusion testing for Huntington's disease in pregnancy with a closely-linked DNA marker. *The Lancet*, **i**, 1281–3.

Quinton, P.M. (1982). Abnormalities in electrolyte secretion in cystic fibrosis sweat glands due to decreased ion permeability. In *Fluid and electrolyte abnormalities in exocrine glands in cystic fibrosis*. (ed. P.M. Quinton, J.R. Martinez, and U. Hopfer). San Francisco Press.

Reardon, W. and Pembrey, M. (1990). The genetics of screening. *Archives of Disease of Childhood*, **65**, 1196–8.

Reeders, S.T., Breuning, M.H., Davies, K.E., Nicholls, R.D., Jarman, A.P., and Higgs, D.R. (1985). A highly polymorphic DNA marker linked to adult polycystic kidney disease on chromosome 16. *Nature*, **317**, 542–4.

Regan, R., Owen, E.J., and Jacobs, H.S. (1990). Hypersecretion of luteinising hormone, infertility, and miscarriage. *The Lancet*, **336**, 1141–4.

Reik, W. (1988). Genomic imprinting: a possible mechanism for the parental origin effect in Huntington's chorea. *Journal of Medical Genetics*, **25**, 805–8.

Ringe, D. and Petsko, G.A. (1990). Cystic fibrosis. A transport problem? *Nature*, **346**, 312–3.

Riordan, J., Rommens, J.M., Kerem, B.-S., Alon, N., Rozmahel, R., Grzelczak, Z. *et al.* (1989). Identification of the cystic fibrosis gene: cloning and characterisation of complementary DNA. *Science*, **245**, 1066–1071.

Rizza, C.R. and Spooner, J.D. (1983). Treatment of haemophilia and related disorders in Britain and Northern Ireland during 1976–80: report on behalf of the directors of haemophilia centres in the United Kingdom. *British Medical Journal*, **286**, 929-33.

Roberts, G., Stanfield, M., Black, A., and Redmond, A. (1988). Screening for cystic fibrosis: a four-year regional experience. *Archives of Diseases in Childhood*, **63**, 1438-43.

Robinson, H. P. and Caines, J. S. (1977). Sonar evidence of early pregnancy failure in patients with twin conceptus. *British Journal of Obstetrics and Gynaecology*, **87**, 22-5.

Rodeck, C. H. and Nicolaides, K. H. (ed.) (1983). *Prenatal Diagnosis*. Proceedings of the Eleventh Study Group of the Royal College of Obstetricians and Gynaecologists. John Wiley and Sons, Chichester.

Rodeck, C. H., Mibashan, R. S., Abramowicz, J., and Campbell, S. (1982). Selective feticide of the affected twin by fetoscopic air embolism. *Prenatal Diagnosis*, **2**, 189-94.

Rodger, M. W. and Baird, D. T. (1987). Induction of therapeutic abortion in early pregnancy with mifepristone in combination with prostaglandin pessary. *The Lancet*, **ii**, 1415-8.

Rooney, D. E., MacLachlan, N., and Smith, J. (1990). Early amniocentesis: a cytogenetic evaluation. *British Medical Journal*, **299**, 25-7.

Rotter, J. I. and Diamond, J. M. (1987). What maintains frequencies of human genetic diseases? *Nature*, **329**, 289-90.

Rowley, P. T. (1984). Genetic screening: marvel or menace? *Science*, **225**, 138-44.

Rowley, P. T., Loader, S., and Walden, M. (1988). Pregnant women identified as haemoglobinopathy carriers by prenatal screening want genetic counseling and use information provided. *Birth Defects Original Articles Series*, Vol. 23, No 5B, pp. 449-454. March of Dimes Birth Defects Foundation.

Royal College of Obstetricians and Gynaecologists (1984). *Report of the RCOG Working Party on routine ultrasound examination in pregnancy*. Royal College of Obstetricians and Gynaecologists, London.

Royal College of Physicians (1989). *Report on prenatal diagnosis and genetic screening; community and service implications*. The Royal College of Physicians, London.

Saiki, R. K., Scarf, S. J., Faloona, F. *et al.* (1985). Enzymatic amplification of Beta globin genomic sequences and restriction site analysis for diagnosis of sickle cell anaemia. *Science*, **230**, 1350-54.

Saiki, R. K., Bugawan, T. L., Horn, G. T., Mullis, K. B., and Ehrlich, H. A. (1986). Analysis of enzymatically amplified β-globin and HLA-DQa DNA with allele-specific oligonucleotide probes. *Nature*, **324**, 163-8.

Sangani, B., Sukumaran, P. K., Mahadik, C., Yagnik, H., Telang, S., Vas, F. *et al.* (1990). Thalassaemia in Bombay: the role of medical genetics in developing countries. *Bulletin of the World Health Organization*, **68**, 75-81.

Sanghvi, L. D. (1966). Inbreeding in India. *Eugenics Quarterly*, **13**, 291-301.

Scientific Steering Committee (1991). Risk of fatal coronary heart disease in familial hypercholesterolaemia. *British Medical Journal*, **303**, 893-6.

Schull, W. J. and Neel, J. V. (1956). *The effects of inbreeding on Japanese children*. Harper and Row, New York.

Scriver, C. R., Beaudet, A. L., Sly, W. S., and Valle, D. (ed.) (1989a). *The metabolic basis of inherited disease*, (6th edn). McGraw Hill, New York.

Scriver, C. R., Kaufman, S., and Woo, S. (1989b). The hyperphenylalaninaemias.

In *The metabolic basis of inherited disease*. (6th edn). (ed. C.R. Scriver, A.L. Beaudet, W.S. Sly, and D. Valle). McGraw Hill, New York.

Segal, A.J., Spataro, R.F., and Barbaric, Z.L. (1977). Adult polycystic kidney disease. A review of 100 cases. *The Journal of Urology*, **118**, 711–3.

Sellar, M. and Nevin, N. (1984). Periconceptional vitamin supplementation and the prevention of neural tube defects in south-east England and Northern Ireland. *Journal of Medical Genetics*, **21**, 325–30.

Serjeant, G.R. (1985). *Sickle cell disease*. Oxford University Press.

Shaw, N.J. and Littlewood, J.M. (1987). Misdiagnosis of cystic fibrosis. *Archives of Disease in Childhood*, **62**, 1271–3.

Sibbald, B. and Turner-Warwick, M. (1979). Factors influencing the prevalence of asthma among first degree relatives of extrinsic and intrinsic asthmatics. *Thorax*, **34**, 332–7.

Silman, A.J. (1987). Why do women live longer and is it worth it? *British Medical Journal*, **245**, 1311–2.

Skinner, R., Emery, A.E.H., Scheuerbrandt, G., and Syme, J. (1982). Feasibility of neonatal screening for Duchenne muscular dystrophy. *Journal of Medical Genetics*, **19**, 1–3.

Slack, J. (1969). Risks of ischaemic heart disease in familial hyperlipoproteinaemic states. *The Lancet*, **ii**, 1380–2.

Slack, J. and Evans, K.A. (1966). The increased risk of death from ischaemic heart disease in first degree relatives of 121 men and 96 women with ischaemic heart disease. *Journal of Medical Genetics*, **3**, 239–57.

Smith, R.A., Rogers, M., Bradley, D.M., Sibert, J.R., and Harper, P.S. (1989). Screening for Duchenne muscular dystrophy. *Archives of Diseases in Childhood*, **64**, 1017–21.

Smith, I., Beasley, M.G., and Ades, A.E. (1990). Intelligence and quality of dietary treatment in phenylketonuria. *Archives of Disease in Childhood*, **65**, 472–8.

Smithells, R.W., Sheppard, S., Schorach, C.J., Seller, M.J., Nevin, N.C., Readir *et al.* (1981). Apparent prevention of neural tube defects by periconceptional vitamin supplementation. *Archives of Diseases in Childhood*, **56**, 911–18.

Smithells, R.W., Sheppard, S., Wild, J., and Schorah, C.J. (1989). Prevention of neural tube defect recurrences in Yorkshire: final report. *The Lancet*, **ii**, 498–9.

Smithells, R.W., Sheppard, S., and Holzel, H. (1990). Congenital rubella in Great Britain 1971–1988. *Health Trends*, **22**, 273–6.

Sofaer, J.A. and Emery, E.H. (1981). Genes for super-intelligence? *Journal of Medical Genetics*, **18**, 410–13.

Stagno, S., Pass, R.F., Dworsky, M.E., Henderson, R.E., Moore, E.G., Walton, P.D., and Alford, C.A. (1982). Congenital cytomegalovirus infection. *The New England Journal of Medicine*, **306**, 945–9.

Stein, Z., Kline, J., Susser, E., Shrout, P., Warburton, D., and Susser, M. (1980). Maternal age and spontaneous abortion. In *Human embryonic and fetal death*, (ed. E.B. Hook and I.H. Porter), pp. 107–27. Academic Press, New York.

Stevenson, A.C. (1957). Achondroplasia: an account of the condition in Northern Ireland. *Annals of Human Genetics*, **9**, 81–91.

Stjernfeldt, M., Berglund, K., Lindstein, J., and Ludvigsson, J. (1986). Maternal smoking during pregnancy and risk of childhood cancer. *The Lancet*, **i**, 1350–2.

Stone, M. H., Clarke, N. M. P., Campbell, M. J., Richardson, J. B., and Johnson, P. A. (1990). Comparison of audible sound transmission with ultrasound in screening for congenital dislocation of the hip. *The Lancet*, **336**, 421–2.

Strom, C. M., Verlinsky, Y., Milayeva, S. Euesikov, S., Cieslak, J., Lifchez, A. (1990). Preconception genetic diagnosis of cystic fibrosis. *The Lancet*, **ii**, 306–7.

Sykes, B. (1987). Genetics cracks bone disease. *Nature*, **330**, 607–8.

Tabor, A., Philip, J., Madsen, M., Bang, J., Obel, E. B., and Norgaard-Pedersen, B. (1986). Randomised controlled trial of genetic amniocentesis in 4606 low-risk women. *The Lancet*, **i**, 1287–93.

Tan, S. L., Steer, C., Royston, P., Rizk, P., Mason, B. A., and Campbell S. (1990). Conception rates and in-vitro fertilisation. *The Lancet*, **i**, 299.

Tanner, J. M. (1989). *Foetus into man. Physical growth from conception to maturity*, (2nd edn). Castlemead Publications, Ware, UK.

Terry, G. H., Ho-Terry, L., Warren, R. C., Rodeck, C. H., Cohen, A., and Rees, M. (1986). First trimester prenatal diagnosis of congenital rubella: a laboratory investigation. *British Medical Journal*, **292**, 930–33.

Thomas, D. F. M., Irving, H. C., and Arthur, R. J. (1985). Prenatal diagnosis: how useful is it? *British Journal of Urology*, **57**, 784–787.

Thompson. A. J., Smith, I., Brenton, D., Youl, B. D., Rylance, G., Davidson, D. C., *et al.* (1990). Neurological deterioration in young adults with phenylketonuria. *The Lancet*, **336**, 602–5.

Tibben, A., van der Vlis, M. V., Niermeijer, M. F., van der Kamp, J. J. P., Roos, R. A. C., Rooimans, H. G. M. *et al.* (1990). Testing for Huntington's disease with support of all parties. *The Lancet*, **i**, 553.

Trichopoulos, D., Zavitsanos, X., Koutis, C., Drogari, P., Proukakis, C., and Petridou, E. (1987). The victims of Chernobyl in Greece: induced abortions after the accident. *British Medical Journal*, **295**, 1100.

Trimble, B. K. and Doughty, J. H. (1974). The amount of hereditary disease in human populations. *Annals of Human Genetics*, **38**, 199–223.

United Nations Demographic Yearbook Series. United Nations, New York.

Veenema, H., Veenema, T., and Geraedts, J. P. M. (1987). The fragile X syndrome in a large family. II. Psychological investigations. *Journal of Medical Genetics*, **24**, 32–8.

Vejerslev, L. O., Dissing, J., Hansen, H. E., and Poulsen, H. (1987). Hydatidiform mole: genetic origin in polyploid conceptuses. *Human Genetics*, **76**, 11–19.

Vitez, M., Koranyi, G., Goncy, E., Rudas, T., and Czeizel, A. (1984). A semiquantitative score system for epidemiologic studies of fetal alcohol syndrome. *American Journal of Epidemiology*, **119**, 301–8.

Vlaanderen, W. and Treffers, P. E. (1987). Prognosis of subsequent pregnancies after recurrent spontaneous abortion in first trimester. *British Medical Journal*, **295**, 92–3.

Vogel, F. and Motulsky, A. G. (1986). *Human genetics: problems and approaches*. (Second edn). Springer-Verlag, Berlin.

Vogel, F. and Rathenberg, R. (1975). Spontaneous mutation in man. *Advances in Human Genetics*, **5**, 221–318.

Vullo, C. and Modell, B. (1990) *What is thalassaemia?* The Cooley's Anemia Foundation, New York. (For the Thalassaemia International Federation).

Wagner, M.G. and St Clair, P.A. (1989). Are in-vitro fertilisation and embryo transfer of benefit to all?. *The Lancet*, **ii**, 1027.

Walbot, V. and Holder, N. (1987). *Developmental biology*. Random House, New York.

Wald, N.J. (ed.) (1984). *Antenatal and neonatal screening*. Oxford University Press.

Wald, N.J., Cuckle, H.S., Densem, J.W., Nanchahal, K., Royston, P., Chard, T. *et al.* (1988). Maternal serum screening for Down's syndrome in early pregnancy. *British Medical Journal*, **297**, 883-7.

Waldenstrom, U., Axelsson, O., Nilsson, S. Eklund, G., Fall, O., Lindeberg, S. (1988). Effects of routine one-stage ultrasound screening in pregnancy: a randomised controlled trial. *The Lancet*, **ii**, 585-8.

Walker, D.A., Harper, P.S., Newcombe, R.G., and Davies, K. (1983). Huntington's chorea in South Wales: mutation, fertility and genetic fitness. *Journal of Medical Genetics*, **20**, 12-17.

Ward, R.H.T., Fairweather, D.V.I., Whyley, G.A., Shirley, I.M., and Lucas, M. (1981). Four years' experience of maternal serum alpha-fetoprotein screening and its effect on the pattern of antenatal care. *Prenatal diagnosis*, **1**, 91-101.

Watson, J.D., Hopkins, N.H., Roberts, J.W., Steitz, J.A., and Weiner, A.M. (1988). *The molecular biology of the gene*, (4th edn). The Benjamin Cummings Publishing Company Inc., California.

Weatherall, D.J. (1991). *The new genetics and clinical practice*. (3rd edn). Oxford University Press (for the Nuffield Provincial Hospitals Trust).

Weatherall, D.J. and Clegg, J.B. (1981). *The thalassaemia syndromes*, (3rd edn). Blackwell Scientific Publications, Oxford.

WHO (1982). Hereditary anaemias: genetic basis, clinical features, diagnosis and treatment. *Bulletin of the World Health Organization*, **60**, 643-60.

WHO (1983). Community control of hereditary anaemias: Memorandum from a WHO meeting. (1983). *Bulletin of the World Health Organization*, **61**, 63-80.

WHO (1985a). *Update of the progress of haemoglobinopathies control*. Unpublished Report of the WHO: HMG/WG/85.8. May be obtained free of charge from: The Hereditary Diseases Programme, WHO, Geneva, Switzerland.

WHO (1985b). *Community approaches to the control of hereditary diseases*. Unpublished WHO document HMG/WG/85.4. May be obtained free of charge from: The Hereditary Diseases Programme, WHO, Geneva, Switzerland.

WHO (1986). Prevention of avoidable mutational disease: Memorandum from a WHO meeting. (1986). *Bulletin of the World Health Organization*, **64**, 205-16.

WHO (1987). Smoking—worldwide trends and their implications. *World Health Statistics Annual* (1986) pp. 16-19. WHO, Geneva.

WHO (1988). *The Haemoglobinopathies in Europe*. WHO Regional Office for Europe, Document IPC/MCH 110. May be obtained free of charge from: Maternal and Child Health Division, WHO Regional Office for Europe, 8 Scherfigsvej, Dk-2100, Copenhagen, Denmark.

WHO (1989). Glucose-6-Phosphate Dehydrogenase Deficiency. Report of a WHO Working Group. *Bulletin of the World Health Organization*, **67**, 601-11.

WHO (1990). *Hereditary anaemias: alpha thalassaemia*. Unpublished WHO document HDP/WG/HA/87.5. May be obtained free of charge from: The Hereditary Diseases Programme, WHO, Geneva, Switzerland.

Wild, J., Read, A. P., Sheppard, S., Seller, M., Smithells, R. W., Nevin, N. C. *et al.* (1986). Recurrent neural tube defects, risk factors and vitamins. *Archives of Disease in Childhood*, **61**, 440–4.

Wilkinson, J. A, (1985). *Congenital dislocation of the hip joint*. Springer-Verlag, Berlin.

Williams, C., Weber, L., Williamson, R., and Hjelm, M. (1988). Guthrie spots for DNA-based carrier testing in cystic fibrosis. *The Lancet*, **ii**, 639.

Wilson Cox, D. (1989). Alpha-1 antitrypsin deficiency. In *The metabolic basis of inherited disease*, (6th edn). (ed. C. R. Scriver, A. L. Beaudet, W. S. Sly, and D. Valle.) McGraw Hill, New York.

Winter, R. M. and Pembrey, M. E. (1982). Does unequal crossing over contribute to the mutation-rate in Duchenne muscular dystrophy? *American Journal of Medical Genetics*, **12**, 437–41.

Woodrow J. (1970). *Rh immunization and its prevention*. Series hematologica III, Vol. 3. Munksgaard, Copenhagen.

World Health Organization European Collaborative Group. (1986). European collaborative trial of multifactorial prevention of coronary heart disease: final report of the 6-year results. *The Lancet*, **i**, 869–72.

Ylinen, K., Aula, P., and Stenman, U-H. (1984). Risk of minor and major malformations in diabetes with high haemoglobin A_2 values in early pregnancy. *British Medical Journal*, **289**, 345–6.

Zeesman, S., Clow, C. L., Cartier, L., and Scriver, C. R. (1984). A private view of heterozygosity: eight-year follow-up study on carriers of the Tay–Sachs gene detected by high-school screening in Montreal. *American Journal of Medical Genetics*, **18**, 769–78.

Zerres, K. and Stephan, M. (1986). Attitudes to early diagnosis of polycystic kidney disease. *The Lancet*, **ii**, 1395.

Zurlo, M. G., De Stefano, P., Borgna-Pignatti, C., Di Palma, A., Piga, A., Melevendi, C., *et al.* (1989). Survival and causes of death in thalassaemia major. *The Lancet*, **ii**, 27–30.

Index